Food Microbiology

Third Edition

Food Microbiology

Third Edition

Martin R. Adams and Maurice O. Moss
University of Surrey, Guildford, UK

RSCPublishing

ISBN 978-0-85404-284-5

A catalogue record for this book is available from the British Library.

Published by The Royal Society of Chemistry,
Thomas Graham House, Science Park, Milton Road, Cambridge CB4 0WF, UK

Registered Charity No. 207890

For further information see our web site at www.rsc.org

Printed by Cromwell Press Limited, Trowbridge, Wiltshire

Preface to the First Edition

In writing this book we have tried to present an account of modern food microbiology that is both thorough and accessible. Since our subject is broad, covering a diversity of topics from viruses to helminths (by way of the bacteria) and from pathogenicity to physical chemistry, this can make presentation of a coherent treatment difficult; but it is also part of what makes food microbiology such an interesting and challenging subject.

The book is directed primarily at students of Microbiology, Food Science and related subjects up to Master's level and assumes some knowledge of basic microbiology. We have chosen not to burden the text with references to the primary literature in order to preserve what we hope is a reasonable narrative flow. Some suggestions for further reading for each chapter are included in Chapter 12. These are largely review articles and monographs which develop the overview provided and can also give access to the primary literature if required. We have included references that we consider are among the most current or best (not necessarily the same thing) at the time of writing, but have also taken the liberty of including some of the older, classic texts which we feel are well worth revisiting on occasion. By the very nature of current scientific publishing, many of our most recent references may soon become dated themselves. There is a steady stream of research publications and reviews appearing in journals such as *Food Microbiology, Food Technology*, the *International Journal of Food Microbiology*, the *Journal of Applied Bacteriology* and the *Journal of Food Protection* and we recommend that these sources are regularly surveyed to supplement the material provided here.

We are indebted to our numerous colleagues in food microbiology from whose writings and conversation we have learned so much over the years. In particular we would like to acknowledge Peter Bean for looking through the section on heat processing, Ann Dale and Janet Cole for their help with the figures and tables and, finally, our long suffering families of whom we hope to see more in the future.

Preface to the Second Edition

The very positive response *Food Microbiology* has had since it was first published has been extremely gratifying. It has reconfirmed our belief in the value of the original project and has also helped motivate us to produce this second edition. We have taken the opportunity to correct minor errors, improve some of the diagrams and update the text to incorporate new knowledge, recent developments and legislative changes. Much of this has meant numerous small changes and additions spread throughout the book, though perhaps we should point out (for the benefit of reviewers) new sections on stress response, *Mycobacterium* spp. and risk analysis, and updated discussions of predictive microbiology, the pathogenesis of some foodborne illnesses, BSE/vCJD and HACCP.

A number of colleagues have provided advice and information and among these we are particularly indebted to Mike Carter, Paul Cook, Chris Little, Johnjoe McFadden, Bob Mitchell, Yasmine Motarjemi and Simon Park. It is customary for authors to absolve those acknowledged from all responsibility for any errors in the final book. We are happy to follow that convention in the unspoken belief that if any errors have crept through we can always blame each other.

Preface to the Third Edition

In this third edition we have taken the opportunity to update and clarify the text in a number of places, removing a few incipient cobwebs along the way. Mostly this has entailed small changes within the existing text though there are new sections dealing with natamycin, subtyping, emerging pathogens and *Enterobacter sakazakii*.

In addition to all those colleagues who have helped with previous editions we are pleased to acknowledge Janet Corry and Marcel Zwietering whose diligent reading of the second edition revealed the need for some corrections that had previously eluded us. We have also rationalised the index which we decided was excessive and contained too many esoteric or trivial entries. As a consequence, terms such as "trub" have been deleted. Those seeking knowledge on this topic will now have to read the book in its entirety.

Contents

Chapter 6   Food Microbiology and Public Health

Chapter 7   Bacterial Agents of Foodborne Illness

Chapter 9   Fermented and Microbial Foods

Chapter 10   Methods for the Microbiological Examination of Foods

Chapter 11   Controlling the Microbiological Quality of Foods

CHAPTER 1

The Scope of Food Microbiology

Microbiology is the science which includes the study of the occurrence and significance of bacteria, fungi, protozoa and algae which are the beginning and ending of intricate food chains upon which all life depends. Most food chains begin wherever photosynthetic organisms can trap light energy and use it to synthesize large molecules from carbon dioxide, water and mineral salts forming the proteins, fats and carbohydrates which all other living creatures use for food.

Within and on the bodies of all living creatures, as well as in soil and water, micro-organisms build up and change molecules, extracting energy and growth substances. They also help to control population levels of higher animals and plants by parasitism and pathogenicity.

When plants and animals die, their protective antimicrobial systems cease to function so that, sooner or later, decay begins liberating the smaller molecules for re-use by plants. Without human intervention, growth, death, decay and regrowth would form an intricate web of plants, animals and micro-organisms, varying with changes in climate and often showing apparently chaotic fluctuations in populations of individual species, but inherently balanced in numbers between producing, consuming and recycling groups.

In the distant past, these cycles of growth and decay would have been little influenced by the small human population that could be supported by the hunting and gathering of food. From around 10 000 BC however, the deliberate cultivation of plants and herding of animals started in some areas of the world. The increased productivity of the land and the improved nutrition that resulted led to population growth and a probable increase in the average lifespan. The availability of food surpluses also liberated some from daily toil in the fields and stimulated the development of specialized crafts, urban centres, and trade – in short, civilization.

1.1 MICRO-ORGANISMS AND FOOD

The foods that we eat are rarely if ever sterile, they carry microbial associations whose composition depends upon which organisms gain access and how they grow, survive and interact in the food over time. The micro-organisms present will originate from the natural micro-flora of the raw material and those organisms introduced in the course of harvesting/slaughter, processing, storage and distribution (see Chapters 2 and 5). The numerical balance between the various types will be determined by the properties of the food, its storage environment, properties of the organisms themselves and the effects of processing. These factors are discussed in more detail in Chapters 3 and 4.

In most cases this microflora has no discernible effect and the food is consumed without objection and with no adverse consequences. In some instances though, micro-organisms manifest their presence in one of several ways:

    (i) they can cause spoilage;
   (ii) they can cause foodborne illness;
  (iii) they can transform a food's properties in a beneficial way – food fermentation.

1.1.1 Food Spoilage/Preservation

From the earliest times, storage of stable nuts and grains for winter provision is likely to have been a feature shared with many other animals but, with the advent of agriculture, the safe storage of surplus production assumed greater importance if seasonal growth patterns were to be used most effectively. Food preservation techniques based on sound, if then unknown, microbiological principles were developed empirically to arrest or retard the natural processes of decay. The staple foods for most parts of the world were the seeds – rice, wheat, sorghum, millet, maize, oats and barley – which would keep for one or two seasons if adequately dried, and it seems probable that most early methods of food preservation depended largely on water activity reduction in the form of solar drying, salting, storing in concentrated sugar solutions or smoking over a fire.

The industrial revolution which started in Britain in the late 18th century provided a new impetus to the development of food preservation techniques. It produced a massive growth of population in the new industrial centres which had somehow to be fed; a problem which many thought would never be solved satisfactorily. Such views were often based upon the work of the English cleric Thomas Malthus who in his 'Essay on Population' observed that the inevitable consequence of the exponential growth in population and the arithmetic growth in agricultural productivity would be over-population and mass starvation. This in fact proved not to be the case as the 19th century saw the development of substantial food preservation industries based around the use of chilling, canning and freezing and the first large scale importation of foods from distant producers.

To this day, we are not free from concerns about over-population. Globally there is sufficient food to feed the world's current population, estimated to be 6600 million in 2006. World grain production has more than managed to keep pace with the increasing population in recent years and the World Health Organization's Food and Agriculture Panel consider that current and emerging capabilities for the production and preservation of food should ensure an adequate supply of safe and nutritious food up to and beyond the year 2010 when the world's population is projected to rise to more than 7 billion.

There is however little room for complacency. Despite overall sufficiency, it is recognized that a large proportion of the population is malnourished and that 840 million people suffer chronic hunger. The principal cause of this is not insufficiency however, but poverty which leaves an estimated one-fifth of the world's population without the means to meet their daily needs. Any long-term solution to this must lie in improving the economic status of those in the poorest countries and this, in its train, is likely to bring a decrease in population growth rate similar to that seen in recent years in more affluent countries.

In any event, the world's food supply will need to increase to keep pace with population growth and this has its own environmental and social costs in terms of the more intensive exploitation of land and sea resources. One way of mitigating this is to reduce the substantial pre- and post-harvest losses which occur, particularly in developing countries where the problems of food supply are often most acute. It has been estimated that the average losses in cereals and legumes exceed 10% whereas with more perishable products such as starchy staples and vegetables the figure is more than 20% – increasing to an estimated 25% for highly perishable products such as fish. In absolute terms, the US National Academy of Sciences has estimated the losses in cereals and legumes in developing countries as 100 million tonnes, enough to feed 300 million people.

Clearly reduction in such losses can make an important contribution to feeding the world's population. While it is unrealistic to claim that food microbiology offers all the answers, the expertise of the food microbiologist can make an important contribution. In part, this will lie in helping to extend the application of current knowledge and techniques but there is also a recognized need for simple, low-cost, effective methods for improving food storage and preservation in developing countries. Problems for the food microbiologist will not however disappear as a result of successful development programmes. Increasing wealth will lead to changes in patterns of food consumption and changing demands on the food industry. Income increases among the poor have been shown to lead to increased demand for the basic food staples while in the better-off it leads to increased demand for more perishable animal products. To supply an increasingly affluent and expanding urban population will require massive extension of a safe distribution network and will place great demands on the food microbiologist.

1.1.2   Food Safety

In addition to its undoubted value, food has a long association with the transmission of disease. Regulations governing food hygiene can be found in numerous early sources such as the Old Testament, and the writings of Confucius, Hinduism and Islam. Such early writers had at best only a vague conception of the true causes of foodborne illness and many of their prescriptions probably had only a slight effect on its incidence. Even today, despite our increased knowledge, 'Foodborne disease is perhaps the most widespread health problem in the contemporary world and an important cause of reduced economic productivity.' (WHO 1992.) The available evidence clearly indicates that biological contaminants are the major cause. The various ways in which foods can transmit illness, the extent of the problem and the principal causative agents are described in more detail in Chapters 6, 7 and 8.

1.1.3   Fermentation

Microbes can however play a positive role in food. They can be consumed as foods in themselves as in the edible fungi, mycoprotein and algae. They can also effect desirable transformations in a food, changing its properties in a way that is beneficial. The different aspects of this and examples of important fermented food products are discussed in Chapter 9.

1.2   MICROBIOLOGICAL QUALITY ASSURANCE

Food microbiology is unashamedly an applied science and the food microbiologist's principal function is to help assure a supply of wholesome and safe food to the consumer. To do this requires the synthesis and systematic application of our knowledge of the microbial ecology of foods and the effects of processing to the practical problem of producing, economically and consistently, foods which have good keeping qualities and are safe to eat. How we attempt to do this is described in Chapter 11.

CHAPTER 2

Micro-organisms and Food Materials

Foods, by their very nature, need to be nutritious and metabolizable and it should be expected that they will offer suitable substrates for the growth and metabolism of micro-organisms. Before dealing with the details of the factors influencing this microbial activity, and their significance in the safe handling of foods, it is useful to examine the possible sources of micro-organisms in order to understand the ecology of contamination.

2.1 DIVERSITY OF HABITAT

Viable micro-organisms may be found in a very wide range of habitats, from the coldest of brine ponds in the frozen wastes of polar regions, to the almost boiling water of hot springs. Indeed, it is now realized that actively growing bacteria may occur at temperatures in excess of 100 °C in the thermal volcanic vents, at the bottom of the deeper parts of the oceans, where boiling is prevented by the very high hydrostatic pressure (see Section 3.2.5). Micro-organisms may occur in the acidic wastes draining away from mine workings or the alkaline waters of soda lakes. They can be isolated from the black anaerobic silts of estuarine muds or the purest waters of biologically unproductive, or oligotrophic, lakes. In all these, and many other, habitats microbes play an important part in the recycling of organic and inorganic materials through their roles in the carbon, nitrogen and sulfur cycles (Figure 2.1). They thus play an important part in the maintenance of the stability of the biosphere.

The surfaces of plant structures such as leaves, flowers, fruits and especially the roots, as well as the surfaces and the guts of animals all have a rich microflora of bacteria, yeasts and filamentous fungi. This natural, or normal flora may affect the original quality of the raw ingredients used in the manufacture of foods, the kinds of contamination which may occur during processing, and the possibility of food spoilage or food associated illness. Thus, in considering the possible sources of

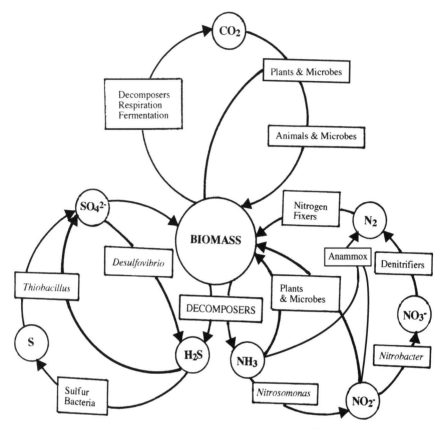

Figure 2.1 *Micro-organisms and the carbon, nitrogen and sulfur cycles*

micro-organisms as agents of food spoilage or food poisoning, it will be necessary to examine the natural flora of the food materials themselves, the flora introduced by processing and handling, and the possibility of chance contamination from the atmosphere, soil or water.

2.2   MICRO-ORGANISMS IN THE ATMOSPHERE

Perhaps one of the most hostile environments for many micro-organisms is the atmosphere. Suspended in the air, the tiny microbial propagule may be subjected to desiccation, to the damaging effects of radiant energy from the sun, and the chemical activity of elemental gaseous oxygen (O_2) to which it will be intimately exposed. Many micro-organisms, especially Gram-negative bacteria, do indeed die very rapidly when suspended in air and yet, although none is able to grow and multiply in the atmosphere, a significant number of microbes are able to survive and use the turbulence of the air as a means of dispersal.

2.2.1   Airborne Bacteria

The quantitative determination of the numbers of viable microbial propagules in the atmosphere is not a simple job, requiring specialized sampling equipment, but a qualitative estimate can be obtained by simply exposing a Petri dish of an appropriate medium solidified with agar to the air for a measured period of time. Such air exposure plates frequently show a diverse range of colonies including a significant number which are pigmented (Figure 2.2).

The bacterial flora can be shown to be dominated by Gram-positive rods and cocci unless there has been a very recent contamination of the air by an aerosol generated from an animal or human source, or from water. The pigmented colonies will often be of micrococci or corynebacteria and the large white-to-cream coloured colonies will frequently be of aerobic sporeforming rods of the genus *Bacillus*. There may also be small raised, tough colonies of the filamentous bacteria belonging to *Streptomyces* or a related genus of actinomycetes. The possession of pigments may protect micro-organisms from damage by both visible and ultraviolet radiation of sunlight and the relatively simple, thick cell walls of Gram-positive bacteria may afford protection from desiccation. The endospores of *Bacillus* and the conidiospores of *Streptomyces* are especially resistant to the potentially damaging effects of suspension in the air.

The effects of radiation and desiccation are enhanced by another phenomenon, the 'open air factor' which causes even more rapid death

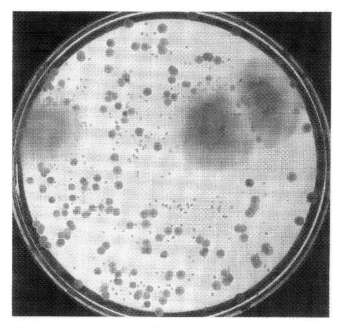

Figure 2.2 *Exposure plate showing air flora*

rates of sensitive Gram-negative organisms such as *Escherichia coli*. It can be shown that these organisms may die more rapidly in outdoor air at night time than they do during the day, in spite of reduced light damage to the cells. It is possible that light may destroy this 'open-air factor', or that other more complex interactions may occur. Phenomena such as this, alert us to the possibility that it can be very difficult to predict how long micro-organisms survive in the air and routine monitoring of air quality may be desirable within a food factory, or storage area, where measures to reduce airborne microbial contamination can have a marked effect on food quality and shelf-life. This would be particularly true for those food products such as bakery goods that are subject to spoilage by organisms that survive well in the air.

Bacteria have no active mechanisms for becoming airborne. They are dispersed on dust particles disturbed by physical agencies, in minute droplets of water generated by any process which leads to the formation of an aerosol, and on minute rafts of skin continuously shed by many animals including humans. The most obvious mechanisms for generating aerosols are coughing and sneezing but many other processes generate minute droplets of water. The bursting of bubbles, the impaction of a stream of liquid onto a surface, or taking a wet stopper out of a bottle are among the many activities that can generate aerosols, the droplets of which may carry viable micro-organisms for a while.

One group of bacteria has become particularly well adapted for air dispersal. Many actinomycetes, especially those in the genus *Streptomyces*, produce minute dry spores which survive well in the atmosphere. Although they do not have any mechanisms for active air dispersal, the spores are produced in chains on the end of a specialized aerial structure so that any physical disturbance dislodges them into the turbulent layers of the atmosphere. The air of farmyard barns may contain many millions of spores of actinomycetes per cubic metre and some species, such as *Thermoactinomyces vulgaris* and *Micropolyspora faeni*, can cause the disabling disease known as farmer's lung where individuals have become allergic to the spores. Actinomycetes are rarely implicated in food spoilage but geosmin-producing strains of *Streptomyces* may be responsible for earthy odours and off-flavours in potable water, and geosmin (Figure 2.3) may impart earthy taints to such foods as shellfish.

2.2.2 Airborne Fungi

It is possible to regard the evolution of many of the terrestrial filamentous fungi (the moulds) as the development of increasingly sophisticated mechanisms for the air dispersal of their reproductive propagules. Some of the most important moulds in food microbiology do not have active spore dispersal mechanisms but produce large numbers of small

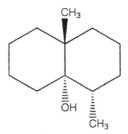

Figure 2.3 *Geosmin*

(a) (b)

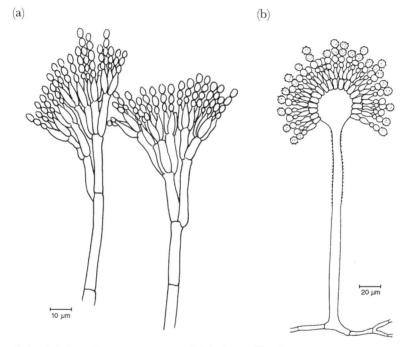

Figure 2.4 *(a) Pencillium expansum and (b) Aspergillus flavus*

unwettable spores which are resistant to desiccation and light damage. They become airborne in the same way as fine dry dust particles by physical disturbance and wind. Spores of *Penicillium* and *Aspergillus* (Figure 2.4) seem to get everywhere in this passive manner and species of these two genera are responsible for a great deal of food spoilage. The individual spores of *Penicillium* are only 2–3 μm in diameter, spherical to sub-globose (*i.e.* oval), and so are small and light enough to be efficiently dispersed in turbulent air.

Some fungi, such as *Fusarium* (Figure 2.5), produce easily wettable spores which are dispersed into the atmosphere in the tiny droplets of water which splash away from the point of impact of a rain drop and so may become very widely distributed in field crops during wet weather.

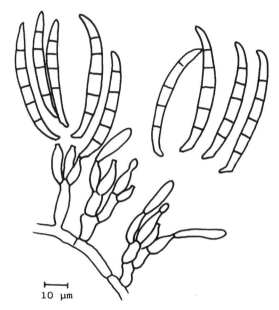

10 μm

Figure 2.5 *Fusarium graminearum*

Such spores rarely become an established part of the long-term air spora and this mechanism has evolved as an effective means for the short-term dispersal of plant pathogens.

As the relative humidity of the atmosphere decreases with the change from night to day, the sporophores of fungi such as *Cladosporium* (Figure 2.6) react by twisting and collapsing, throwing their easily detached spores into the atmosphere. At some times of the year, especially during the middle of the day, the spores of *Cladosporium* may be the most common spores in the air spora. Species such as *Cladosporium herbarum* grow well at refrigeration temperatures and may form unsightly black colonies on the surface of commodities such as chilled meat.

Many fungi have evolved mechanisms for actively firing their spores into the atmosphere (Figure 2.7), a process which usually requires a high relative humidity. Thus the ballistospores of the mirror yeasts, which are frequently a part of the normal microbial flora of the leaf surfaces of plants, are usually present in highest numbers in the atmosphere in the middle of the night when the relative humidity is at its highest.

The evolutionary pressure to produce macroscopic fruiting bodies, which is seen in the mushrooms and toadstools, has produced a structure which provides its own microclimate of high relative humidity so that these fungi can go on firing their spores into the air even in the middle of a dry day.

In our everyday lives we are perhaps less aware of the presence of micro-organisms in the atmosphere than anywhere else, unless we

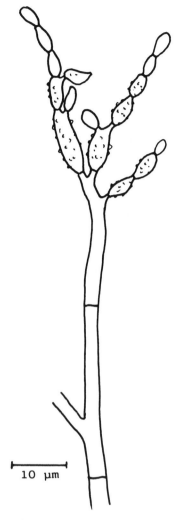

Figure 2.6 *Cladosporium cladosporioides*

happen to suffer from an allergy to the spores of moulds or act-inomycetes, but, although they cannot grow in it, the atmosphere forms an important vehicle for the spread of many micro-organisms, and the subsequent contamination of foods.

2.3 MICRO-ORGANISMS OF SOIL

The soil environment is extremely complex and different soils have their own diverse flora of bacteria, fungi, protozoa and algae. The soil is such a rich reservoir of micro-organisms (Figure 2.8) that it has provided many of the strains used for the industrial production of antibiotics, enzymes, amino acids, vitamins and other products used in both the

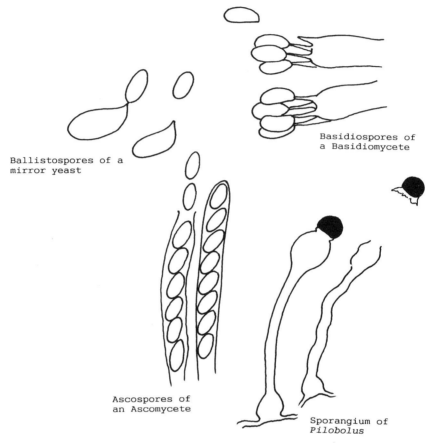

Ballistospores of a
mirror yeast

Basidiospores of
a Basidiomycete

Ascospores of
an Ascomycete

Sporangium of
Pilobolus

Figure 2.7   *Mechanisms for active dispersal of fungal propagules*

pharmaceutical and food industries. Soil micro-organisms participate in
the recycling of organic and nitrogenous compounds which is essential if
the soil is to support the active growth of plants, but this ability to
degrade complex organic materials makes these same micro-organisms
potent spoilage organisms if they are present on foods. Thus the commonly accepted practice of protecting food from 'dirt' is justified in
reducing the likelihood of inoculating the food with potential spoilage
organisms.

   The soil is also a very competitive environment and one in which the
physico-chemical parameters can change very rapidly. In response to
this, many soil bacteria and fungi produce resistant structures, such as
the endospores of *Bacillus* and *Clostridium*, and chlamydospores and
sclerotia of many fungi, which can withstand desiccation and a wide
range of temperature fluctuations. Bacterial endospores are especially
resistant to elevated temperatures, indeed their subsequent germination
is frequently triggered by exposure to such temperatures, and their

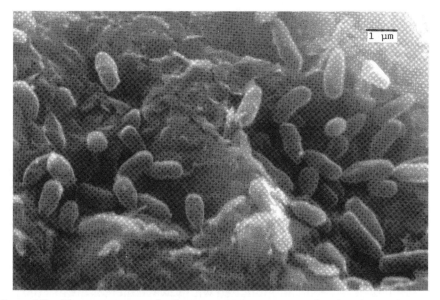

Figure 2.8 *Electron micrograph of micro-organisms associated with soil particles*

common occurrence in soil makes this a potent source of spoilage and food poisoning bacilli and clostridia.

2.4 MICRO-ORGANISMS OF WATER

The aquatic environment represents in area and volume the largest part of the biosphere and both fresh water and the sea contain many species of micro-organisms adapted to these particular habitats. The bacteria isolated from the waters of the open oceans often have a physiological requirement for salt, grow best at the relatively low temperatures of the oceans and are nutritionally adapted to the relatively low concentrations of available organic and nitrogenous compounds. Thus, from the point of view of a laboratory routinely handling bacteria from environments directly associated with man, marine bacteria are usually described as oligotrophic psychrophiles with a requirement for sodium chloride for optimum growth.

The surfaces of fish caught from cold water in the open sea will have a bacterial flora which reflects their environment and will thus contain predominantly psychrophilic and psychrotrophic species. Many of these organisms can break down macromolecules, such as proteins, poly-saccharides and lipids, and they may have doubling times as short as ten hours at refrigeration temperatures of 0–7 °C. Thus, in ten days, *i.e.* 240 hours, one organism could have become 2^{24} or between 10^7 and 10^8 under such conditions. Once a flora has reached these numbers it could be responsible for the production of off-odours and hence spoilage. Of course, during the handling of a commodity such as fish, the natural flora of the environment will be contaminated with organisms associated with man, such as members of the Enterobacteriaceae and *Staphylococcus*, which can grow well at 30–37 °C. It is readily possible to distinguish the environmental flora from the 'handling' flora by comparing the numbers of colonies obtained by plating-out samples on nutrient agar and incubating at 37 °C with those from plates of sea water agar, containing a lower concentration of organic nutrients, and incubated at 20 °C.

The seas around the coasts are influenced by inputs of terrestrial and freshwater micro-organisms and, perhaps more importantly, by human activities. The sea has become a convenient dump for sewage and other waste products and, although it is true that the seas have an enormous capacity to disperse such materials and render them harmless, the scale of human activity has had a detrimental effect on coastal waters. Many shellfish used for food grow in these polluted coastal waters and the majority feed by filtering out particles from large volumes of sea water. If these waters have been contaminated with sewage there is always the risk that enteric organisms from infected individuals may be present and will be concentrated by the filter feeding activities of shellfish. Severe diseases such as hepatitis or typhoid fever, and milder illnesses such as gastroenteritis have been caused by eating contaminated oysters and mussels which seem to be perfectly normal in taste and appearance. In warmer seas even unpolluted water may contain significant numbers of *Vibrio parahaemolyticus* and these may also be concentrated by filter-feeding shellfish, indeed they may form a stable part of the natural enteric flora of some shellfish. This organism may be responsible for outbreaks of food poisoning especially associated with sea foods.

The fresh waters of rivers and lakes also have a complex flora of micro-organisms which will include genuinely aquatic species as well as components introduced from terrestrial, animal and plant sources. As with the coastal waters of the seas, fresh water may also act as a vehicle for bacteria, protozoa and viruses causing disease through contamination with sewage effluent containing human faecal material. These organisms do not usually multiply in river and lake water and may be present in very low, but nonetheless significant, numbers making it difficult to demonstrate their presence by direct methods. It is usual to infer the possibility of the presence of such organisms by actually looking for a species of bacterium which is always present in large numbers in human faeces, is unlikely to grow in fresh water, but will survive at least as long as any pathogen. Such an organism is known as an 'indicator organism' and the species usually chosen in temperate climates is *Escherichia coli*.

Fungi are also present in both marine and fresh waters but they do not have the same level of significance in food microbiology as other micro-organisms. There are groups of truly aquatic fungi including some which are serious pathogens of molluscs and fish. There are fungi which have certainly evolved from terrestrial forms but have become morphologically and physiologically well adapted to fresh water or marine habitats. They include members of all the major groups of terrestrial fungi, the ascomycetes, basidiomycetes, zygomycetes and deuteromycetes and there is the possibility that some species from this diverse flora could be responsible for spoilage of a specialized food commodity associated with water such as a salad crop cultivated with overhead irrigation from a river or lake, but this is speculation.

Of the aquatic photosynthetic micro-organisms, the cyanobacteria, or blue-green algae, amongst the prokaryotes and the dinoflagellates amongst the eukaryotes, have certainly had an impact on food quality and safety. Both these groups of micro-organisms can produce very toxic metabolites which may become concentrated in shellfish without apparently causing them any harm. When consumed by humans, however, they can cause a very nasty illness such as paralytic shellfish poisoning (see Chapter 8).

2.5 MICRO-ORGANISMS OF PLANTS

All plant surfaces have a natural flora of micro-organisms which may be sufficiently specialized to be referred to as the phylloplane flora, for that of the leaf surface, and the rhizoplane flora for the surface of the roots. The numbers of organisms on the surfaces of healthy, young plant leaves may be quite low but the species which do occur are well adapted for this highly specialized environment. Moulds such as *Cladosporium* and the so-called black yeast, *Aureobasidium pullulans*, are frequently present. Indeed, if the plant is secreting a sugary exudate, these moulds may be present in such large numbers that they cover the leaf surface with a black sooty deposit. In the late summer, the leaves of such trees as oak and lime may look as though they are suffering from some form of industrial pollution, so thick is the covering of black moulds. *Aureobasidium* behaves like a yeast in laboratory culture but develops into a filamentous mould-like organism as the culture matures.

There are frequently true yeasts of the genera *Sporobolomyces* and *Bullera* on plant leaf surfaces. These two genera are referred to as mirror yeasts because, if a leaf is attached to the inner surface of the lid of a Petri dish containing malt extract agar, the yeasts produce spores which they actively fire away from the leaf surface. These ballistospores hit the agar surface and germinate to eventually produce visible colonies in a pattern which forms a mirror image of the leaf. An even richer yeast flora is found in association with the nectaries of flowers and the surfaces of fruits and the presence of some of these is important in the spontaneous fermentation of fruit juices, such as that of the grape in the production of wine. The bacterial flora of aerial plant surfaces which is most readily detected is made up predominantly of Gram-negative rods, such as *Pectobacterium, Erwinia, Pseudomonas* and *Xanthomonas* but there is usually also present a numerically smaller flora of fermentative Gram-positive bacteria such as *Lactobacillus* and *Leuconostoc* which may become important in the production of such fermented vegetable products as sauerkraut (see Chapter 9).

The specialized moulds, yeasts and bacteria living as harmless commensals on healthy, young plant surfaces are not usually any problem in the spoilage of plant products after harvest. But, as the plant matures, both the bacterial and fungal floras change. The numbers of pectinolytic bacteria increase as the vegetable tissues mature and a large number of mould species are able to colonize senescent plant material. In the natural cycling of organic matter these organisms would help to break down the complex plant materials and so bring about the return of carbon, nitrogen and other elements as nutrients for the next round of plant growth. But, when humans break into this cycle and harvest plant products such as fruits, vegetables, cereals, pulses, oilseeds and root crops, these same organisms may cause spoilage problems during prolonged periods of storage and transport.

Plants have evolved several mechanisms for resisting infection by micro-organisms but there are many species of fungi and bacteria which overcome this resistance and cause disease in plants and some of these may also cause spoilage problems after harvesting and storage. Amongst the bacteria, *Pectobacterium caratovorum* subsp. *caratovorum* (previously known as *Erwinia carotovora* var. *atroseptica*) is a pathogen of the potato plant causing blackleg disease of the developing plant. The organism can survive in the soil when the haulms of diseased plants fall to the ground and, under the right conditions of soil moisture and temperature, it may then infect healthy potato tubers causing a severe soft rot during storage. One of the conditions required for such infection is a film of moisture on the tuber for this species can only infect the mature tuber through a wound or via a lenticel in the skin of the potato. This process may be unwittingly aided by washing potatoes and marketing them in plastic bags so that, the combination of minor damage and moisture trapped in the bag, favours the development of *Pectobacterium* soft rot.

Amongst the fungi, *Botrytis cinerea* (Figure 2.9) is a relatively weak pathogen of plants such as the strawberry plant where it may infect the flower. However, this low pathogenicity is often followed by a change to an aggressive invasion of the harvested fruit, usually through the calyx into the fruit tissue. Once this 'grey mould' has developed on one fruit, which may have been damaged and infected during growth before harvest, the large mass of spores and actively growing mould readily infects neighbouring fruit even though they may be completely sound.

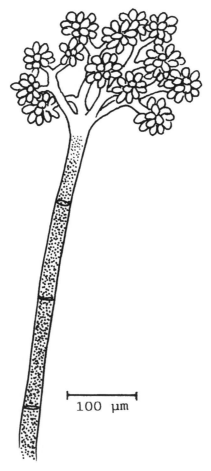

100 μm

Figure 2.9 *Botrytis cinerea*

The cereals are a group of plant commodities in which there is a pronounced and significant change in the microbial flora following harvesting. In the field the senescent plant structures carrying the cereal grain may become infected by a group of fungi, referred to as the field fungi, which includes such genera as *Cladosporium, Alternaria, Helminthosporium* and *Chaetomium*. After harvest and reduction of the moisture content of the grain, the components of the field flora decrease in numbers and are replaced by a storage flora which characteristically includes species of the genera *Penicillium* and *Aspergillus*. Some genera of fungi, such as *Fusarium*, contain a spectrum of species, some of which are specialized plant pathogens, others saprophytic field fungi and others capable of growth during the initial stages of storage. Indeed, the more that is learnt about the detailed ecology of individual species, the more it is realized that it may be misleading to try to pigeon hole them into simple categories such as field fungi and storage fungi. Thus it is now known that *Aspergillus flavus*, a very important species because of its ability to produce the carcinogenic metabolite known as aflatoxin, is not just a storage mould as was once believed, but may infect the growing plant in the field and produce its toxic metabolites before harvesting and storage (see Chapter 8). Indeed, it is now recognized that many plants carry fungal endophytes in their naturally healthy state.

2.6 MICRO-ORGANISMS OF ANIMAL ORIGIN

All healthy animals carry a complex microbial flora, part of which may be very specialized and adapted to growth and survival on its host, and part of which may be transient, reflecting the immediate interactions of the animal with its environment. From a topological point of view, the gut is also part of the external surface of an animal but it offers a very specialized environment and the importance of the human gut flora will be dealt with in Chapter 6.

2.6.1 The Skin

The surfaces of humans and other animals are exposed to air, soil and water and there will always be the possibility of contamination of foods and food handling equipment and surfaces with these environmental microbes by direct contact with the animal surface. However, the surface of the skin is not a favourable place for most micro-organisms since it is usually dry and has a low pH due to the presence of organic acids secreted from some of the pores of the skin. This unfavourable environment ensures that most micro-organisms reaching the skin do not multiply and frequently die quite quickly. Such organisms are only 'transients' and would not be regularly isolated from the cleaned skin surface.

Nevertheless, the micro-environments of the hair follicles, sebaceous glands and the skin surface have selected a specialized flora exquisitely adapted to each environment. The bacteria and yeasts making up this 'normal' flora are rarely found in other habitats and are acquired by the host when very young, usually from the mother. The micro-organisms are characteristic for each species of animal and, in humans, the normal skin flora is dominated by Gram-positive bacteria from the genera *Staphylococcus, Corynebacterium* and *Propionibacterium*. For animals which are killed for meat, the hide may be one of the most important sources of spoilage organisms while, in poultry, the micro-organisms associated with feathers and the exposed follicles, once feathers are removed, may affect the microbial quality and potential shelf-life of the carcass.

2.6.2 The Nose and Throat

The nose and throat with the mucous membranes which line them represent even more specialized environments and are colonized by a different group of micro-organisms. They are usually harmless but may have the potential to cause disease, especially following extremes of temperature, starvation, overcrowding or other stresses which lower the resistance of the host and make the spread of disease more likely in both humans and other animals. *Staphylococcus aureus* is carried on the mucous membranes of the nose by a significant percentage of the human population and some strains of this species can produce a powerful toxin capable of eliciting a vomiting response. The food poisoning caused by this organism will be dealt with in Chapter 7.

2.7 CONCLUSIONS

In this chapter we have described some of the major sources of micro-organisms which may contaminate food and cause problems of spoilage or create health risks when the food is consumed. It can be seen that most foods cannot be sterile but have a natural flora and acquire a transient flora derived from their environment. To ensure that food is safe and can be stored in a satisfactory state, it is necessary to either destroy the micro-organisms present, or manipulate the food so that growth is prevented or hindered. The manner in which environmental and nutritional factors influence the growth and survival of micro-organisms will be considered in the next chapter. The way in which this knowledge can be used to control microbial activity in foods will be considered in Chapter 4.

CHAPTER 3

Factors Affecting the Growth and Survival of Micro-organisms in Foods

3.1 MICROBIAL GROWTH

Microbial growth is an autocatalytic process: no growth will occur without the presence of at least one viable cell and the rate of growth will increase with the amount of viable biomass present. This can be represented mathematically by the expression:

$$dx/dt = \mu x \qquad (3.1)$$

where dx/dt is the rate of change of biomass, or numbers x with time t, and μ is a constant known as the specific growth rate.

The same exponential growth rule applies to filamentous fungi which grow by hyphal extension and branching since the rate of branching normally increases with hyphal length.

Integration of Equation (3.1) gives:

$$x = x_0 \, e^{\mu t} \qquad (3.2)$$

or, taking natural logarithms and re-arranging:

$$\ln(x/x_0) = \mu t \qquad (3.3)$$

where x_0 is the biomass present when $t = 0$.

The doubling or generation time of an organism τ can be obtained by substituting $x = 2x_0$ in Equation 3.3. Thus:

$$\tau = \ln 2/\mu = 0.693/\mu \qquad (3.4)$$

An alternative way of representing exponential growth in terms of the doubling time is:

$$x = x_0 \, 2^{t/\tau} \qquad (3.5)$$

This can be simply illustrated by considering the case of a bacterial cell dividing by fission to produce two daughter cells. In time τ, a single cell will divide to produce two cells; after a further doubling time has elapsed four cells will be present; after another, eight, and so on. Thus, the rate of increase as well as the total cell number is doubling with every doubling time that passes.

If, however, we perform the experiment measuring microbial numbers with time and then plot log x against time, we obtain the curve shown as Figure 3.1 in which exponential growth occurs for only a part of the time.

A simple analysis of this curve can distinguish three major phases. In the first, the lag-phase, there is no apparent growth while the inoculum adjusts to the new environment, synthesizes the enzymes required for its exploitation and repairs any lesions resulting from earlier injury, *e.g.* freezing, drying, heating. The exponential or logarithmic phase which follows is characterized by an increase in cell numbers following the simple growth law equation. Accordingly, the slope of this portion of the curve will equal the organism's specific growth rate μ, which itself will depend on a variety of factors (see below). Finally, changes in the medium as a result of exponential growth bring this phase to an end

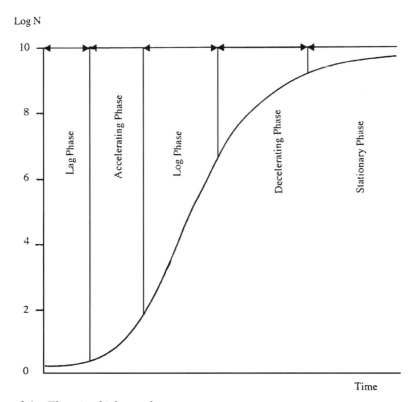

Figure 3.1 *The microbial growth curve*

as key nutrients become depleted, or inhibitory metabolites accumulate, and the culture moves into the stationary phase.

One way of representing this overall process mathematically is to modify the basic growth Equation (3.1) so that the growth rate decreases as the population density increases. An equation that does this and gives us a closer approximation to the observed microbial growth curve is the logistic equation:

$$\mathrm{d}x/\mathrm{d}t = (\mu_m - \mu_m x/K)x \tag{3.6}$$

where K is the carrying capacity of the environment (the stationary phase population) and μ_m, the maximum specific growth rate. As x increases and approaches K, the growth rate falls to zero. Or, in its integrated form:

$$x = Kc/(c + e^{-\mu_m t}) \tag{3.7}$$

where $c = x_0/(K-x_0)$.

The significance of exponential growth for food processing hardly needs emphasizing. A single bacterium with a doubling time of 20 minutes ($\mu = 2.1\mathrm{hr}^{-1}$) growing in a food, or pockets of food trapped in equipment, can produce a population of greater than 10^7 cells in the course of an 8-hour working day. It is therefore, a prime concern of the food microbiologist to understand what influences microbial growth in foods with a view to controlling it.

The situation is complicated by the fact that the microflora is unlikely to consist of a single pure culture. In the course of growth, harvesting/slaughter, processing and storage, food is subject to contamination from a range of sources (Chapter 2). Some of the micro-organisms introduced will be unable to grow under the conditions prevailing, while others will grow together in what is known as an association, the composition of which will change with time.

The factors that affect microbial growth in foods, and consequently the associations that develop, also determine the nature of spoilage and any health risks posed. For convenience they can be divided into four groups along the lines suggested more than 50 years ago in a seminal review by Mossel and Ingram (Table 3.1) – physico-chemical properties of the food itself (intrinsic factors); conditions of the storage environment (extrinsic factors); properties and interactions of the micro-organisms present (implicit factors); and processing factors. This last group of factors (subsumed under intrinsic properties by Mossel and Ingram) usually exert their effect in one of two ways: either they change an intrinsic or extrinsic property, for example slicing a product will damage antimicrobial structures and increase nutrient availability and redox potential, or they eliminate a proportion of the product microflora as would occur in washing, pasteurization or irradiation.

Table 3.1 *Factors affecting the development of microbial associations in food*

Intrinsic Factors
Nutrients
pH and buffering capacity
Redox potential
Water activity
Antimicrobial constituents
Antimicrobial structures

Environmental factors
Relative humidity
Temperature
Gaseous atmosphere

Implicit factors
Specific growth rate
Mutualism
Antagonism
Commensalism

Processing factors
Slicing
Washing
Packing
Irradiation
Pasteurization

The microbiological effects of different processing factors applied in the food industry will be discussed as they arise elsewhere in the text, principally in Chapter 4.

Although it is often convenient to examine the factors affecting microbial growth individually, some interact strongly, as in the relationships between relative humidity and water activity a_w, and gaseous atmosphere and redox potential. For this reason, in the following discussion, we have not been over zealous in discussing individual factors in complete isolation.

3.2 INTRINSIC FACTORS (SUBSTRATE LIMITATIONS)

3.2.1 Nutrient Content

Like us, micro-organisms can use foods as a source of nutrients and energy. From them, they derive the chemical elements that constitute microbial biomass, those molecules essential for growth that the organism cannot synthesize, and a substrate that can be used as an energy source. The widespread use of food products such as meat or casein digests (peptone and tryptone), meat infusions, tomato juice, malt extract, sugar and starch in microbiological media bears eloquent testimony to their suitability for this purpose.

The inability of an organism to utilize a major component of a food material will limit its growth and put it at a competitive disadvantage compared with those that can. Thus, the ability to synthesize amylolytic (starch degrading) enzymes will favour the growth of an organism on cereals and other farinaceous products. The addition of fruits containing sucrose and other sugars to yoghurt increases the range of carbohydrates available and allows the development of a more diverse spoilage microflora of yeasts.

The concentration of key nutrients can, to some extent, determine the rate of microbial growth. The relationship between the two, known as the Monod equation, is mathematically identical to the Michaelis–Menten equation of enzyme kinetics, reflecting the dependence of microbial growth on rate-limiting enzyme reactions:

$$\mu = \frac{\mu_m S}{S + K_s} \tag{3.8}$$

where μ is the specific growth rate; μ_m the maximum specific growth rate; S the concentration of limiting nutrient; and K_s the saturation constant.

When $S \gg K_s$, a micro-organism will grow at a rate approaching its maximum, but as S falls to values approaching K_s, so too will the growth rate. Values for K_s have been measured experimentally for a range of organisms and nutrients; generally they are extremely low, of the order of 10^{-5} M for carbon and energy sources, suggesting that in most cases, nutrient scarcity is unlikely to be rate-limiting. Exceptions occur in some foods, particularly highly structured ones where local microenvironments may be deficient in essential nutrients, or where nutrient limitation is used as a defence against microbial infection, for example the white of the hen's egg (Section 3.2.4).

3.2.2   pH and Buffering Capacity

As measured with the glass electrode, pH is equal to the negative logarithm of the hydrogen ion activity. Activity is proportional to concentration and the proportionality constant, the activity coefficient, approaches unity as the solution becomes more dilute. Thus:

$$pH = -\log(a_H) = \log 1/(a_H) \approx \log 1/[H^+] \tag{3.9}$$

where (a_H) is the hydrogen ion activity and $[H^+]$ the hydrogen ion concentration.

For aqueous solutions, pH 7 corresponds to neutrality (since $[H^+][OH^-] = 10^{-14}$ for water), pH values below 7 are acidic and those above 7 indicate an alkaline environment. It is worth remembering that since pH is a logarithmic scale differences in pH of 1, 2 and 3 units correspond to 10-, 100- and 1000-fold differences in the hydrogen ion concentration.

The acidity or alkalinity of an environment has a profound effect on the activity and stability of macromolecules such as enzymes, so it is not surprising that the growth and metabolism of micro-organisms are influenced by pH. Plotting microbial growth rate against pH produces an approximately symmetrical bell-shaped curve spanning 2–5 pH units, with a maximum rate exhibited over a range of 1–2 units.

In general, bacteria grow fastest in the pH range 6.0–8.0, yeasts 4.5–6.0 and filamentous fungi 3.5–4.0. As with all generalizations there are exceptions, particularly among those bacteria that produce quantities of acids as a result of their energy-yielding metabolism. Examples important in food microbiology are the lactobacilli and acetic acid bacteria with optima usually between pH 5.0 and 6.0.

Most foods are at least slightly acidic, since materials with an alkaline pH generally have a rather unpleasant taste (Table 3.2). Egg white, where the pH increases to around 9.2 as CO_2 is lost from the egg after laying, is a commonplace exception to this. A somewhat more esoteric example, which many would take as convincing evidence of the inedibility of

Table 3.2 *Approximate pH ranges of some common food commodities*

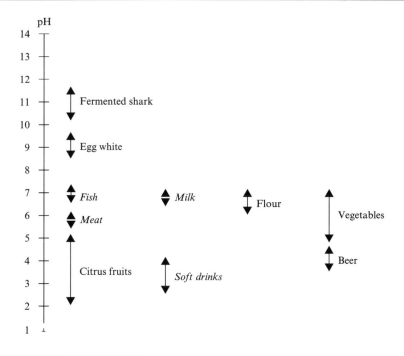

alkaline foods, is fermented shark, produced in Iceland and known as hakar, which has a pH of 10–12.

The acidity of a product can have important implications for its microbial ecology and the rate and character of its spoilage. For example, plant products classed as vegetables generally have a moderately acid pH and soft-rot producing bacteria such as *Pectobacterium carotovorum* and pseudomonads play a significant role in their spoilage. In fruits, however, a lower pH prevents bacterial growth and spoilage is dominated by yeasts and moulds.

As a rule, fish spoil more rapidly than meat under chill conditions. The pH of post-rigor mammalian muscle, around 5.6, is lower than that of fish (6.2–6.5) and this contributes to the longer storage life of meat. The pH-sensitive genus *Shewanella* (formerly *Alteromonas*) plays a significant role in fish spoilage but has not been reported in normal meat (pH < 6.0). Those fish that have a naturally low pH such as halibut (pH ≈ 5.6) have better keeping qualities than other fish.

The ability of low pH to restrict microbial growth has been deliberately employed since the earliest times in the preservation of foods with acetic and lactic acids (see Chapters 4 and 9).

With the exception of those soft drinks that contain phosphoric acid, most foods owe their acidity to the presence of weak organic acids. These do not dissociate completely into protons and conjugate base in solution but establish an equilibrium:

$$HA \rightleftharpoons H^+ + A^- \tag{3.10}$$

The equilibrium constant for this process, K_a, is given by

$$K_a = \frac{[H^+][A^-]}{[HA]} \tag{3.11}$$

where [ ] denotes concentration.

This expression can be rearranged:

$$\frac{1}{[H^+]} = \frac{1}{K_a} \frac{[A^-]}{[HA]} \tag{3.12}$$

If we take logarithms to the base 10 we get:

$$pH = pK_a + \log \frac{[A^-]}{[HA]} \tag{3.13}$$

Equation (3.13) is known as the Henderson–Hasselbalch equation and describes the relationship between the pH of a solution, the strength of the acid present and its degree of dissociation. When the pH is equal to an acid's pK_a, then half of the acid present will be undissociated. If the pH is increased then dissociation of the acid will increase as well, so that

Table 3.3 *pK$_a$ values of some common food acids*

Acid	pK$_a$
Acetic (ethanoic)	4.75
Propionic	4.87
Lactic	3.86
Sorbic	4.75
Citric	3.14, 4.77, 6.39
Benzoic	4.19
Parabens	8.5
Phosphoric	2.12, 7.12, 12.67
Carbonic	6.37, 10.25
Nitrous	3.37
Sulfurous	1.81, 6.91

when pH = pK_a+1 there will be 10 times as much dissociated acid as undissociated. Similarly as the pH is decreased below the pK_a the proportion of undissociated acid increases. Table 3.3 presents a list of some common food-associated acids and their pK_a values.

This partial dissociation of weak acids, such as acetic acid, plays an important part in their ability to inhibit microbial growth. It is well established that, although addition of strong acids has a more profound effect on pH *pro rata*, they are less inhibitory than weak lipophilic acids at the same pH. This is because microbial inhibition by weak acids is not solely due to the creation of a high extracellular proton concentration, but is also directly related to the concentration of undissociated acid.

Many essential cell functions such as ATP synthesis in bacteria, active transport of nutrients and cytoplasmic regulation occur at the cell membrane and are dependent on potential energy stored in the membrane in the form of a proton motive force. This force is an electro-chemical potential produced by the active translocation of protons from the cell interior to the external environment. Unlike protons and other charged molecules, undissociated lipophilic acid molecules can pass freely through the membrane; in doing so they pass from an external environment of low pH where the equilibrium favours the undissociated molecule to the high pH of the cytoplasm (around 7.5 in neutrophiles). At this higher pH, the equilibrium shifts in favour of the dissociated molecule, so the acid ionizes producing protons which will tend to acidify the cytoplasm and break down the pH component of the proton motive force. The cell will try to maintain its internal pH by neutralizing or expelling the protons leaking in but this will slow growth as it diverts energy from growth-related functions. If the external pH is sufficiently low and the extracellular concentration of acid high, the burden on the cell becomes too great, the cytoplasmic pH drops to a level where growth is no longer possible and the cell eventually dies (Figure 3.2).

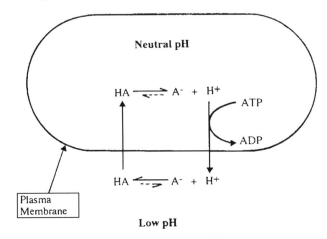

Figure 3.2 *Microbial inhibition by weak organic acids*

3.2.3   Redox Potential, E_h

An oxidation–reduction (redox) reaction occurs as the result of a transfer of electrons between atoms or molecules. In the equation below, this is represented in its most general form to include the many redox reactions which also involve protons and have the overall effect of transferring hydrogen atoms.

$$[\text{Oxidant}] + H^+ + ne \rightleftharpoons [\text{Reductant}] \qquad (3.14)$$

where n is the number of electrons, e, transferred.

In living cells an ordered sequence of both electron and hydrogen transfer reactions is an essential feature of the electron transport chain and energy generation by oxidative phosphorylation.

The tendency of a medium to accept or donate electrons, to oxidize or reduce, is termed its redox potential (E_h) and is measured against an external reference by an inert metal electrode, usually platinum, immersed in the medium. If the balance of the various redox couples present favours the oxidized state then there will be a tendency to accept electrons from the electrode creating a positive potential which signifies an oxidizing environment. If the balance is reversed, the sample will tend to donate electrons to the electrode which will then register a negative potential – a reducing environment. The redox potential we measure in a food is the result of several factors summarized in Table 3.4.

The tendency of an atom or molecule to accept or donate electrons is expressed as its standard redox potential, E_0'. A large positive E_0' indicates that the oxidized species of the couple is a strong oxidizing agent and the reduced form only weakly reducing. A large negative E_0' indicates the reverse. Some redox couples typically encountered in food

Table 3.4 *Factors influencing the measured E_h of foods*

Redox couples present
Ratio of oxidant to reductant
pH
Poising capacity
Availability of oxygen (physical state, packing)
Microbial activity

Table 3.5 *Some important redox couples and their standard redox potential*

Couple	$E_0(mV)$
1/2 O_2/H_2O	+820
Fe^{3+}/Fe^{2+}	+760
Cytochrome C ox/red	+250
Dehydroascorbic acid/ascorbic acid	+80
Methylene blue ox/red	+11
Pyruvate/lactate	−190
Glutathione oxid./Glutathione red.	−230
$NAD^+/NADH$	−320

materials and their E_0' values are shown in Table 3.5. The measured E_h will also be influenced by the relative proportions of oxidized and reduced species present. This relationship for a single couple is expressed by the Nernst equation:

$$E_h = E_0' + \frac{RT}{nF} \ln \frac{[\text{Oxidant}][H^+]}{[\text{Reductant}]} \qquad (3.15)$$

where E_h and E_0' are both measured at pH 7; R is the gas constant; T, the absolute temperature; n, the number of electrons transferred in the process and F is the Faraday constant.

Thus, if there is a preponderance of the oxidant over its corresponding reductant, then this will tend to increase the redox potential and the oxidizing nature of the medium.

With the notable exception of oxygen, most of the couples present in foods, *e.g.* glutathione and cysteine in meats, and to a lesser extent, ascorbic acid and reducing sugars in plant products, would on their own tend to establish reducing conditions. From the Nernst equation, it is clear that the hydrogen ion concentration will affect the E_h, and for every unit decrease in the pH the E_h increases by 58 mV. The high positive E_h values registered by fruit juices (see Table 3.6) are largely a reflection of their low pH.

As redox conditions change there will be some resistance to change in a food's redox potential, known as poising. This is analogous to buffering of a medium against pH changes and is, like buffering, a 'capacity'

Table 3.6   *Redox potentials of some food materials*

	E (mV)	pH
Raw meat (post-rigor)	−200	5.7
Raw minced meat	+225	5.9
Cooked sausages and canned meats	−20 to −150	Ca. 6.5
Wheat (whole grain)	−320 to −360	6.0
Barley (ground grain)	+225	7.0
Potato tuber	Ca. −150	Ca. 6.0
Spinach	+74	6.2
Pear	+436	4.2
Grape	+409	3.9
Lemon	+383	2.2

effect dependent on, and increasing with, the concentration of the couple. Also, like buffering, poising is greatest when the two components of a redox couple are present in equal amounts.

Oxygen, which is present in the air at a level of around 21%, is usually the most influential redox couple in food systems. It has a high E_0' and is a powerful oxidizing agent; if sufficient air is present in a food, a high positive potential will result and most other redox couples present will, if allowed to equilibrate, be largely in the oxidized state. Hence the intrinsic factor of redox potential is inextricably linked with the extrinsic factor of storage atmosphere. Increasing the access of air to a food material by chopping, grinding, or mincing will increase its E_h. This can be seen by comparing the values recorded for raw meat and minced meat, and for whole grain and ground grain in Table 3.6. Similarly, exclusion of air as in vacuum packing or canning will reduce the E_h.

Microbial growth in a food reduces its E_h. This is usually ascribed to a combination of oxygen depletion and the production of reducing compounds such as hydrogen by the micro-organisms. Oxygen depletion appears to be the principal mechanism; as the oxygen content of the medium decreases, so the redox potential declines from a value of around 400 mV at air saturation by about 60 mV for each tenfold reduction in the partial pressure of oxygen.

The decrease in E_h as a result of microbial activity is the basis of some long-established rapid tests applied to foods, particularly dairy products. Redox dyes such as methylene blue or resazurin are sometimes used to indicate changes in E_h which are correlated with microbial levels. Methylene blue is also used to determine the proportion of viable cells in the yeast used in brewing. A cell suspension stained with methylene blue is examined under the microscope and viable cells with a reducing cytoplasm appear colourless. Non-viable cells fail to reduce the dye and appear blue.

Redox potential exerts an important elective effect on the microflora of a food. Although microbial growth can occur over a wide spectrum of redox potential, individual micro-organisms are conveniently classified into one of several physiological groups on the basis of the redox range over which they can grow and their response to oxygen.

Obligate or strict aerobes are those organisms that are respiratory, generating most of their energy from oxidative phosphorylation using oxygen as the terminal electron acceptor in the process. Consequently they have a requirement for oxygen and a high E_h and will predominate at food surfaces exposed to air or where air is readily available. For example, pseudomonads, such as *Pseudomonas fluorescens*, which grows at an E_h of +100 to +500 mV, and other oxidative Gram-negative rods produce slime and off-odours at meat surfaces. *Bacillus subtilis* (E_h −100 to +135 mV) produces rope in the open texture of bread and *Acetobacter* species growing on the surface of alcoholic beverages, oxidize ethanol to acetic acid to produce either spoilage or vinegar.

Obligate anaerobes tend only to grow at low or negative redox potentials and often require oxygen to be absent. Anaerobic metabolism gives the organism a lower yield of utilizable energy than aerobic respiration, so a reducing environment that minimizes the loss of valuable reducing power from the microbial cell is favoured. The presence or absence of oxygen can naturally affect this, but for many anaerobes, oxygen exerts a specific toxic effect of its own. For example, it has been observed that *Clostridium acetobutylicum* can grow at an E_h as high as +370 mV maintained by ferricyanide, but would not grow at +110 mV in an aerated culture. This effect is linked to the inability of obligate or aero-intolerant anaerobes to scavenge and destroy toxic products of molecular oxygen such as hydrogen peroxide and, more importantly, the superoxide anion radical ($O_2^{-\bullet}$) produced by a one-electron reduction of molecular oxygen. They lack the enzymes catalase and superoxide dismutase, which catalyse the breakdown of these species as outlined below.

$$2\,O_2^{-\bullet} + 2H \xrightarrow[\text{superoxide dismutase}]{} H_2O_2 + O_2 \qquad (3.16)$$

$$2\,H_2O_2 \xrightarrow[\text{catalase}]{} 2\,H_2O_2 + O_2 \qquad (3.17)$$

Obligate anaerobes, such as clostridia, are of great importance in food microbiology. They have the potential to grow wherever conditions are anaerobic such as deep in meat tissues and stews, in vacuum packs and canned foods causing spoilage and, in the case of *C. botulinum*, the major public health concern: botulism.

Aerotolerant anaerobes are incapable of aerobic respiration, but can nevertheless grow in the presence of air. Many lactic acid bacteria fall into this category; they can only generate energy by fermentation and lack both catalase and superoxide dismutase, but are able to grow in the presence of oxygen because they have a mechanism for destroying superoxide based on the accumulation of millimolar concentrations of manganese.

3.2.4 Antimicrobial Barriers and Constituents

All foods were at some stage part of living organisms and, as such, have been equipped through the course of evolution with ways in which potentially damaging microbial infections might be prevented or at least limited.

The first of these is the integument: a physical barrier to infection such as the skin, shell, husk or rind of a product. It is usually composed of macromolecules relatively resistant to degradation and provides an inhospitable environment for micro-organisms by having a low water activity, a shortage of readily available nutrients and, often, antimicrobial compounds such as short chain fatty acids (on animal skin) or essential oils (on plant surfaces).

The value of these physical barriers can be clearly seen when they are breached in some way. Physical damage to the integument allows microbial invasion of the underlying nutrient-rich tissues and it is a commonplace observation that damaged fruits and vegetables deteriorate more rapidly than entire products, and that this process is initiated at the site of injury. Consequently it is important to the farmer and food processor that harvesting and transport maintain these barriers intact as far as possible.

As a second line of defence, the product tissues may contain antimicrobial components, the local concentration of which often increases as a result of physical damage. In plants, injury can rupture storage cells containing essential oils or may bring together an enzyme and substrate which were separated in the intact tissue. The latter occurs in plants such as mustard, horseradish, watercress, cabbage and other brassicas to produce antimicrobial isothiocyanates (mustard oils) and in *Allium* species (garlic, onions and leeks) to produce thiosulfinates such as allicin (Figure 3.3). A class of antimicrobials known collectively as phytoalexins are produced by many plants in response to microbial invasion, for example the antifungal compound phaseollin produced in green beans.

Many natural constituents of plant tissues such as pigments, alkaloids and resins have antimicrobial properties, but limited practical use is made of these. Benzoic and sorbic acids found in cranberries and mountain ash berries respectively are notable exceptions that are used in their pure forms as food preservatives. Considerable attention has been directed to the antimicrobial properties of those plants used as herbs and spices to flavour food (Table 3.7). Analysis of their volatile flavour and odour fractions, known as essential oils, has frequently

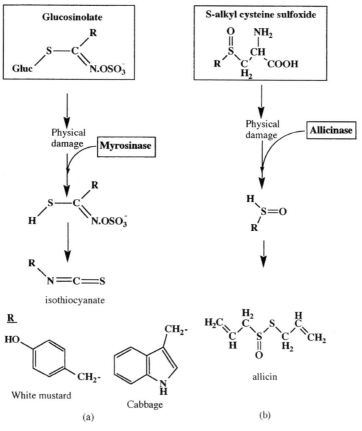

Figure 3.3 *Production of plant antimicrobials as a result of physical damage: (a) isothiocyanate and (b) allicin*

Table 3.7   *Plants used for flavouring food that also possess antimicrobial activity*

Achiote	Dill	Oregano
Allspice (pimento)	Elecampane	Paprika
Almond (bitter)	Fennel	Parsley
Angelica	Garlic	Pennyroyal
Basil (sweet)	Ginger	Pepper
Bay (laurel)	Lemon	Peppermint
Bergamot	Liquorice	Rosemary
Calmus	Lime	Sage
Cananga	Mace	Sassafras
Caraway	Mandarin	Spearmint
Cardamon	Marjoram	Star anise
Celery	Musky bugle	Tarragon (estragon)
Cinnamon	Mustard	Thyme
Citronella	Nutmeg	Turmeric
Clove	Onion	Verbena
Coriander	Orange	Wintergreen

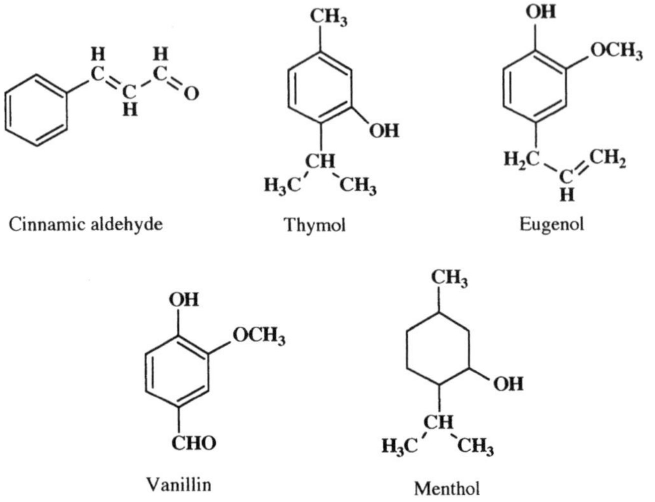

Figure 3.4   *Essential oil components with antimicrobial activity*

identified compounds such as allicin in garlic, eugenol from allspice (pimento), cloves and cinnamon, thymol from thyme and oregano, and cinnamic aldehyde from cinnamon and cassia which have significant antimicrobial activity (Figure 3.4). As a consequence, herbs and spices may contribute to the microbiological stability of foods in which they are used. It has, for example, been claimed that inclusion of cinnamon in raisin bread retards mould spoilage. Usually, however, their role in preservation is likely to be minor and, in some cases, they can be a source of microbial contamination leading to spoilage or public health problems. Outbreaks of botulism associated with crushed garlic in oil and home canned peppers demonstrate that even in relatively high concentrations plant antimicrobials are not a complete guarantee of safety.

Antimicrobial components differ in their spectrum of activity and potency, they are present at varying concentrations in the natural product, and are frequently at levels too low to have any effect. Hops and their extracts are ubiquitous ingredients in beer. Humulones contained in the hop resin and isomers produced during processing, impart the characteristic bitterness of the product but have also been shown to possess activity against the common beer spoilage organisms, lactic acid bacteria. When first introduced into brewing, hops probably contributed to microbiological stability, but this is less likely nowadays with the relatively low hopping rates used. In fact the ability of lactic acid bacteria to acquire resistance to hop resins means that the brewery environment probably acts as a very efficient natural enrichment culture for humulone-tolerant bacteria, thus negating any beneficial effects.

A rather different example of the importance of plant antimicrobials is provided by oleuropein, the bitter principle of green olives. In the production of Spanish-style green olives, it is removed by an alkali extraction process, primarily for reasons of flavour. However oleuropein and its aglycone are also thought to be inhibitory to lactic acid bacteria; if not removed at this early stage, they would prevent the necessary fermentation occurring subsequently.

Animal products too, have a range of non-specific antimicrobial constituents. Probably the supreme example of this is the white or albumen of the hen's egg which possesses a whole battery of inhibitory components. Many of the same or similar factors can also be found in milk where they are present in lower concentrations and are thus less effective.

Both products contain the enzyme lysozyme which catalyses the hydrolysis of glycosidic linkages in peptidoglycan, the structural polymer responsible for the strength and rigidity of the bacterial cell wall. Destruction or weakening of this layer causes the cell to rupture (lyse) under osmotic pressure. Lysozyme is most active against Gram-positive bacteria, where the peptidoglycan is more readily accessible, but it can also kill Gram-negatives if their protective outer membrane is damaged in some way.

Other components limit microbial growth by restricting the availability of key nutrients. Ovotransferrin in egg white and lactoferrin in milk are proteins that scavenge iron from the medium. Iron is an essential nutrient for all bacteria and many have evolved means of overcoming iron limitation by producing their own iron-binding compounds known as siderophores.

In addition, egg white has powerful cofactor-binding proteins such as avidin and ovoflavoprotein which sequester biotin and riboflavin restricting the growth of those bacteria for which they are essential nutrients, see Table 3.8.

Table 3.8 *Antimicrobials in hen's egg albumen and milk*

Albumen	Milk
Nutrient Status	
High pH	Moderate pH
Low levels of available nitrogen	High levels of protein, carbohydrate and fat
Antimicrobials	
Ovotransferrin (conalbumin) (12% of solids)	Lactoferrin
Lysozyme (3.5% of solids)	Lysozyme
Avidin – (0.05% of solids)	–
Ovoflavoprotein –(0.8% of solids)	–
Ovomucoid & ovoinhibitor – (protease inhibitors) (11% of solids)	–
	Lactoperoxidase $(30\,\mathrm{mg\,l^{-1}})$
	Immunoglobulin $(300\,\mathrm{mg\,l^{-1}})$

Milk also has the capacity to generate antimicrobials in the presence of hydrogen peroxide. The enzyme lactoperoxidase constitutes about 0.5% of whey proteins and catalyses the oxidation of thiocyanate by hydrogen peroxide. Thiocyanate is naturally present in milk and its level can be boosted by consumption of brassicas which are rich in thiocyanate precursors. Hydrogen peroxide can be generated by endogenous enzyme activity or by the aerobic metabolism of lactic acid bacteria. The reaction produces short lived oxidation products such as hypothiocyanate which can kill Gram negative bacteria and inhibit Gram positives, possibly by damaging the bacterial cytoplasmic membrane (Figure 3.5).

3.2.5　Water Activity

Water is a remarkable compound. Considered as a hydride of oxygen (H_2O) it has quite exceptional properties when compared with the hydrides of neighbouring elements in the periodic table such as ammonia (NH_3), methane (CH_4), hydrogen sulfide (H_2S), and hydrofluoric acid (HF), see Table 3.9. Life as we know it is totally dependent on the presence of water in its liquid state. The reactions which take place in the cytoplasm do so in an aqueous environment and the cytoplasm is surrounded by a membrane which is generally permeable to water molecules which may pass freely from the cytoplasm to the environment and from the environment to the cytoplasm. This dynamic two way flow

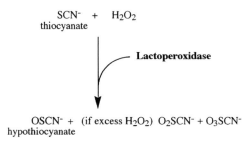

Figure 3.5　*The lactoperoxidase system*

Table 3.9　*The boiling points ($^{\circ}C$) of hydrides of elements surrounding oxygen in the periodic table*

B_2H_6	CH_4	NH_3	H_2O	HF
−92.5	−161.7	−33	100	19.4
	SiH_4	PH_3	H_2S	HCl
	−112	−87	−60	−85
			H_2Se	
			−42	

of water molecules is normally in a steady state and a living organism will only be stressed if there is a net flow out of the cytoplasm, leading to plasmolysis, or a net flow into the cell leading to rupture of the membrane, and the latter is normally prevented by the presence of a cell wall in the bacteria and fungi.

In our everyday lives we think of water as existing in its liquid state between its freezing point (0 °C) and boiling point (100 °C) and we might expect that this would limit the minimum and maximum temperatures at which growth could possibly occur. But, of course, the freezing point of water can be depressed by the presence of solutes and there are a number of micro-organisms which can actively grow at subzero temperatures because their cytoplasm contains one or more compounds, such as a polyol, which act as an antifreeze. Similarly the boiling point of water can be elevated by increased hydrostatic pressure and, in nature, very high pressures exist at the bottom of the deep oceans. Under these circumstances the temperature of liquid water may be well above 100 °C and the relatively recent exploration of submarine volcanic vents has uncovered some remarkable bacteria which can indeed grow at such high temperatures.

Although the cytoplasm must be in the liquid phase for active growth (and it is important not to confuse growth and survival, for many microorganisms can survive but not grow when their cytoplasm has been completely dried), water in the environment of the living organism may be present, not only in the liquid phase as pure water or a solution, but also in the atmosphere in the gaseous phase, or associated with what would be described macroscopically as the solid phase (Figure 3.6).

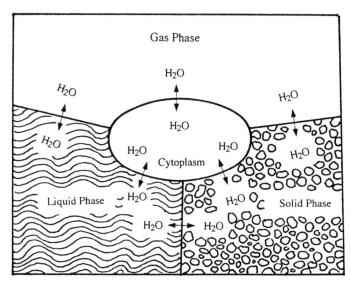

Figure 3.6 *A cell in equilibrium with liquid, solid and gaseous phases, each of these being in equilibrium with each other*

A useful parameter which helps us to understand the movement of water from the environment to the cytoplasm or from the cytoplasm to the environment is water activity, a_w. The water activity of a substrate is most conveniently defined as the ratio of the partial pressure of water in the atmosphere in equilibrium with the substrate, P, compared with the partial pressure of the atmosphere in equilibrium with pure water at the same temperature, P_0. This is numerically equal to the equilibrium relative humidity (ERH) expressed as a fraction rather than as a percentage:

$$a_w = \frac{P}{P_0} = \frac{1}{100} \text{ERH} \tag{3.18}$$

This has important implications for the storage of low a_w foods (see Section 3.3.1).

In 1886 François Marie Raoult described the behaviour of an ideal solution by an equation which has since then been known as Raoult's law:

$$P_A = X_A P_{A_0} \tag{3.19}$$

where P_A is the partial vapour pressure of A above a solution, in which X_A is the mole fraction of the solvent A, and PA_0 is the vapour pressure of pure liquid A at the same temperature. If the solvent A is water then Equations (3.18) and (3.19) can be combined to give:

$$a_w = X_{\text{water}} \tag{3.20}$$

Thus for an aqueous solution the water activity is approximately given by the ratio of the number of moles of water to the total number of moles (*i.e.* water + solute), *i.e.*:

$$a_w = N_w/(N_w + N_s) \tag{3.21}$$

It should be noted that water activity is a colligative property, that is to say it depends on the number of molecules or ions present in solution, rather than their size. Thus a compound like sodium chloride, which dissociates into two ions in solution, is more effective at reducing the water activity than a compound like sucrose on a mole-to-mole basis.

Physical chemists would prefer to work with the chemical potential of water (μ_w), which is a complex parameter made up of a reference state, a water activity term, a pressure term and a gravitational term:

$$\mu_w = \mu_w^* + RT \ln a_w + V_m P + mgh \tag{3.22}$$

which can be rearranged to give a new parameter, ψ, known as the water potential having the same dimensions as pressure:

$$\psi = \frac{\mu_w - \mu_w^*}{V_m} = \frac{RT}{V_m} \ln a_w + P + \frac{mgh}{V_m} \tag{3.23}$$

For situations associated with everyday life on the surface of the Earth it is possible to ignore the pressure and gravity terms and a good approximation of the relationship between the water potential and water activity is given by Equation (3.24):

$$\psi = \frac{RT}{V_m}\ln a_w \qquad (3.24)$$

where R (the gas constant) $= 0.08205\,\text{dm}^3\,\text{atm}\,\text{K}^{-1}\,\text{mol}^{-1}$; and V_m (the molar volume of water) $= 0.018\,\text{dm}^3\,\text{mol}^{-1}$.

Thus at 25 °C (298 °K) a water activity of 0.9 would correspond to a water potential of -143 atm or -14.5 MPa.

Water potential may contain both an osmotic component, associated with the effect of solutes in solution, and a matric component, associated with the interaction of water molecules with surfaces, which can be clearly demonstrated by the rise of water in a capillary tube. The latter might be particularly important in discussions about the availability of water in a complex matrix such as cake.

A parameter related to water activity is osmotic pressure which can be thought of as the force per unit area required to stop the net flow of water molecules from a region of high to one of low water activity. Cytoplasm is an aqueous solution and so must have a lower water activity than pure water; thus a micro-organism in an environment of pure water will experience a net flow of water molecules into the cytoplasm. If it cannot control this it will increase in size and burst. Bacteria, fungi and algae cope by having a rigid strong wall capable of withstanding the osmotic pressure of the cytoplasm which may be as high as 30 atm (*ca.* 3 MPa) in a Gram-positive bacterium or as little as 5 atm (*ca.* 0.5 MPa) in a Gram-negative species. Freshwater protozoa, on the other hand, cope with the net flow of water into the cell by actively excreting it out again with a contractile vacuole.

As water activity is decreased, or osmotic pressure is increased, in the environment it is essential that the water activity of the cytoplasm is even lower, or its osmotic pressure even higher. This is achieved by the production of increasing concentrations of solutes which must not interfere with cytoplasmic function. They are thus known as compatible solutes and include such compounds as the polyols glycerol, arabitol and mannitol in the fungi and amino acids or amino acid derivatives in the bacteria.

With a reduction of water activity in their environment the number of groups of micro-organisms capable of active growth decreases (Table 3.10). The exact range of water activities allowing growth is influenced by other physico-chemical and nutritional conditions but Figure 3.7 illustrates the range for a number of individual species of micro-organisms

Table 3.10   *Minimum water activities at which active growth can occur*

Group of micro-organism	Minimum a_w
Most Gram-negative bacteria	0.97
Most Gram-positive bacteria	0.90
Most yeasts	0.88
Most filamentous fungi	0.80
Halophilic bacteria	0.75
Xerophilic fungi	0.61

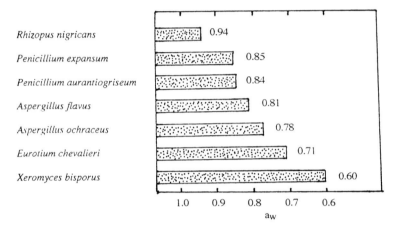

Figure 3.7   *Range of a_w values allowing growth of a number of species of micro-organisms*

and Figure 3.8 demonstrates the interaction between temperature and water activity for *Aspergillus flavus* and *Penicillium expansum*.

Figure 3.9 shows the range of a_w values associated with a number of different food commodities. Because low water activities are associated with three distinct types of food three terms are used to describe the micro-organisms especially associated with these foods:

(i) *halotolerant* – able to grow in the presence of high concentrations of salt
(ii) *osmotolerant* – able to grow in the presence of high concentrations of unionized organic compounds such as sugars
(iii) *xerotolerant* – able to grow on dry foods.

These terms do not describe rigidly exclusive groups of micro-organisms but are useful in the context of studies of particular food commodities. Some micro-organisms actually grow better at reduced a_w and may be described as halophilic, osmophilic or xerophilic, indeed the halobacteria are obligately halophilic and cannot grow in the absence of high

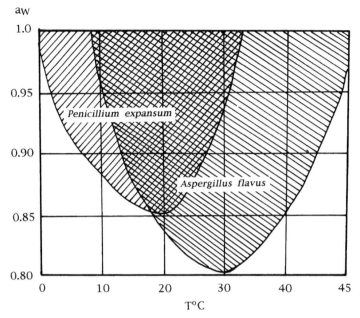

Figure 3.8 *Temperature water activity combinations allowing the growth of* Aspergillus flavus *and* Penicillium expansum

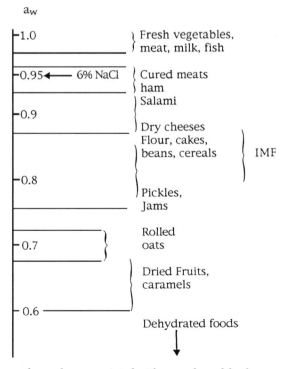

Figure 3.9 *Range of a_w values associated with a number of food commodities*

concentrations of salt. This group of bacteria, which includes such genera as *Halobacterium* and *Halococcus*, belong to the Archaebacteria and accumulate potassium chloride as their compatible solute. They are obligately halophilic because the integrity of their outer wall depends on a high concentration of sodium chloride in their environment. They are usually associated with salt lakes or salt pans where solar salt is being made and may cause the proteolytic spoilage of dried, salted fish.

The limiting value of water activity for the growth of any micro-organism is about 0.6 and below this value the spoilage of foods is not microbiological but may be due to insect damage or chemical reactions such as oxidation. At a water activity of 0.6, corresponding to a water potential of -68 MPa, the cytoplasm would need to contain very high concentrations of an appropriate compatible solute and it is probable that macromolecules such as DNA would no longer function properly and active growth must cease. However, it is important to note that, even if active growth is impossible, survival may still occur and many micro-organisms can survive at very low water activities and are frequently stored in culture collections in this form.

It is a relatively simple matter to determine the water content of a food commodity by drying to constant weight under defined conditions. The water content, however, may not give a good indication of how available that water is, *i.e.* what the water activity is, unless the relationship between these two properties has been established. Thus, oil-rich nuts with a water content of 4–9%, protein rich legumes with 9–13% water content and sucrose rich dried fruits with a water content of 18–25% could all have the same water activity of about 0.7 and would thus be acceptably stable to spoilage by most micro-organisms.

The relationship between water activity and water content is very sensitive to temperature and may seem to depend on whether water is being added or removed from a substrate. An example of a water sorption isotherm is shown in Figure 3.10. In this example the material has been allowed to equilibrate effectively at a known water activity before measuring the water content but Figure 3.11 demonstrates the differences which may be observed depending on whether a given water content is achieved by adding water to a dry commodity or removing it from a wet commodity. The same water content seems to be associated with a higher a_w in the former case than in the latter. This hysteresis phenomenon is a reflection of the long time that it may take for water to equilibrate with the constituents of a complex food matrix.

The measurement of water activity can thus be achieved by measuring the water content if the shape of the isotherm has been determined. Water activity can be measured by measuring the equilibrium relative humidity of the atmosphere in contact with the sample. This can be done by the dew point method or with a hair hygrometer. There are a number

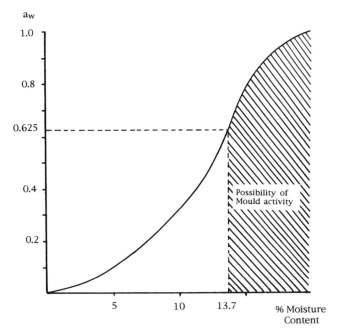

Figure 3.10 *Water sorption isotherm for wheat at 25 °C*

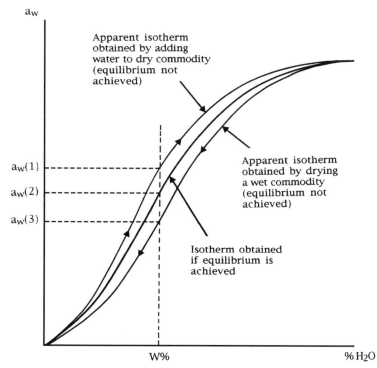

Figure 3.11 *Hysteresis associated with the relationship between apparent water activity and water content*

of instruments which measure relative humidity through its effect on the electrical properties, such as conductivity or resistivity, of materials. Thus the resistance of lithium chloride, or the capacitance of anodized aluminium, changes with changes of relative humidity.

A method known as the Landrock–Proctor method depends on gravimetrically measuring changes in water content of samples of the material after equilibration with atmospheres of known relative humidity which can be obtained using saturated solutions of a number of inorganic salts. If the sample has a lower a_w than the atmosphere then it will gain weight, if it has a higher a_w then it will lose weight. By carrying out measurements of weight change over a range of relative humidities it is possible to extrapolate to the relative humidity which would cause no weight loss and thus corresponds to the a_w of the sample. Figure 3.12 shows the result of such an experiment with samples of Madeira cake and Table 3.11 shows the water activities of a variety of saturated salt

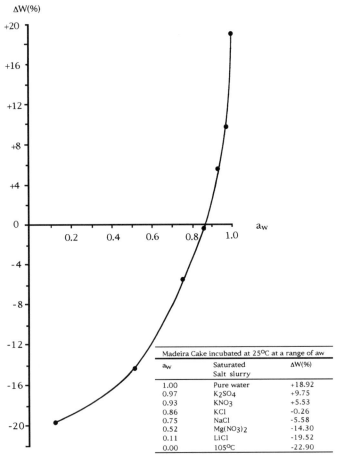

Madeira Cake incubated at 25°C at a range of aw		
a_w	Saturated Salt slurry	$\Delta W(\%)$
1.00	Pure water	+18.92
0.97	K_2SO_4	+9.75
0.93	KNO_3	+5.53
0.86	KCl	-0.26
0.75	NaCl	-5.58
0.52	$Mg(NO_3)_2$	-14.30
0.11	LiCl	-19.52
0.00	105°C	-22.90

Figure 3.12 *Weight changes of samples of madeira cake at different ERH values*

Table 3.11 *Water activities of saturated salts solution at 25 °C*

Salt	a_w
Lithium chloride	0.11
Zinc nitrate	0.31
Magnesium chloride	0.33
Potassium carbonate	0.43
Magnesium nitrate	0.52
Sodium bromide	0.57
Lithium acetate	0.68
Sodium chloride	0.75
Potassium chloride	0.86
Potassium nitrate	0.93
Pure water	1.00

solutions at 25 °C. Some of these salt solutions have large temperature coefficients and so the temperature needs to be very carefully controlled.

3.3 EXTRINSIC FACTORS (ENVIRONMENTAL LIMITATIONS)

3.3.1 Relative Humidity

It has already been seen in Section 3.2.5. that relative humidity and water activity are interrelated, thus relative humidity is essentially a measure of the water activity of the gas phase. When food commodities having a low water activity are stored in an atmosphere of high relative humidity water will transfer from the gas phase to the food. It may take a very long time for the bulk of the commodity to increase in water activity, but condensation may occur on surfaces giving rise to localized regions of high water activity. It is in such regions that propagules which have remained viable, but unable to grow, may now germinate and grow. Once micro-organisms have started to grow and become physiologically active they usually produce water as an end product of respiration. Thus they increase the water activity of their own immediate environment so that eventually micro-organisms requiring a high a_w are able to grow and spoil a food which was initially considered to be microbiologically stable.

Such a situation can occur in grain silos or in tanks in which concentrates and syrups are stored. Another problem in large-scale storage units such as grain silos occurs because the relative humidity of air is very sensitive to temperature. If one side of a silo heats up during the day due to exposure to the sun then the relative humidity on that side is reduced and there is a net migration of water molecules from the cooler side to re-equilibrate the relative humidity. When that same side cools down again the relative humidity increases and, although water molecules migrate back again, the temporary increase in relative humidity may be sufficient to cause local condensation onto the grain with a localized increase in a_w sufficient to allow germination of fungal spores and subsequent spoilage of the grain. This type of phenomenon can often account for localized caking of grain which had apparently been stored at a 'safe' water content.

The storage of fresh fruit and vegetables requires very careful control of relative humidity. If it is too low then many vegetables will lose water and become flaccid. If it is too high then condensation may occur and microbial spoilage may be initiated.

3.3.2   Temperature

Microbial growth can occur over a temperature range from about $-8\,^{\circ}\text{C}$ up to $100\,^{\circ}\text{C}$ at atmospheric pressure. The most important requirement is that water should be present in the liquid state and thus available to support growth (see Section 3.2.5). No single organism is capable of growth over the whole of this range; bacteria are normally limited to a temperature span of around $35\,^{\circ}\text{C}$ and moulds rather less, about $30\,^{\circ}\text{C}$.

A graph showing the variation of growth rate with temperature illustrates several important features of this relationship (Figure 3.13). Firstly, each organism exhibits a minimum, optimum and maximum temperature at which growth can occur. These are known as cardinal temperatures and are, to a large extent, characteristic of an organism, although they are influenced by other environmental factors such as nutrient availability,

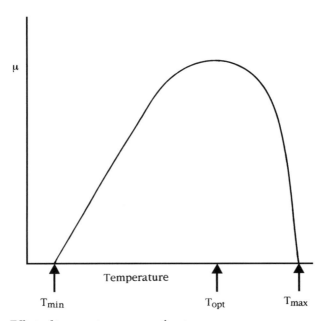

Figure 3.13   *Effect of temperature on growth rate*

Table 3.12 *Cardinal temperatures for microbial growth*

Group	Temperature (°C)		
	Minimum	*Optimum*	*Maximum*
Thermophiles	40–45	55–75	60–90
Mesophiles	5–15	30–40	40–47
Psychrophiles (obligate psychrophiles)	−5 to +5	12–15	15–20
Psychrotrophs (facultative psychrophiles)	−5 to +5	25–30	30–35

(Adapted from ICMSF 1980)

pH and a_w. Micro-organisms can be classified into several physiological groups based on their cardinal temperatures. This is a useful, if rather arbitrary, convention, since the distribution of micro-organisms through the growth temperature range is continuous. To take account of this and the effect of other factors, it is more appropriate to define cardinal temperatures as ranges rather than single values (Table 3.12).

In food microbiology mesophilic and psychrotrophic organisms are generally of greatest importance. Mesophiles, with temperature optima around 37 °C, are frequently of human or animal origin and include many of the more common foodborne pathogens such as *Salmonella*, *Staphylococcus aureus* and *Clostridium perfringens*.

As a rule mesophiles grow more quickly at their optima than psychrotrophs and so spoilage of perishable products stored in the mesophilic growth range is more rapid than spoilage under chill conditions. Because of the different groups of organisms involved, it can also be different in character.

Among the organisms capable of growth at low temperatures, two groups can be distinguished: the true or strict psychrophiles ('cold loving') have optima of 12–15 °C and will not grow above about 20 °C. As a result of this sensitivity to quite moderate temperatures, psychrophiles are largely confined to polar regions and the marine environment. Psychrotrophs or facultative psychrophiles will grow down to the same low temperatures as strict psychrophiles but have higher optimum and maximum growth temperatures. This tolerance of a wider range of temperatures means that psychrotrophs are found in a more diverse range of habitats and consequently are of greater importance in the spoilage of chilled foods.

Thermophiles are generally of far less importance in food microbiology, although thermophilic spore formers such as certain *Bacillus* and *Clostridium* species do pose problems in a restricted number of situations (see Chapter 4).

Another feature evident from Figure 3.13 is that the curve is asymmetric – growth declines more rapidly above the optimum temperature than below it. As the temperature is decreased from the optimum the growth rate slows, partly as a result of the slowing of enzymic reactions within the cell. If this were the complete explanation however, then the change in growth rate with temperature below the optimum might be expected to follow the Arrhenius Law which describes the relationship between the rate of a chemical reaction and the temperature. The fact that this is not observed in practice is, on reflection, hardly surprising since microbial growth results from the activity of a network of inter-acting and interregulating reactions and represents a far higher order of complexity than simple individual reactions.

A most important contribution to the slowing and eventual cessation of microbial growth at low temperatures is now considered to be changes in membrane structure that affect the uptake and supply of nutrients to enzyme systems within the cell. It has been shown that many micro-organisms respond to growth at lower temperatures by increasing the proportion of unsaturated and/or shorter chain fatty acids in their membranes and that psychrotrophs generally have higher levels of these acids than mesophiles. Increasing the degree of unsaturation or decreas-ing the carbon chain length of a fatty acid decreases its melting point so that membranes containing these will remain fluid and hence functional at lower temperatures.

As the temperature increases above the optimum, the growth rate declines much more sharply as a result of the irreversible denaturation of proteins and the thermal breakdown of the cell's plasma membrane. At temperatures above the maximum for growth, these changes are suffi-cient to kill the organism – the rate at which this occurs increasing with increasing temperature. The kinetics of this process and its importance in food preservation are discussed in Chapter 4.

3.3.3 Gaseous Atmosphere

Oxygen comprises 21% of the earth's atmosphere and is the most important gas in contact with food under normal circumstances. Its presence and its influence on redox potential are important determinants of the microbial associations that develop and their rate of growth. Since this topic has already been discussed in some detail under redox potential (Section 3.2.3), this section will be confined to the microbiological effects of other gases commonly encountered in food processing.

The inhibitory effect of carbon dioxide (CO_2) on microbial growth is applied in modified-atmosphere packing of food and is an advantageous consequence of its use at elevated pressures (hyperbaric) in carbonated mineral waters and soft drinks.

Carbon dioxide is not uniform in its effect on micro-organisms. Moulds and oxidative Gram-negative bacteria are most sensitive and the Gram-positive bacteria, particularly the lactobacilli, tend to be most resistant. Some yeasts such as *Brettanomyces* spp. also show considerable tolerance of high CO_2 levels and dominate the spoilage microflora of carbonated beverages. Growth inhibition is usually greater under aerobic conditions than anaerobic and the inhibitory effect increases with decrease of temperature, presumably due to the increased solubility of CO_2 at lower temperatures. Some micro-organisms are killed by prolonged exposure to CO_2 but usually its effect is bacteriostatic.

The mechanism of CO_2 inhibition is a combination of several processes whose precise individual contributions are yet to be determined. One factor often identified is the effect of CO_2 on pH. Carbon dioxide dissolves in water to produce carbonic acid which partially dissociates into bicarbonate anions and protons. Carbonic acid is a weak dibasic acid (pK_a 6.37 and 10.25); in an unbuffered solution it can produce an appreciable drop in pH, distilled water in equilibrium with the CO_2 in the normal atmosphere will have a pH of about 5, but the effect will be less pronounced in buffered food media so that equilibration of milk with 1 atmosphere pCO_2 decreased the pH from 6.6 to 6.0. Probably of more importance than its effect on the growth medium is the ability of CO_2 to act in the same way as weak organic acids (see Section 3.2.2), penetrating the plasma membrane and acidifying the cell's interior.

Other contributory factors are thought to include changes in the physical properties of the plasma membrane adversely affecting solute transport; inhibition of key enzymes, particularly those involving carboxylation /decarboxylation reactions in which CO_2 is a reactant; and reaction with protein amino groups causing changes in their properties and activity.

3.4 IMPLICIT FACTORS

A third set of factors that are important in determining the nature of microbial associations found in foods are described as *implicit factors* – properties of the organisms themselves, how they respond to their environment and interact with one another.

At its simplest, an organism's specific growth rate can determine its importance in a food's microflora; those with the highest specific growth rate are likely to dominate over time. This will of course depend upon the conditions prevailing; many moulds can grow perfectly well on fresh foods such as meat, but they grow more slowly than bacteria and are therefore out-competed. In foods where the faster growing bacteria are inhibited by factors such as reduced pH or a_w, moulds assume an important role in spoilage. Alternatively, two organisms may have similar maximum specific growth rates but differ in their affinity (K_s) for a growth limiting substrate (see Equation 3.8). If the level of that substrate is sufficiently low that it becomes limiting, then the organism with the lower K_s (higher affinity) will outgrow the other.

In Sections 3.2 and 3.3 we described how microbial growth and survival are influenced by a number of factors and how micro-organ-isms respond to changes in some of these. This response does however depend on the physiological state of the organism. Exponential phase cells are almost always killed more easily by heat, low pH or antimicrobials than stationary phase cells and often the faster their growth rate the more readily they are killed. This makes sense intuitively; the consequences of a car crash are invariably more serious the faster the car is travelling at the time. At higher growth rates, where cell activity is greater and more finely balanced, the damage caused by a slight jolt to the system will be more severe than the same perturbation in cells growing very slowly or not at all. The precise mechanism leading to cell death is almost certainly very complex. One proposal is that lethal damage is largely a result of an oxidative burst, the production of large numbers of damaging free radicals within the cell in response to the physical or chemical stress applied. This would mean that cell death is in fact a function of the organism's response to a stress rather than a direct effect of the stress itself.

A cell's sensitivity to potentially lethal treatments can also be affected by its previous history. Generally, some form of pre-adaptation will decrease the damaging effect of adverse conditions. Growth or holding organisms such as *Salmonella* at higher temperatures has been shown to increase their heat resistance. Pre-exposure to moderately low pH can increase an organism's subsequent resistance to a more severe acid challenge. Growth at progressively lower temperatures can reduce the minimum temperature at which an organism would otherwise grow.

Some reaction to stress can be apparent very soon after exposure as existing enzymes and membrane proteins sense and react to the change. Other responses occur more slowly since they involve gene transcription and the production of proteins. The most extensively studied of this type of response is the production of heat shock proteins; proteins produced following exposure to elevated temperatures and which protect the cell from heat damage. Some heat shock proteins, described as chaperones or chaperonins, interact with unfolded or partially unfolded proteins and assist them in reaching their correct conformation. Chaperonins are present in normal cells but obviously far more will be needed during processes such as heating which increase the rate at which cellular proteins denature.

Heat shock proteins are encoded by genes which have a specific sigma factor, sigma 32 also known as RpoH, for transcription. Sigma factors are proteins which bind to DNA-dependent RNA polymerase, the enzyme which transcribes DNA into messenger RNA. When bound to the polymerase they confer specificity for certain classes of promoter on the DNA and thus help determine which regions of the genome are transcribed. Another alternative sigma factor RpoS, also known as the stationary phase sigma factor, has been identified in a number of Gram-negative bacteria and a similar regulon sigma B operates in Gram-positive bacteria. RpoS is produced in cells throughout growth but is rapidly degraded in exponential phase cells. As growth slows at the end of exponential phase it accumulates and directs the transcription of a battery of genes associated with the stationary phase, many of which are protective.

It is now clear that RpoS is a general stress response regulator and also accumulates in response to environmental stresses such as low pH and osmotic stress. Since the RpoS response confers resistance to a range of stresses, exposure to one factor such as low pH can confer increased resistance to other stresses such as heat. Of equal concern is the observation that RpoS also plays a role in regulating expression of genes associated with virulence in some food borne pathogens and that virulence factors expressed as the cells enter stationary phase can also be induced by stress. The implications of this for food microbiology are considerable, for not only do they suggest that stresses micro-organisms encounter during food processing may increase resistance to other stresses, but that they could also increase the virulence of any pathogens present.

Until now we have dealt with micro-organisms largely as isolated individuals and have not considered any effects they might have on each other. Cell to cell communication has however been shown to play a part in the induction of stress responses. Molecules such as acylhomoserine lactones and proteins secreted by cells in response to a stress have been shown to produce a stress response in others, implying that cells in the vicinity which have not necessarily been directly exposed to the stress may also increase in resistance.

Ecologists have identified a number of different ways in which organisms can interact and several of these can be seen in the microbial ecology of food systems. Mutualism, when growth of one organism stimulates the growth of another, is well illustrated by the interaction of the starter cultures in yoghurt fermentation (see Section 9.5.1). Similar stimulatory effects can be seen in spoilage associations or in sequences of spoilage organisms seen when growth of one organism paves the way for others. For example, a grain's water activity may be sufficiently low to prevent the growth of all but a few fungi, once these begin to grow however water produced by their respiration increases the local water activity allowing less xerophilic moulds to take over. Alternatively, one organism might increase the availability of nutrients to others by degrading a food component such as starch or protein into more readily assimilable compounds. Some micro-organisms may remove an inhibitory component and thereby permit the growth of others. This last example has had safety implications in mould-ripened cheeses where mould growth increases the pH allowing less acid tolerant organisms such as *Listeria monocytogenes* to grow.

Alternatively, micro-organisms may be antagonistic towards one another producing inhibitory compounds or sequestering essential nutrients such as iron. The best practical examples of this in food microbiology are the lactic fermented foods which are discussed in some detail in Chapter 9.

3.5 PREDICTIVE FOOD MICROBIOLOGY

Understanding how different properties of a food, its environment and its history can influence the microflora that develops on storage is an important first step towards being able to make predictions concerning shelf-life, spoilage and safety. The food industry is continually creating new microbial habitats, either by design in developing new products and reformulating traditional ones, or by chance, as a result of variations in the composition of raw materials or in a production process. To be able to predict microbial behaviour in each new situation and determine its consequences for food safety and quality, we must first describe accurately the food environment and then determine how this will affect microbial growth and survival.

Characterization of a habitat in terms of its chemical and physical properties is generally straightforward, although problems can arise if a property is not uniformly distributed throughout the product. This can be a particular problem with solid foods, for example the local salt concentration may vary considerably within a ham or a block of cheese and we have seen in Section 3.3.1 how water can migrate through a mass of food.

A considerable amount of data is available on how factors such as pH, a_w, and temperature affect the growth and survival of microorganisms and some of this was described in Sections 3.2 and 3.3. Much of this information was however acquired when only one or two factors had been changed and all the others were optimal or near-optimal. In many foods a very different situation applies, micro-organisms experience a whole battery of sub-optimal factors which collectively determine the food's characteristics as a medium for microbial growth.

Leistner described this situation as the 'hurdle effect', where each inhibitory factor can be visualized as a hurdle contributing to a food's overall stability and safety (Figure 3.14). The analogy of a hurdle race, while vivid, has been criticized on the grounds that it can lead to the misapprehension that micro-organisms confront each hurdle in turn. In reality they are usually faced with the aggregate effect of all the barriers at once and it is this which determines whether the organisms can grow

Figure 3.14   *The hurdle effect*

or how fast they can grow. Perhaps a better way to visualise it is as a protective wall; each adverse condition contributes one or more layers of bricks to the wall. The overall height of the wall will determine which organisms are able to climb it and how fast they will grow once they have done so. Scientific description of this multi-factorial technique for preserving foods may be relatively recent but the concept has been applied empirically since antiquity in numerous traditional products such as cheese, cured meats, smoked fish and fruit preserves, all of which rely on a number of contributing factors for their stability and safety.

When confronted with this situation there are three basic approaches to predicting the fate of particular organisms. The first is to seek an expert judgement, based on the individual expertise of a food microbiologist and their interpretation of the published literature. While this can be useful qualitatively, it rarely provides reliable quantitative data.

To its credit, the food industry has generally not placed too great a reliance on this sort of approach but has resorted to the challenge trial. In this, the organism of concern is inoculated into the food material and its fate followed through simulated conditions of processing, storage, distribution, temperature abuse, or whatever is required. Though it provides reliable data, the challenge trial is expensive, time consuming and labour intensive to perform properly. It also has extremely limited predictive value since its predictions hold only for the precise set of conditions tested. Any change in formulation or conditions of processing or storage will invalidate the predictions and necessitate a fresh challenge trial under the new set of conditions.

The third and increasingly popular approach is the use of mathematical models. A model is simply an object or concept that is used to represent something else and a mathematical model is one constructed using mathematical concepts such as constants, variables, functions, equations, *etc.*

Mathematical models are not entirely new to food microbiology having been used with great success since the 1920s for predicting the probability of *Clostridium botulinum* spores surviving a particular heat process and enabling the design of heat processes for low acid canned foods with an acceptable safety margin (see Section 4.1.5).

The log-linear *C. botulinum* model is an inactivation model, describing microbial survival, but models predicting the potential for microbial growth to occur under a range of conditions can also be constructed. These are generally more complex but their development has been facilitated by the availability and accessability of powerful modern computers.

There are four essential steps in developing a model.

(1) *Planning.* This requires a clear definition of the problem:

> (i) are we interested in spoilage or safety, which organisms are our main concern?
> (ii) what is the appropriate response or dependent variable, *e.g.* growth rate, toxin production, time to spoilage?
> (iii) what are the relevant explanatory or independent variables, *e.g.* temperature, pH, a_w?

(2) *Data collection.* The response variable identified in the planning stage is measured for various levels of the explanatory variables. These should cover the full range in which we may be interested since the predictive value of the model is limited to situations where unknown values can be interpolated. Extrapolation into areas where there are no data points will not yield valid predictions.

(3) *Model fitting.* Different models which relate the response variable to the explanatory variables are tested to see how well they fit the experimental data.

(4) *Model validation.* The model is evaluated using experimental data not used in building the model.

A number of different types of model are commonly used. Probabilistic models give a quantitative assessment of the chance that a particular microbiological event will occur within a given time and are most suited to situations where the hazard is severe. The event most often described in such models is the probability of toxin formation (*i.e.* growth) by *C. botulinum*. The work was initially prompted by the perceived need to reduce nitrite levels in cured meats such as hams and to assess quantitatively the relative importance of factors contributing to their safety.

In the original work, the probability of toxin production, p, (the proportion of samples containing toxin within each treatment combination) was fitted to a logistic model to describe the relationship between the probability of toxin production and the level of factors/variables present (Figure 3.15). Any factor which tends to decrease μ in Figure 3.15 reduces the probability of toxin production. Of the different factors included in the model, it can be seen that nitrite, incubation temperature and salt are more important in preventing toxin production than the others and that they are acting independently; there is no evidence of synergistic interactions between them.

Here the logistic equation is being used simply as a regression equation, a common practice in modelling situations where there are two possible outcomes to an event, *e.g.* pass/fail, toxin production/no toxin production. Its use in this context should not be confused with its use to represent the microbial growth curve (Section 3.1).

Probability of toxin production (P)

$$P = 1/(1 + e^{-\mu})$$

---

where $\mu =$

4.679

- (1.47 x N)	where $N = \text{NaNO}_2$, µg/g x 10^2
- (1.104 x S)	where $S = \text{NaCl}$, % w/v on water
+ (0.1299 x T)	where $T = $ storage temperature, °C
- 2.09 + (0.67 x N)	if 500 µg/g nitrate added
- 6.238 + (0.8264 x S)	if 1000 µg/g isoascorbate added
- 1.7049 +(0.3987 x N)	if heat treatment is high (80°C/7 min + 70°C/60 min)
- (0.01937 x N x T) - 1.2824	if nitrate and polyphosphate added
+ 0.99	if nitrate added and heat treatment high

Figure 3.15 *Logistic model for probability of toxin production by* Clostridium botulinum *types A and B in pasteurized pork slurry in the pH range 5.5 to 6.3*

One disadvantage of probabilistic models is that they do not give us much information about the rate at which changes occur. Models that predict times to a particular event such as growth to a certain level or detectable toxin production are termed response surface models. One such model for the growth of *Yersinia enterocolitica* at sub-optimal pH and temperature is described by the equation:

$$\text{LTG} = 423.8 - 2.54(T) - 10.97(\text{pH}) + 0.0041(T)^2 + 0.52(\text{pH})^2 \\ + 0.0129(\text{pH})(T)^* \tag{3.25}$$

where LTG is the natural logarithm of the time for a 100-fold increase in numbers, T is temperature, pH is pH with acetic acid as acidulant. Terms marked with an asterisk have an insignificant contribution at the 5% confidence level.

Such models are derived by analysing the data for growth under different (known) conditions for a least squares fit to a quadratic equation. For many, the practical implications of equations are not immediately obvious and a graphical representation as a three dimensional response surface has more impact (Figure 3.16).

Although the model is simply a fitted curve and is not based on any assumptions about microbial growth, an interesting consequence of the *Yersinia* model is that the cross-product term is not significant. This means that the two preservative factors, temperature and pH appear to be acting independently; a fact that is also apparent from the graph where there is little or no curvature in the response surface.

Kinetic models take parameters which describe how fast a micro-organism will grow, such as duration of lag phase and generation time, and model these as the response variable for different conditions of pH, temperature, a_w, etc. This is described as a second level growth model since it is used to predict lag phase and generation time for a given set of conditions. The predictions are then used in a primary level growth model, one which describes the microbial growth curve, to predict the effect of the chosen conditions on how microbial numbers change over time. This approach is more precise than response surface methods since individual parts of the growth curve may respond differently to changing conditions.

In building the model, experimental values for lag phase and growth rate are derived by fitting microbial count data to a mathematical function, the primary model, which describes the microbial growth curve. Some have used the logistic equation for this, but more commonly the Gompertz equation was used:

$$y = a \exp[-\exp(b - ct)] \tag{3.26}$$

where y is bacterial concentration, a, b and c are constants, and t is time.

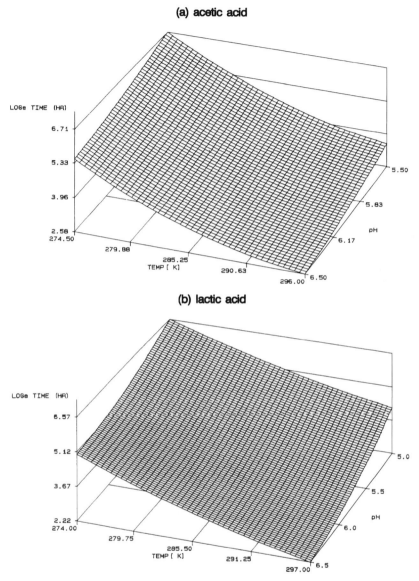

Figure 3.16 *Response surface plot describing the combined effect of temperature and pH on the time for a 2 $\log_{10}$ cycle growth increase of* Yersinia enterocolitica. *(a) Acetic acid and (b) lactic acid*
(C. Little)

This equation was originally developed in the 19th century to describe mortality as a function of age and is used for actuarial purposes, but was found to give a good fit to microbial growth data. More recently the Baranyi equation has found favour and, although considerably more complex, has the advantage of having been developed specifically to describe microbial growth in foods. Once a large number of such values

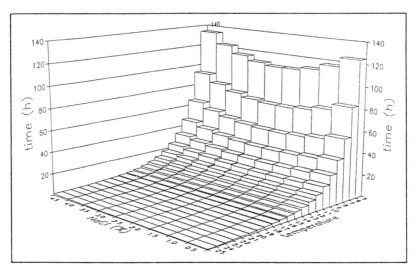

Figure 3.17   *3-D graph showing predicted generation time of salmonellae at fixed pH*
(Reprinted from Food Technology International 1990 with permission
from Sterling Publications Ltd)

of lag phase and growth rate have been obtained under a variety of
environmental conditions, their variation with factors such as temperature, salt, pH, *etc.* can be modelled (the secondary model) using response
surface techniques to give a polynomial equation, usually of degree 2 or
3, *i.e.* a quadratic or cubic polynomial (Figure 3.17). This is the approach
used in two software packages, Food Micromodel developed in the UK
as a result of a Ministry of Agriculture funded research programme, and
the United States Department of Agriculture Pathogen Modeling
Programme. Food Micromodel has now been superseded by ComBase
Predictor based on the same data set and available at http://www.
combase.cc/. The latest version of the Pathogen Modelling Programme
is available at http://ars.usda.gov/main/site_main.htm?modecode=
19353000. ComBase is a collaborative project involving researchers in
the UK, USA, Australia and elsewhere which is combining the databases
of numerous groups with a view to producing an integrated set of models
to be known as ComBase-PMP: Combined Database and Predictive
Microbiology Programme. This is due to be available in 2007.

Some models have started off as attempts to model the effect of
temperature on microbial growth and have been refined to incorporate
other factors such as pH and a_w. The classical Arrhenius equation relates
the rate constant (k) of a chemical reaction to absolute temperature T:

$$k = A \exp\left(-E/RT\right) \tag{3.27}$$

where E is the activation energy, A is the collision factor and R is the
universal gas constant.

If we assume that microbial growth is governed by a single rate limiting enzyme, then we can interpret k as the specific growth rate constant and E as a temperature characteristic. If this is the case and A and E are constant with temperature, then a plot of $\ln k$ against $1/T$ (the absolute temperature) would give a straight line. In fact a concave downward curve is obtained indicating that the activation energy E increases with decreasing temperature.

To improve the fit with observed behaviour, the basic equation has been modified by Davey to include a quadratic term:

$$\ln k = C_0 + C_1/T + C_2/T^2 \qquad (3.28)$$

This can be further modified to include other parameters affecting k such as pH and a_w. For example:

$$\ln k = C_0 + C_1/T + C_2/T^2 + C_3 a_w + C_4 a_w^2 \qquad (3.29)$$

The Schoolfield equation is another variation of the Arrhenius model where additional terms have been added to the basic equation to account for the effects of high- and low-temperature inactivation on growth rate. Terms describing the effect of a_w and pH can also be incorporated here to give a considerably more complex equation.

An alternative, rather simpler, approach which has met with some success is the square root model to describe growth at sub-optimal temperatures:

$$\sqrt{k} = b(T - T_{\min}) \qquad (3.30)$$

where k is the rate of growth, T the absolute temperature (K), and $T_{\min}$ is a conceptual minimum temperature of no physiological significance since it is usually below the freezing point of microbiological media.

Application of this expression to describe microbial growth was first described by Ratkowsky, although it is now recognized as a special form of the Bělehrádek power function originally described nearly 70 years ago.

A plot of $\sqrt{k}$ against T should give a straight line with an intercept on the T axis at $T_{\min}$ and this has been observed and reported by a number of authors monitoring growth in both laboratory media and foods (Figure 3.18).

To include the effects of other constraints on growth the square root equation has been extended separately to give similar equations including an a_w term and a pH term.

$$\sqrt{k} = c\sqrt{(a_w - a_{w_{\min}})}(T - T_{\min}) \qquad (3.31)$$

and

$$\sqrt{k} = d\sqrt{(\mathrm{pH} - \mathrm{pH}_{\min})}(T - T_{\min}) \qquad (3.32)$$

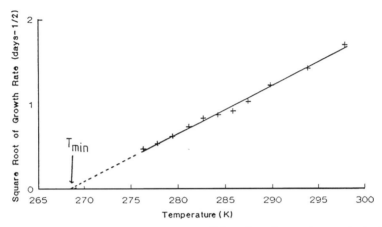

Figure 3.18   *Growth data for* Yersinia enterocolitica *plotted according to the Ratkowsky square root model*

The fact that T_{min} is not affected by a_w or by pH over the ranges tested indicate that these factors act independently of temperature. These two models have been combined to describe the growth of *Listeria monocytogenes* at sub-optimal pH, a_w and temperature using an equation of the form:

$$\sqrt{k} = e\sqrt{(a_w - a_{w_{min}})(\mathrm{pH} - \mathrm{pH}_{min})}(T - T_{min}) \tag{3.33}$$

Mathematical models of growth are not simply tools for use in development laboratories. For instance, by being able to predict accurately the response of microbial growth rate to temperature, the effect of a fluctuating temperature environment on microbial numbers throughout a distribution chain can be predicted. The value of the technique is illustrated by Figure 3.19 where what might appear as slightly different temperature histories between depot and supermarket can have a dramatic effect on microbial numbers.

Time–temperature function integrators are available which integrate the temperature history of a batch of product and express it as time at some reference temperature. If the temperature of the product remains at the reference temperature, say 0 °C, then they run as clocks recording real time. If the temperature fluctuates, then they speed up or slow down depending on whether the temperature deviates above or below the reference temperature. The relationship between rate and temperature used is the same as that between microbial growth rate and temperature. So quality loss as a result of microbial growth in a fluctuating temperature environment can be known with some accuracy and without the need for microbiological testing.

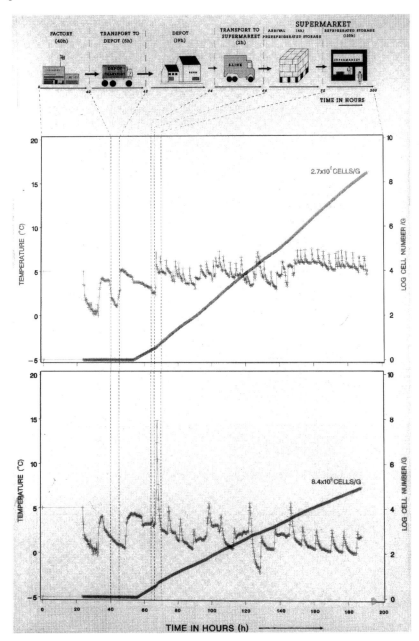

Figure 3.19 *Predicted growth of spoilage organisms during two chill distribution chains* (C. Adair, Unilever Research)

Mathematical models may play a part in the development of computer-based expert systems in food microbiology. The expert systems would provide advice and interpretation of results provided by mathematical models in the same way as human experts would but by embodying their expertise in the form of rules a computer can apply. One expert system, developed at the UK's Flour Milling and Baking Research Association, predicts the mould-free shelf life of bakery products. Here a_w and temperature are the principal determinants of shelf-life and previous storage trials have shown that, at a given temperature, there is a linear relationship between the logarithm of the mould-free shelf-life and the a_w, expressed as equilibrium relative humidity (ERH). For example, at 27 °C:

$$\log_{10} \text{mould-free shelf-life} = 6.42 - (0.0647 \times \text{ERH}) \qquad (3.34)$$

The user is led through a series of screen menus, to choose a product type, and input the ingredients, their relative amounts, the weight loss during processing and the storage temperature. The programme then calculates the ERH of the product and uses the appropriate isotherm to calculate the mould-free shelf-life.

CHAPTER 4

The Microbiology of Food Preservation

In Chapter 3 we outlined the physical and chemical factors that influence microbial growth and survival in foods. We have also seen how knowledge of these helps us to form a qualitative picture of a food's microflora and how mathematical models can be formulated which give a quantitative description of microbial growth under differing conditions. Manipulation of the factors affecting microbial behaviour is the basis of food preservation (Table 4.1). In this chapter we will survey the principal techniques of food preservation, with the notable exception of fermentation which is discussed separately in Chapter 9. Since our main concern here is the effect of preservation treatments on micro-organisms, technological features will only occasionally be touched on. For more detail on these aspects, readers are referred to more specialized texts on food technology.

4.1 HEAT PROCESSING

4.1.1 Pasteurization and Appertization

Foods are subject to thermal processes in a number of different contexts (Table 4.2). Often, their main objective is not destruction of micro-organisms in the product, although this is an inevitable and frequently useful side effect.

Credit for discovering the value of heat as a preservative agent goes to the French chef, distiller and confectioner, Nicolas Appert. In 1795 the French Directory offered a prize of 12 000 francs to anyone who could develop a new method of preserving food. Appert won this prize in 1810 after he had experimented for a number of years to develop a technique based on packing foods in glass bottles, sealing them, and then heating them in boiling water. He described his technique in detail in 1811 in a book called the ' . . . Art of Conserving all kinds of Animal and Vegetable Matter for several Years'. A similar technique was used by the

Table 4.1 *Mechanisms of principal food preservation procedures*

Procedure	Factor influencing growth or survival
Cooling, chill distribution and storage	Low temperature to retard growth
Freezing, frozen distribution and storage	Low temperature and reduction of water activity to prevent growth
Drying, curing and conserving	Reduction in water activity sufficient to delay or prevent growth
Vacuum and oxygen-free 'modified atmosphere' packaging	Low oxygen tension to inhibit strict aerobes and delay growth of facultative anaerobes
Carbon dioxide-enriched 'modified atmosphere'packaging	Specific inhibition of some micro-organisms by carbon dioxide
Addition of acids	Reduction of pH value and sometimes additional inhibition by the particular acid
Lactic fermentation	Reduction of pH value *in situ* by microbial action and sometimes additional inhibition by the lactic and acetic acids formed and by other microbial products, *e.g.* ethanol, bacteriocins
Emulsification	Compartmentalization and nutrient limitation within the aqueous droplets in water-in-oil emulsion foods
Addition of preservatives	Inhibition of specific groups of micro-organisms
Pasteurization and appertization	Delivery of heat sufficient to inactivate target micro-organisms to the desired extent
Radurization, radicidation and radappertization	Delivery of ionizing radiation at a dose sufficient to inactivate target micro-organisms to the desired extent
Application of high hydrostatic pressure Pascalization	Pressure-inactivation of vegetative bacteria, yeasts and moulds

Adapted from Gould (1989)

Table 4.2 *Heat processes applied to foods*

Heat process	Temperature	Objective
Cooking baking boiling frying grilling	$\leqslant 100\,°C$	Improvement of digestibility, *e.g.* starch gelatinization, collagen breakdown during cooking of meat. Improvement of flavour. Destruction of pathogenic micro-organisms
Blanching	$< 100\,°C$	Expulsion of oxygen from tissues. Inactivation of enzymes
Drying/ Concentration	$< 100\,°C$	Removal of water to enhance keeping quality
Pasteurization	$60–80\,°C$	Elimination of key pathogens and spoilage organisms
Appertization	$> 100\,°C$	Elimination of micro-organisms to achieve 'commercial sterility'

Englishman Saddington in 1807 to preserve fruits and for which he too received a prize, this time of five guineas, from the Royal Society of Arts. British patents describing the use of iron or metal containers were issued to Durand and de Heine in 1810 and the firm of Donkin and Hall established a factory for the production of canned foods in Bermondsey, London around 1812.

Appert held the view that the cause of food spoilage was contact with air and that the success of his technique was due to the exclusion of air from the product. This view persisted with sometimes disastrous consequences for another 50 years until Pasteur's work established the relationship between microbial activity and putrefaction. Today, the two types of heat process employed to destroy micro-organisms in food, pasteurization and appertization, bear the names of these eminent figures.

Pasteurization, the term given to heat processes typically in the range 60–80 °C and applied for up to a few minutes, is used for two purposes. First is the elimination of a specific pathogen or pathogens associated with a product. This type of pasteurization is often a legal requirement introduced as a public health measure when a product has been frequently implicated as a vehicle of illness. Notable examples are milk, bulk liquid egg (see Section 7.10.5) and ice cream mix, all of which have a much improved safety record as a result of pasteurization. The second reason for pasteurizing a product is to eliminate a large proportion of potential spoilage organisms, thus extending its shelf-life. This is normally the objective when acidic products such as beers, fruit juices, pickles, and sauces are pasteurized.

Where pasteurization is introduced to improve safety, its effect can be doubly beneficial. The process cannot discriminate between the target pathogen(s) and other organisms with similar heat sensitivity so a pasteurization which destroys say *Salmonella* will also improve shelf-life. The converse does not normally apply since products pasteurized to improve keeping quality are often considered as being intrinsically safe due to other factors such as low pH. This may be less true than was previously thought following several food poisoning outbreaks associated with unpasteurized fruit juices (see Section 7.7.5).

On its own, the contribution of pasteurization to extension of shelf-life can be quite small, particularly if the pasteurized food lacks other preservative factors such as low pH or a_w. Thermoduric organisms such as spore formers and some Gram positive vegetative species in the genera *Enterococcus*, *Microbacterium* and *Arthrobacter* can survive pasteurization temperatures. They can also grow and spoil a product quite rapidly at ambient temperatures, so refrigerated storage is often an additional requirement for an acceptable shelf-life.

Appertization refers to processes where the only organisms that survive processing are non-pathogenic and incapable of developing within the product under normal conditions of storage. As a result, appertized products have a long shelf-life even when stored at ambient temperatures. The term was coined as an alternative to the still widely used description *commercially sterile* which was objected to on the grounds that sterility is not a relative concept; a material is either sterile or it is not. An appertized or commercially sterile food is not necessarily sterile – completely free from viable organisms. It is however free from organisms capable of growing in the product under normal storage conditions. Thus for a canned food in temperate climates, it is not a matter of concern if viable spores of a thermophile are present as the organism will not grow at the prevailing ambient temperature.

4.1.2 Quantifying the Thermal Death of Micro-organisms: D and z Values

In Chapter 3 we described how a micro-organism can grow only over a restricted range of temperature, defined by three cardinal temperatures. When the temperature is increased above the maximum for growth, cells are injured and killed as key cellular components are destroyed and cannot be replaced. This occurs at an increasing rate as the temperature increases.

The generally accepted view is that thermal death is a first order process, that is to say, at a given lethal temperature, the rate of death depends upon the number of viable cells present. We can express this mathematically as:

$$dN/dt = -cN \qquad (4.1)$$

where dN/dt is the rate of death, N is the number of viable cells present and c is a proportionality constant. The minus sign signifies that N is decreasing.

To obtain information about the number of cells surviving after different periods of heating, this equation can be integrated between time zero and time t to give:

$$\log_e(N/N_0) = -ct \qquad (4.2)$$

or

$$N = N_0 e^{-ct} \qquad (4.3)$$

where N and N_0 are the numbers of viable cells present at times t and 0 respectively.

It is more convenient to represent Equation (4.2) in terms of logarithms to the base 10 as:

$$\log_{10}(N/N_0) = -kt \qquad (4.4)$$

where $k = c/\log_e 10 = c/2.303$.

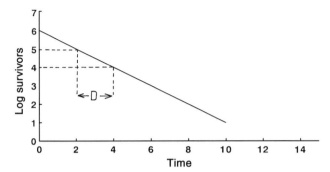

Figure 4.1 *The D value*

From Equation (4.4), it is clear that a plot of the log of the number of surviving cells at a given temperature against time should give a straight line with negative slope, k (Figure 4.1). As the temperature increases, so the slope of the survivor curve increases.

From this relationship we can derive a measure of an organism's heat resistance that is useful in calculating the lethality of heat processes. The D value or decimal reduction time is defined *as the time at a given temperature for the surviving population to be reduced by 1 log cycle, i.e.* 90%. The temperature at which a D value applies is indicated by a subscript, *e.g.* D_{65}.

A D value can be obtained from a plot of $\log_{10}$ survivors *versus* time, where it is the reciprocal of the slope, $1/k$. [You can confirm this by substituting $N_0 = 10N$ and $t = D$ in Equation (4.4).]

Alternatively, it can be calculated from:

$$D = (t_2 - t_1)/(\log N_1 - \log N_2) \tag{4.5}$$

where N_1 and N_2 are survivors at times t_1 and t_2 respectively.

One consequence of Equation (4.5) is that one can never predict with certainty how many decimal reductions a heat process must achieve (its lethality) for a product to be sterile since there is no log N_2 for $N_2 = 0$. When the initial microbial population in a batch of product is 10^n and a heat process producing n decimal reductions (nD) is applied, there will be one surviving organism in the product (log $1 = 0$). If you apply a more severe heat process, $(n+1)D$, $(n+2)D$ or $(n+4)D$ say, then the number of survivors will be $10^{-1}, 10^{-2}$ or 10^{-4} respectively. Physically, it is meaningless to talk of a fraction of an organism surviving, so these figures are interpreted as a probability of survival, corresponding to there being a 1 in 10, 1 in 100 and a 1 in 10 000 chance respectively that an organism will survive the heat process.

To give an example: if the D_{72} of *Salmonella* Senftenberg 775 W (the most heat-resistant salmonella) in milk is 1.5 s, then HTST pasteurization (15 s at 72 °C) will produce a 10D reduction in viable numbers. If we

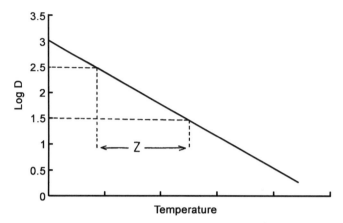

Figure 4.2   *The z-value*

assume the incidence of salmonella in the raw milk to be 1 colony-forming unit (cfu) 1^{-1}, then after pasteurization it will be reduced to 10^{-10} cfu 1^{-1} or 1 cfu $(10^{10}\ 1)^{-1}$. This means that if the milk is packed in 1 litre containers, one pack in 10 billion (10^{10}) would contain salmonella. If however the level of salmonella contamination was higher, say 10^{4} cfu 1^{-1}, then the same heat treatment would result in contamination of one in a million packs. Such simple calculations underestimate the true lethal effect since they assume instantaneous heating and cooling.

As the temperature is increased so the D value decreases. This is an exponential process over the range of temperatures used in the heat processing of food so that plotting log D against temperature gives a straight line. From this we can derive another important parameter in heat processing, *z*: *the temperature change which results in a tenfold (1 log) change in D* (Figure 4.2).

$$z = (T_2 - T_1)/(\log D_1 - \log D_2) \tag{4.6}$$

Knowledge of an organism's *z* value is important if we are to take into account the lethal effect of the different temperatures experienced during a heat process.

4.1.3   Heat Sensitivity of Micro-organisms

The heat sensitivity of various micro-organisms is illustrated by Table 4.3 which shows their D values. Generally psychrotrophs are less heat resistant than mesophiles, which are less heat resistant than thermophiles; and Gram-positives are more heat resistant than Gram-negatives. Most vegetative cells are killed almost instantaneously at 100 °C and their D values are measured and expressed at temperatures appropriate to pasteurization.

Table 4.3 *Microbial heat resistance*

Vegetative organisms $(z \sim 5\,°C)$	*D* (mins)	
Salmonella sp.	D_{65}	0.02–0.25
Salmonella Senftenberg	D_{65}	0.8–1.0
Staphylococcus aureus	D_{65}	0.2–2.0
Escherichia coli	D_{65}	0.1
Yeasts and moulds	D_{65}	0.5–3.0
Listeria monocytogenes	D_{60}	5.0–8.3
Campylobacter jejuni	D_{55}	1.1
Bacterial Endospores	D_{121}	
$(z \sim 10\,°C)$		
B. stearothermophilus		4–5
C. thermosaccharolyticum		3–4
Desulfotomaculum nigrificans		2–3
B. coagulans		0.1
C. botulinum types A & B		0.1–0.2
C. sporogenes		0.1–1.5
C. botulinum type E	D_{80}	0.1–3.0
	D_{110}	<1 second

Bacterial spores are usually far more heat resistant than vegetative cells; thermophiles produce the most heat resistant spores while those of psychrotrophs and psychrophiles are most heat sensitive. Since spore inactivation is the principal concern in producing appertized foods, much higher temperatures are used in appertization processes and in the measurement of spore D values.

Yeast ascospores and the asexual spores of moulds are only slightly more heat resistant than the vegetative cells and will normally be killed by temperatures at or below 100 °C, *e.g.* in the baking of bread. Ascospores of the mould *Byssochlamys fulva*, and a few other ascomycetes do show a more marked heat resistance and can be an occasional cause of problems in canned fruits which receive a relatively mild heat process (see Section 5.5.4).

The heat resistance exhibited by the bacterial endospore is due mainly to its ability to maintain a very low water content in the central DNA-containing protoplast; spores with a higher water content have a lower heat resistance. The relative dehydration of the protoplast is maintained by the spore cortex, a surrounding layer of electronegative peptidoglycan which is also responsible for the spore's refractile nature. The exact mechanism by which it does this is not known, although it may be some combination of physical compression of the protoplast by the cortex and osmotic extraction of the water. As the cortex is dissolved during germination and the protoplast rehydrates, so the spore's heat resistance declines. Suspension of a germinated spore population in a strong solution of a non-permeant solute such as sucrose will reverse this process of rehydration and restore the spore's heat resistance. The total picture is probably more complex than this however, since other features of the spore such as its high content of divalent cations, particularly calcium, are thought to make some contribution to heat resistance.

Thermal sensitivity as measured by the D value can vary with factors other than the intrinsic heat sensitivity of the organism concerned. This is most pronounced with vegetative cells where the growth conditions and the stage of growth of the cells can have an important influence. For example, stationary phase cells are generally more heat resistant than log phase cells. Heat sensitivity is also dependent on the composition of the heating menstruum; cells tend to show greater heat sensitivity as the pH is increased above 8 or decreased below 6. Fat enhances heat resistance as does decreasing a_w through drying or the addition of solutes such as sucrose. The practical implications of this can be seen in the more severe pasteurization conditions used for high sugar or high fat products such as ice cream mix and cream compared with that used for milk. This effect is quite dramatic in the instance of milk chocolate where the D_{70} value of *Salmonella* Senftenberg 775 W has been measured as between 6 and 8 hours compared with only a few seconds in milk. A more specialized example of medium effects on heat sensitivity occurs in brewing where the ethanol content of beer has been shown to have a profound effect on the heat sensitivity of a spoilage *Lactobacillus*; an observation that has implications for the pasteurization of low-alcohol beers (Figure 4.3).

At present all thermal process calculations are based on the assumption that the death of micro-organisms follows the log-linear kinetics described by Equation 4.4. Though this is often the case, deviations from log-linear behaviour are also often observed (Figure 4.4). Sometimes these deviations can be rationalized on the basis of some special property of the organism. For example, an apparent increase in viable numbers of organisms or a lag at the start of heating may be ascribed to heat activation of spores so that in the first moments of heating the number of spores being activated equals or exceeds the number being destroyed. Alternatively a lag phase may reflect the presence of clumps of cells, all of which require to be inactivated before that colony forming unit is destroyed. The frequently observed tailing of the curves, which has greater practical significance, may be due to sub-populations of cells that are more heat resistant. These deviations from the accepted model tend to be observed more often when studying the thermal death of vegetative organisms and in some cases may reflect inadequacy of the logarithmic death concept in this situation.

The primary assumption which gives rise to log-linear kinetics is that at a constant temperature each cell has an equal chance of inactivation at any instant. This can be explained on theoretical grounds if there is a

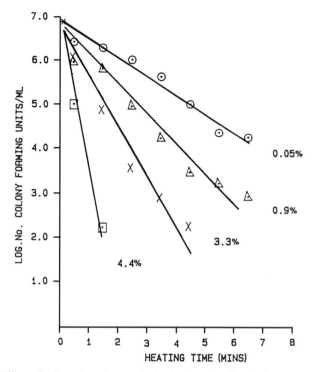

Figure 4.3 *Effect of ethanol on the survival of a lactobacillus in beer at 60 °C*

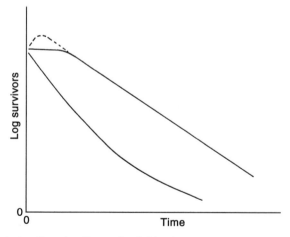

Figure 4.4 *Deviation from log-linear death kinetics*

single target molecule in each cell whose inactivation causes death. If this is not the case and there is, say, a large number of the same target molecules present or a large number of different targets, then several inactivation events would be required for cell death and a lag would be observed in the thermal death curve.

Damage to DNA has been identified as the probable key lethal event in both spores and vegetative cells. In spores, however, inactivation of germination mechanisms is also important. If this inactivation can be bypassed in some way, then apparently dead spores may be cultured. This has been demonstrated by the inclusion of lysozyme in recovery media where the enzyme hydrolyses the spore cortex, replacing the spore's own inactivated germination system.

Deviations from log-linear kinetics in the thermal death of vegetative cells probably reflect a greater multiplicity of target sites for thermal inactivation such as the cytoplasmic membrane, key enzymes, RNA and the ribosomes. This type of damage can be cumulative rather than instantly lethal. Individual inactivation events may not kill the cell but will inflict sub-lethal injury making it more vulnerable to other stresses. If however injured cells are allowed time in a non-inhibitory medium, they can repair and recover their full vigour. Examples of sub-lethal damage can be seen when cells do not grow aerobically but can be cultured anaerobically or in the presence of catalase, or when selective agents such as bile salts or antibiotics, which are normally tolerated by the organism, prove inhibitory.

Two other factors also contribute to deviations from log-linear behaviour in vegetative organisms. Individual cells within a population may exhibit a broader range of heat resistance than is seen with spores and, since vegetative cells are not metabolically inert, they may also respond and adapt to a heating regime modifying their sensitivity.

4.1.4 Describing a Heat Process

Heating processes are neither uniform nor instantaneous. To be able to compare the lethal effect of different processes it is necessary for us to have some common currency to describe them. For appertization processes this is known as the F value; a parameter which expresses the integrated lethal effect of a heat process in terms of minutes at a given temperature indicated by a subscript. A process may have an F_{121} value of say 4, which means that its particular combination of times and temperatures is *equivalent* to instantaneous heating to 121 °C, holding at that temperature for four minutes and then cooling instantly, it does not even necessarily imply that the product ever reaches 121 °C. The F value will depend on the z value of the organism of concern; if $z = 10$ °C then 1 minute at 111 °C has an $F_{121} = 0.1$, if $z = 5$ °C then the F_{121} value will be 0.01. It is therefore necessary to specify both the z value and the temperature when stating F. For spores z is commonly about 10 °C and the F_{121} determined using this value is designated F_0.

To determine the F_0 value required in a particular process one needs to know the D_{121} of the target organism and the number of decimal reductions considered necessary.[†]

$$F_{121} = D_{121}(\log N_0 - \log N) \tag{4.7}$$

In this exercise, the canner will have two objectives, a safe product and a stable product. From the point of view of safety in low acid canned foods (defined as those with a pH > 4.5) *Clostridium botulinum* is the principle concern. The widely accepted minimum lethality for a heat process applied to low-acid canned foods is that it should produce 12 decimal reductions in the number of surviving *C. botulinum* spores ($\log N_0 - \log N = 12$). This is known as the 12D or botulinum cook. If D_{121} of *C. botulinum* is 0.21 minutes then a botulinum cook will have an F_0 of $12 \times 0.21 = 2.52$ min. The effect of applying a process with this F_0 to a product in which every can contains one spore of *C. botulinum* ($N_0 = 1$) will be that a spore will survive in one can out of every 10^{12}.

The canner also has the objective of producing a product which will not spoil at an unacceptably high rate. Since spoilage is a more acceptable form of process failure than survival of *C. botulinum*, the process lethality requirements with respect to spoilage organisms do not need to be so severe. In deciding the heat process to be applied, a number of factors have to be weighed up.

(1) What would be the economic costs of a given rate of spoilage?
(2) What would be the cost of additional processing to reduce the rate of spoilage?
(3) Would this additional processing result in significant losses in product quality?

Most canners would regard an acceptable spoilage rate due to under-processing as something around 1 in 10^5–10^6 cans and this is normally achievable through 5–6 decimal reductions in the number of spores with spoilage potential (the USFDA use 6D as their yardstick). PA3679, *Clostridium sporogenes* is frequently used as an indicator for process spoilage and typically has a D_{121} of about 1 min. This will translate into a process with an F_0 value of 5–6; sufficient to produce about 24–30 decimal reductions in viable *C. botulinum* spores – well in excess of the minimal requirements of the botulinum cook. Some typical F_0 values used in commercial canning are presented in Table 4.4.

[†] For pasteurization treatments, an analogous procedure is used to express processes in terms of their P value or pasteurization units. Once again the reference temperature must be specifed. This can be 60 °C, in which case: $P_{60} = D_{60}(\log N_0 - \log N)$.

Table 4.4   *Typical F_0 values for some canned foods*

Food	F_0 (min)
Asparagus	2–4
Beans in tomato sauce	4–6
Carrots	3–4
Peas	4–6
Milk pudding	4–10
Meats in gravy	8–10
Potatoes	4–10
Mackerel in brine	3–4
Meat loaf	6
Chocolate pudding	6

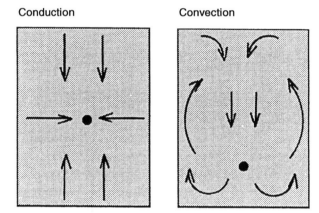

Figure 4.5   *Conduction and convection heating in cans.* ● *Denotes slowest heating point*

Having decided the *F* value required, it is necessary to ensure that the F_0 value actually delivered by a particular heating regime achieves this target value. To do this, the thermal history of the product during processing is determined using special cans fitted with thermocouples to monitor the product temperature. These must be situated at the slowest heating point in the pack where the F_0 value will be at a minimum. The precise location of the slowest heating point and the rate at which its temperature increases depend on the physical characteristics of the can contents. Heat transfer in solid foods such as meats is largely by conduction which is a slow process and the slowest heating point is the geometric centre of the can (Figure 4.5). When fluid movement is possible in the can, heating is more rapid because convection currents are set up which transfer heat more effectively. In this case the slowest heating point lies on the can's central axis but nearer the base.

The slowest heating point is not always easy to predict. It may change during processing as in products which undergo a sol–gel transition during heating, producing a broken heating curve which shows a phase of convection heating followed by one of conduction heating. In most cases heating is by conduction but some can contents show neither pure convection nor pure conduction heating and the slowest heating point must be determined experimentally.

Movement of material within the can improves heat transfer and will reduce the process time. This is exploited in some types of canning retort which agitate the cans during processing to promote turbulence in the product.

The F value can be computed from the thermal history of a product by assigning a lethal rate to each temperature on the heating curve. The lethal rate, L_R, at a particular temperature is the ratio of the microbial death rate at that temperature to the death rate at the lethal rate reference temperature. For example, using 121 °C as the reference temperature:

$$L_R = D_{121}/D_T \tag{4.8}$$

where L_R is the lethal rate at 121 °C.

Since

$$z = (T_2 - T_1)/(\log D_1 - \log D_2) \tag{4.9}$$

and substituting $T_2 = 121$ °C; $T_1 = T$; $D_2 = D_{121}$ and $D_1 = D_T$

$$L_R = 1/10^{(121-T)/z} \tag{4.10}$$

Lethal rates calculated in this way can be obtained from published tables where the L_R can be read off for each temperature (from about 90 °C and above) and for a number of different z values (Table 4.5). Nowadays though this is unnecessary since the whole process of F value calculation tends to be computerized.

Total lethality is the sum of the individual lethal rates over the whole process; for example 2 minutes at a temperature whose L_R is 0.1 contributes 0.2 to the F_0 value, 2 minutes at a L_R of 0.2 contributes a further 0.4, and so on. Another way of expressing this is that the area

Table 4.5 *Selected lethal rate values* ($F_{121.1}$ *min*$^{-1}$)

Temp. (°C)	z value					
	7	8	9	10	11	12
100.0	0.001	0.002	0.005	0.008	0.012	0.017
101.0	0.001	0.003	0.006	0.010	0.015	0.021
102.0	0.002	0.004	0.008	0.012	0.018	0.026
103.0	0.003	0.005	0.010	0.015	0.023	0.031
104.0	0.004	0.007	0.013	0.019	0.028	0.038
105.0	0.005	0.010	0.016	0.024	0.034	0.045
110.0	0.026	0.041	0.058	0.077	0.098	0.119
115.0	0.134	0.172	0.209	0.245	0.278	0.310
120.0	0.694	0.727	0.753	0.774	0.793	0.808

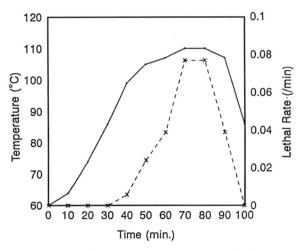

Figure 4.6   *A lethal rate plot.* ● *Product temperature;* × *lethal rate*

under a curve describing a plot of lethal rate against time gives the
overall process lethality, F_0 (Figure 4.6).

$$F_0 = \int L_R \, dt \qquad\qquad (4.11)$$

This procedure has safeguards built into it. If the slowest heating point
receives an appropriate treatment then the lethality of the process
elsewhere in the product will be in excess of this. A further safety margin
is introduced by only considering the heating phase of the process; the
cooling phase, although short, will also have some lethal effect.

Process confirmation can also be achieved by microbiological testing
in which inoculated packs are put through the heat process and the
spoilage/survival rate determined. Heat penetration studies though give
much more precise and useable information since inoculated packs are
subject to culture variations which can affect resistance and also recovery
patterns.

A change in any aspect of the product or its preparation will require
the heat process to be re-validated and failure to do this could have
serious consequences. An early example of this was the scandal in the
mid-nineteenth century when huge quantities of canned meat supplied to
the Royal Navy putrefied leading to the accusation that the meat had
been bad before canning. It transpired that the problem arose because
cans with a capacity of 9–14 lb were being used instead of the original
2–6 lb cans. In these larger cans the centre of the pack took longer to heat
and did not reach a temperature sufficient to kill all the bacteria. More
recently, replacement of sugar with an artificial sweetener in hazelnut
puree meant that spores of *C. botulinum* surviving the mild heat process

given to the product were no longer prevented from growing by the reduced a_w (see Section 7.5.5).

4.1.5 Spoilage of Canned Foods

If a canned food contains viable micro-organisms capable of growing in the product at ambient temperatures, then it will spoil. Organisms may be present as a result of an inadequate heat process, underprocessing, or of post process contamination through container leakage. Spoilage by a single spore former is often diagnostic of underprocessing since rarely would such a failure be so severe that vegetative organisms would survive.

A normal sound can will either be under vacuum with slightly concave ends or have flat ends in those cases where the container is brimful. Spoilage often manifests itself through microbial gas production which causes the ends to distend and a number of different terms are used to describe the extent to which this has occurred (Table 4.6). The spore-forming anaerobes *Clostridium* can be either predominantly proteolytic or saccharolytic but both activities are normally accompanied by gas production causing the can to swell. Cans may sometimes swell as a result of chemical action. Defects in the protective lacquer on the inside of the can may allow the contents to attack the metal releasing hydrogen. These hydrogen swells can often be distinguished from microbiological spoilage since the appearance of swelling occurs after long periods of storage and the rate at which the can swells is usually very slow.

In cases where microbial growth occurs without gas production, spoilage will only be apparent once the pack has been opened. *Bacillus* species, with the exceptions of *B. macerans* and *B. polymyxa*, usually break down carbohydrates to produce acid but no gas giving a type of spoilage known as a 'flat sour', which describes the characteristics of both the can and the food.

The heat process a product receives is determined largely by its acidity: the more acidic a product is, the milder the heat process applied.

Table 4.6 *Description of blown cans*

Name	Description
Flat	No evidence of swelling.
Hard swell	Both ends of the can are permanently and firmly bulged and do not yield readily to thumb pressure.
Soft swell	Both ends bulged but not tightly; they yield to thumb pressure.
Springer	One end flat, the other bulged. When the bulged end is pressed in then the flat one springs out.
Flipper	A can with a normal appearance which when brought down sharply on a flat surface causes a flat end to flip out. The bulged end can be forced back by very slight pressure.

Although more complex schemes have been described, the essential classification of canned foods is into low acid (pH > 4.5, or 4.6 in the United States) and acid foods (pH < 4.5 or 4.6). We have already seen how this is applied to assure safety with the requirement that products with a pH > 4.5 must undergo a botulinum cook to ensure 12 decimal reductions of *C. botulinum* spores. This is not a concern in acid foods as *C. botulinum* cannot grow and the F_0 applied to products with a pH in the range 4.0–4.5 such as canned tomatoes and some canned fruits is generally 0.5–3.0. In higher acidity products such as canned citrus fruits (pH < 3.7) the heat process is equivalent only to a pasteurization.

A product's acidity also determines the type of spoilage that may result from underprocessing since it can prevent the growth of some spoilage organisms. At normal ambient temperatures (< 38 °C) only mesophilic species will grow. Typical examples would be *C. botulinum*, *C. sporogenes* and *B. subtilis* in low acid products and *C. butyricum* and *C. pasteurianum* in products with a pH below 4.5.

Cans are cooled rapidly after processing to prevent spoilage by thermophiles. Thermophilic spores are more likely to survive the normal heat process but would not normally pose a problem. If however a large assemblage of cans is allowed to cool down naturally after retorting, the process will be slow and the cans will spend some time passing through the thermophilic growth range. Under these conditions surviving thermophilic spores may be able to germinate and grow, spoiling the product before it cools. This may also occur if cans are stored at abnormally high ambient temperatures (> 40 °C) and canned foods destined for very hot climates may receive a more stringent process to reduce thermophilic spoilage.

Thermophilic organisms commonly associated with spoilage of low acid canned foods are the saccharolytic organism *C. thermosaccharolyticum*, *B. stearothermophilus* and *Desulfotomaculum nigrificans*. The last of these causes a type of spoilage known as 'sulfur stinker'. It produces hydrogen sulfide which does not usually distend the can but does give the product an objectionable smell and reacts with iron from the can to cause blackening.

Leakage is the most common cause of microbiological spoilage in canned foods. Cans are the most common containers used for retorted products, although glass jars, rigid plastic containers and soft pouches are also sometimes used. Cans are usually made of two or three parts: the three-part can consists of a base, body and lid while in two part cans the body and base are made from a single piece of metal. In a three-part can the body seam is electrically welded but the lid on all cans is held in place by a double seam (Figure 4.7). The correct formation and integrity of this seam are crucial to preventing leakage and monitoring seam integrity is an important aspect of quality control procedures in canning.

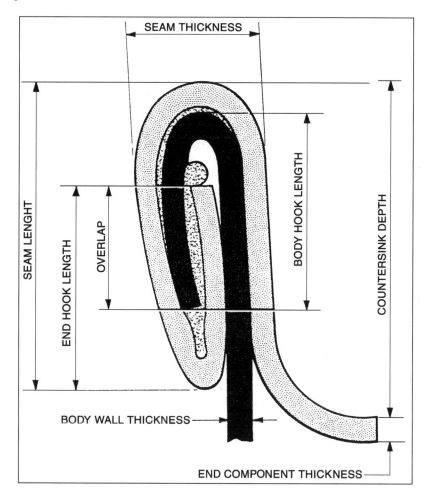

DOUBLE SEAM DIMENSIONAL TERMINOLOGY

Figure 4.7 *The can double seam (Carnaud Metalbox)*

During processing cans are subjected to extreme stress, particularly when the hot can is cooled down rapidly from processing temperature. The negative pressure created in the can under these conditions could lead to micro-organisms on the container's surface or in the cooling water being sucked inside through a small defect in the seam. The defect in the hot can that allowed leakage to occur may seal up and be undetectable when the can is cool so leaker spoilage can cause cans to blow. Since the micro-organisms enter the can after processing there is no restriction on the type of organism capable of causing leaker spoilage, therefore the presence of a mixed culture or non-sporing organisms is almost certainly a result of can leakage.

To prevent leaker spoilage it is essential that the outside of cans is clean and uncontaminated and that chlorinated water is used to cool them. Failures in this respect have been the cause of a large typhoid outbreak in Aberdeen, Scotland where cans of corned beef made in the Argentine had been cooled with river water contaminated with *Salmonella* Typhi and in an outbreak of botulism associated with canned salmon where the *C. botulinum* type E spores which were associated with the raw product contaminated the outside of the cans after processing and were sucked into one can during cooling.

There have been occasional reports of pre-process spoilage in canned foods where there was an unacceptable delay between preparing the product and heat processing. During this time spoilage may occur although the organisms responsible will have been killed by the heat process.

4.1.6   Aseptic Packaging

Up until now in our consideration of appertized foods we have discussed only retorted products; those which are hermetically sealed into containers, usually cans, and then subjected to an appertizing heat process in-pack. While this has been hugely successful as a long-term method of food preservation, it does require extended heating periods in which a food's functional and chemical properties can be adversely affected.

In UHT processing the food is heat processed before it is packed and then sealed into sterilized containers in a sterile environment. This approach allows more rapid heating of the product, the use of higher temperatures than those employed in canning, typically 130–140 °C, and processing times of seconds rather than minutes. The advantage of using higher temperatures is that the z value for chemical reactions such as vitamin loss, browning reactions and enzyme inactivation is typically 25–40 °C compared with 10 °C for spore inactivation. This means that they are less temperature sensitive so that higher temperatures will increase the microbial death rate more than they increase the loss of food quality associated with thermal reactions.

F_0 values for UHT processes can be estimated from the holding temperature (T) and the residence time of the fastest moving stream of product, t.

$$F_0 = t.10^{(T-121)/10} \qquad (4.12)$$

Initially UHT processing and aseptic packaging were confined to liquid products such as milk, fruit juices and some soups which would heat up very quickly due to convective heat transfer. If a food contained solid particles larger than about 5 mm diameter it was unsuited to the rapid processing times due to the slower conductive heating of the particulate phase. Scraped surface heat exchangers have been used to process products containing particles up to 25 mm in diameter but at the cost of overprocessing the liquid phase. To avoid this, one system processes the liquid and solid phase separately. A promising alternative is the use of ohmic heating in which a food stream is passed down a tube which contains a series of electrodes. An alternating voltage is applied across the electrodes and the food's resistance causes it to heat up rapidly. Most of the energy supplied is transformed into heat and the rate at which different components heat up is determined by their conductivities rather than heat transfer.

A common packing system used in conjunction with UHT processing is a form/fill/seal operation in which the container is formed in the packaging machine from a reel of plastic or laminate material, although some systems use preformed containers. Packaging material is generally refractory to microbial growth and the level of contamination on it is usually very low. Nevertheless to obtain commercial sterility it is given a bactericidal treatment, usually with hydrogen peroxide, sometimes coupled with UV irradiation.

4.2 IRRADIATION

Electromagnetic (e.m.) radiation is a way in which energy can be propagated through space. It is characterized in terms of its wavelength λ, or its frequency v, and the product of these two properties gives the speed, c, at which it travels (3×10^8 m sec^{-1} in a vacuum).

$$\lambda v = c \tag{4.13}$$

The range of frequencies (or wavelengths) that e.m. radiation can have is known as the electromagnetic spectrum and is grouped into a number of regions, visible light being only one small region (Figure 4.8).

The energy carried by e.m. radiation is not continuous but is transmitted in discrete packets or quanta; the energy, E, contained in each quantum being given by the expression:

$$E = hv \tag{4.14}$$

where h is a constant (6.6×10^{-27} ergs sec^{-1}) known as Planck's constant. Thus, the higher the frequency of the radiation the higher its quantum energy.

As far as food microbiology is concerned, only three areas of the e.m. spectrum concern us; microwaves, the UV region and gamma rays. We will now consider each of these in turn.

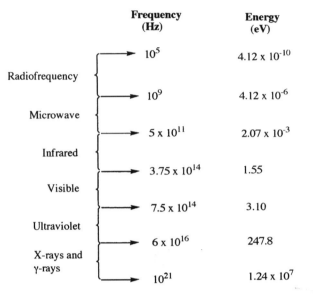

Figure 4.8   *The electromagnetic spectrum*

4.2.1   Microwave Radiation

The microwave region of the e.m. spectrum occupies frequencies between 10^9 Hz up to 10^{12} Hz and so has a relatively low quantum energy. For the two frequencies used in food processing, 2450 MHz and 915 MHz, this is around 10^{-18} ergs or 10^{-6} eV. Domestic microwave ovens use 2450 MHz which is less penetrating than the lower frequency.

Unlike the other forms of radiation we will discuss, microwaves act indirectly on micro-organisms through the generation of heat. When a food containing water is placed in a microwave field, the dipolar water molecules align themselves with the field. As the field reverses its polarity about 2.5×10^9 times each second the water molecules are continually oscillating. This kinetic energy is transmitted to neighbouring molecules leading to a rapid rise in temperature throughout the product. In foods with a high salt content, surface heating due to ions acquiring kinetic energy from the microwave field can also contribute, but this is generally of minor importance.

Microwaves are generated using a magnetron, a device first developed in the UK during research into radar during the Second World War. Magnetrons are used both commercially and domestically, but their biggest impact has been in the domestic microwave oven and in catering where their speed and convenience have enormous advantages. The principal problem associated with the domestic use of microwaves is non-uniform heating of foods, due to the presence of cold spots in the oven, and the non-uniform dielectric properties of the food. These can lead to cold spots in some microwaved foods and concern over the risks associated with consumption of inadequately heated meals has led to more explicit instructions on microwaveable foods. These often specify a tempering period after heating to allow the temperature to equilibrate.

Microwaves have been slow to find industrial applications in food processing, although they are used in a number of areas. Microwaves have been used to defrost frozen blocks of meat prior to their processing into products such as burgers and pies thus reducing wear and tear on machinery. There has also been a limited application of microwaves in the blanching of fruits and vegetables and in the pasteurization of soft bakery goods and moist (30% H_2O) pasta to destroy yeasts and moulds. In Japan, microwaves have been used to pasteurize high-acid foods, such as fruits in syrup, intended for distribution at ambient temperature. These are packed before processing and have an indefinite microbiological shelf-life because of the heat process and their low pH. However, the modest oxygen barrier properties of the pack has meant that their biochemical shelf-life is limited to a few months.

4.2.2 UV Radiation

UV radiation has wavelengths below 450 nm ($v \simeq 10^{15}$ Hz) and a quantum energy of 3–5 eV (10^{-12} ergs). The quanta contain energy sufficient to excite electrons in molecules from their ground state into higher energy orbitals making the molecules more reactive. Chemical reactions thus induced in micro-organisms can cause the failure of critical metabolic processes leading to injury or death.

Only quanta providing energy sufficient to induce these photochemical reactions will inhibit micro-organisms, so those wavelengths that are most effective give us an indication of the sensitive chemical targets within the cell. The greatest lethality is shown by wavelengths around 260 nm which correspond to a strong absorption by nucleic acid bases. The pyrimidine bases appear particularly sensitive, and UV light at this wavelength will, among other things, induce the formation of covalently linked dimers between adjacent thymine bases in DNA (Figure 4.9). If left intact these will prevent transcription and DNA replication in affected cells.

The resistance of micro-organisms to UV is largely determined by their ability to repair such damage, although some organisms such as micrococci also synthesize protective pigments. Generally, the resistance to UV irradiation follows the pattern:

Gram-negatives < Gram-positives ≈ yeasts < bacterial spores
< mould spores < < viruses.

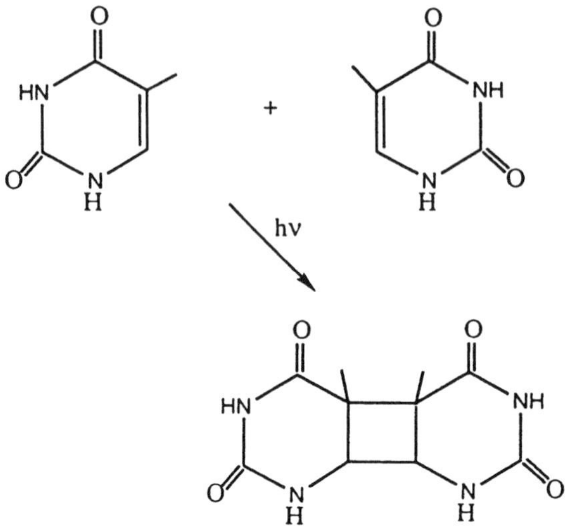

Figure 4.9   *The photochemical dimerization of thymine*

Table 4.7   *UV resistance of some selected food-borne micro-organisms*

Species	D (ergs $\times 10^2$)
E. coli	3–4
Proteus vulgaris	3–4
Serratia marcescens	3–4
Shigella flexneri	3–4
Pseudomonas fluorescens	3–4
Bacillus subtilis (vegetative cells)	6–8
Bacillus subtilis (spores)	8–10
Micrococcus luteus	10–20
Staph. aureus	3–4
Aspergillus flavus	50–100
Mucor racemosus	20–50
Penicillium roquefortii	20–50
Rhizopus nigrificans	> 200
Saccharomyces cerevisiae	3–10

Data from 'Microbial Ecology of Foods'. Vol. 1.ICMSF.

Death of a population of UV-irradiated cells demonstrates log-linear kinetics similar to thermal death and, in an analogous way, D values can be determined. These give the dose required to produce a tenfold reduction in surviving numbers where the dose, expressed in ergs or μWs, is the product of the intensity of the radiation and the time for which it is applied. Some published D values are presented in Table 4.7.

Determination of UV D values is not usually a straightforward affair since the incident radiation can be absorbed by other medium components and has very low penetration. Passage through 5 cm of clear water will reduce the intensity of UV radiation by two-thirds. This effect increases with the concentration of solutes and suspended material so that in milk 90% of the incident energy will be absorbed by a layer only 0.1 mm thick. This low penetrability limits application of UV radiation in the food industry to disinfection of air and surfaces.

Low-pressure mercury vapour discharge lamps are used: 80% of their UV emission is at a wavelength of 254 nm which has 85% of the biological activity of 260 nm. Wavelengths below 200 nm are screened out by surrounding the lamp with an absorbent glass since these wavelengths are absorbed by oxygen in the air producing ozone which is harmful. The output of these lamps falls off over time and they need to be monitored regularly.

Air disinfection is only useful when the organisms suspended in air can make a significant contribution to the product's microflora and are likely to harm the product; for example, in the control of mould spores in bakeries. UV lamps have also been mounted in the head space of tanks storing concentrates, the stability of which depends on their low a_w. Fluctuations in temperature can cause condensation to form inside the tank. If this contacts the product, then areas of locally high a_w can form where previously dormant organisms can grow, spoiling the product. Process water can be disinfected by UV; this avoids the risk of tainting sometimes associated with chlorination, although the treated water will not have the residual antimicrobial properties of chlorinated water. UV radiation is commonly used in the depuration of shellfish to disinfect the water recirculated through the depuration tanks. Chlorination would not be suitable in this situation since residual chlorine would cause the shellfish to stop feeding thus stopping the depuration process.

Surfaces can be disinfected by UV, although protection of microorganisms by organic material such as fat can reduce its efficacy. Food containers are sometimes treated in this way and some meat chill store rooms have UV lamps to retard surface growth. UV can however induce spoilage of products containing unsaturated fatty acids where it accelerates the development of rancidity. Process workers must also be protected from UV since the wavelengths used can cause burning of the skin and eye disorders.

4.2.3 Ionizing Radiation

Ionizing radiation has frequencies greater than 10^{18} Hz and carries sufficient energy to eject electrons from molecules it encounters. In practice three different types are used.

(1) *High-energy electrons.* in the form of β particles produced by radioactive decay or machine generated electrons. Strictly

speaking they are particles rather than electromagnetic radiation, although in some of their behaviour they do exhibit the properties of waves. Because of their mass and charge, electrons tend to be less penetrating than ionizing e.m. radiation; for example, 5 MeV β particles will normally penetrate food materials to a depth of about 2.5 cm.

(2) *X-rays* generated by impinging high energy electrons on a suitable target.

(3) *Gamma γ rays* produced by the decay of radioactive isotopes. The most commonly used isotope cobalt 60, ^{60}Co, is produced by bombarding non-radioactive cobalt, ^{59}Co, with neutrons in a nuclear reactor. It emits high-energy γ-rays (1.1 MeV) which can penetrate food up to a depth of 20 cm (*cf.* β particles). An isotope of caesium, ^{137}Cs, which is extracted from spent nuclear fuel rods, has also been used but is less favoured for a number of reasons.

Ionizing radiation can affect micro-organisms directly by interacting with key molecules within the microbial cell, or indirectly through the inhibitory effects of free radicals produced by the radiolysis of water (Figure 4.10). These indirect effects play the more important role since in the absence of water, doses 2–3 times higher are required to obtain the same lethality. Removal of oxygen also increases microbial resistance 2–4 fold and it is thought that this may be due to the ability of oxygen to participate in free radical reactions and prevent the repair of radiation induced lesions. As with UV irradiation, the main site of damage in cells is the chromosome. Hydroxyl radicals cause single- and double-strand breaks in the DNA molecule as a result of hydrogen abstraction from deoxyribose followed by β-elimination of phosphate which cleaves the molecule. They can also hydroxylate purine and pyrimidine bases.

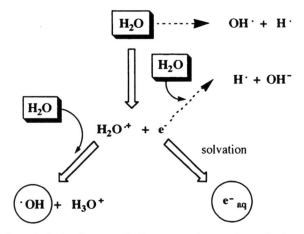

Figure 4.10   *The radiolysis of water. $\bigcirc$ Denotes main reactive radicals*

Resistance to ionizing radiation depends on the ability of the organism to repair the damage caused. Inactivation kinetics are generally logarithmic, although survival curves often appear sigmoidal exhibiting a shoulder and a tail to the phase of log-linear death. The shoulder is usually very slight but is more pronounced with bacteria which have more efficient repair mechanisms where substantially more damage can be accumulated before death ensues.

D values can be derived from the linear portion of these curves and Table 4.8 presents 6D values (the dose to produce a millionfold reduction) reported for a number of foodborne organisms. These are expressed in terms of the absorbed dose of ionizing radiation which is measured in Grays (1 Gy = 1 joule kg^{-1}). Resistance generally follows the sequence:

Gram-negative < Gram-positive ≈ moulds < spores ≈ yeasts < viruses.

Food-associated organisms do not generally display exceptional resistance, although spores of some strains of *Clostridium botulinum* type A have the most radiation resistant spores. Since studies on food irradiation started, a number of bacteria which are highly resistant to radiation have been isolated and described. Although one of these, *Deinococcus radiodurans*, was first isolated from meat, their role in foods is not significant in the normal course of events.

Although patents describing the use of ionizing radiation in the treatment of food appeared soon after the discovery of radioactivity at the turn of the 20th century, it was not until after the Second World War that food irradiation assumed commercial potential. This was largely due to technological advances during the development of nuclear weapons,

Table 4.8 *Radiation resistance of some foodborne micro-organisms*

Species	6D dose (KGy)
E. coli	1.5–3.0
Salmonella Enteritidis	3–5
S. Typhimurium	3–5
Vibrio parahaemolyticus	<0.5–1
Pseudomonas fluorescens	0.5–1
Bacillus cereus	20–30
B. stearothermophilus	10–20
C. botulinum type A	20–30
Lactobacillus spp.	2–7.5
Micrococcus spp.	3–5
Deinococcus radiodurans	>30
Aspergillus flavus	2–3
Penicillium notatum	1.5–2
S. cerevisiae	7.5–10
Viruses	>30

Data from 'Microbial Ecology of Foods'. Vol. 1. ICMSF.

but also to a strong desire to demonstrate that nuclear technology could offer the human race something other than mass destruction. In particular, food irradiation has the advantage of being a much more precisely controlled process than heating, since penetration is deep, instantaneous and uniform. It also retains the fresh character of the product as low level irradiation produces no detectable sensory change in most products.

This failure of low doses of radiation to produce appreciable chemical change in the product has been an obstacle to the development of simple tests to determine whether a food has been irradiated. Although availability of such a test is not essential for the control of irradiation, it is generally accepted that it would facilitate international trade in irradiated food, enhance consumer confidence and help enforce labelling regulations. A number of methods have been developed that are applicable to specific types of food. Free radicals created by irradiation can be detected using electron spin resonance when they are trapped in solid matrices such as bone, seeds and shells. The energy stored in grains of silicate minerals as a result of irradiation can be measured in foods such as herbs and spices using thermoluminescence and long chain volatile hydrocarbons and 2-alkylcyclobutanones produced by irradiation of fatty foods can be detected using gas chromatography. One microbiological test for irradiated food is based on the ratio between an assessment of total microbial numbers using the DEFT technique (see Chapter 10) and a plate count to determine the number of viable bacteria present.

Food irradiation is not without its disadvantages, but a lot of the concerns originally voiced have proved to be unfounded. In 1981 an expert international committee of the FAO/WHO and the International Atomic Energy Authority recommended general acceptance of food irradiation up to a level of 10 kGy. They held the view that it 'constitutes no toxicological risk. Further toxicological examinations of such treated foods are therefore not required'.

It had been thought that irradiation could lead to pathogens becoming more virulent but, apart from one or two exceptions, it has been found that where virulence is affected it is diminished. In the exceptions noted, the effect was slight and not sufficient to compensate for the overall reduction in viable numbers. No example has been found where a non-pathogenic organism has been converted to a pathogen as a result of irradiation. Although it has been reported that spores of some mycotoxigenic moulds which survive irradiation may yield cultures with increased mycotoxin production.

Morphological, biochemical and other changes which may impede isolation and identification and increased radiation resistance have been noted as a result of repeated cyclic irradiation. However, these experiments were performed under the most favourable conditions and for this to occur in practice would require extensive microbial regrowth after each irradiation; a condition that is readily preventable by good hygienic practices and is most unlikely to occur.

The levels of radiation proposed for foods are not sufficient to induce radioactivity in the product and there is no evidence that consumption of irradiated foods is harmful. Food irradiation facilities do require stringent safety standards to protect workers but that is already in place for the irradiation of other materials such as the sterilization of medical supplies and disposables.

By far the greatest obstacle to the more widespread use of food irradiation is not technical but sociological in the form of extensive consumer resistance and distrust. Much of this is based on inadequate information and false propaganda and parallels very closely earlier arguments over the merits of milk pasteurization. Among the same objections raised then were that pasteurization would be used to mask poor quality milk and would promote poor practices in food preparation. While it has to be agreed that those who take the most cynical view of human nature are often proved correct, this did not prove to be the case with milk pasteurization where the production standards and microbiological quality of raw milk are now higher than they have ever been.

Depending on the lethality required, food irradiation can be applied at two different levels. At high levels it can be used to produce a safe shelf stable product in a treatment known as *radappertization*. Though this has been investigated in the context of military rations, it is unlikely to be a commercial reality in the forseeable future. *C. botulinum* spores are the most radiation resistant known, so very high doses are required to achieve the minimum standard of a 12D reduction ($\approx 45\,\text{kGy}$) for low-acid foods. In the event of a process failure, the growth of more resistant, non-pathogenic clostridia would not act as a warning as it can in thermal processing. High radiation doses are also more likely to produce unacceptable sensory changes and the product has to be irradiated in the frozen state to minimize migration of the radiolytic species that cause such changes. These considerations would not apply when the food was inhibitory to the growth of *C. botulinum* as a result of low pH or the presence of agents such as curing salts.

Two terms are used to distinguish different types of radiation pasteurization. *Radicidation* is used to describe processes where the objective is the elimination of a pathogen, as, for example, in the removal of *Salmonella* from meat and poultry. *Radurization* applies to processes aiming to prolong shelf-life. This distinction may be thought a little over elaborate since, as with thermal pasteurization, irradiation treatments are relatively non-discriminating and will invariably improve both safety and shelf-life.

Several potential applications have been identified (Table 4.9) and food irradiation for specific applications is now permitted in more than

Table 4.9 *Applications of food irradiation*

Application	Commodity	Dose (kGy)
Inhibition of sprouting	Potatoes	0.1–3
	Onions	
	Garlic	
	Mushrooms	
Decontamination of food ingredients	Spices	3–10
	Onion powder	
Insect disinfestation	Grains	0.2–7
Destruction of parasites	Meats	0.3–0.5
Inactivation of *Salmonella*	Poultry	3–10
	Eggs	
	Shrimps and frog's legs	
Delay in fruit maturation	Strawberries	2–5
	Mangoes	
	Papayas	
Mould and yeast reduction		1–3

Table 4.10 *Foods which may be treated with ionizing radiation in the UK*

	Maximum permitted dose (kGy)
Fruit and mushrooms	2
Vegetables	1
Cereals	1
Bulbs and tubers	0.2
Spices and condiments	10
Fish and shellfish	3
Poultry	7

50 countries, the USA, South Africa, the Netherlands, Thailand and France being among the leading exponents. In South Africa 1,754 tonnes of herbs and spices were irradiated in 2004 and in the USA in 2003 22,000 tonnes of hamburgers were irradiated. In the UK, the applications listed in Table 4.10 have been permitted since 1991, although consumer resistance and the requirement that irradiated foods are labelled as such have meant that, to-date, only one licence has been granted covering the treatment of some herbs and spices.

4.3 HIGH-PRESSURE PROCESSING–PASCALIZATION

Hite, working at the University of West Virginia Agricultural Experimental Station at the turn of the 20th century, showed that high hydrostatic pressures, around 650 MPa (6500 atm), reduced the microbial load in foods such as milk, meats and fruits. He found that 680 MPa applied for 10 min at room temperature reduced the viable count of milk from 10^7 cfu ml^{-1} to 10^1–10^2 cfu ml^{-1} and that peaches and pears subjected to 410 MPa for 30 min remained in good condition after 5 years storage. He also noted that the microbicidal activity of high pressure is enhanced by low pH or temperatures above and below ambient.

Since then, microbiologists have continued to study the effect of pressure on micro-organisms, although this work has centred on organisms such as those growing in the sea at great depths and pressures. Interest in the application of high pressures in food processing, sometimes called *pascalization*, lapsed until the 1980s when progress in industrial ceramic processing led to the development of pressure equipment capable of processing food on a commercial scale and a resurgence of interest, particularly in Japan.

High hydrostatic pressure acts primarily on non-covalent linkages, such as ionic bonds, hydrogen bonds and hydrophobic interactions, and it promotes reactions in which there is an overall decrease in volume. It can have profound effects on proteins, where such interactions are critical to structure and function, although the effect is variable and depends on individual protein structure. Some proteins such as those of egg, meat and soya form gels and this has been employed to good effect in Japan where high pressure has been used to induce the gelation of fish proteins in the product surimi. Other proteins are relatively unaffected and this can cause problems when they have enzymic activity which limits product shelf-life. Pectin esterase in orange juice, for instance, must be inactivated to stabilize the desired product cloudiness. Nonprotein macromolecules can also be affected by high pressures so that pascalized starch products often taste sweeter due to conformational changes in the starch which allow salivary amylase greater access.

Adverse effects on protein structure and activity obviously contribute to the antimicrobial effect of high pressures, although the cell membrane also appears to be an important target. Membrane lipid bilayers have been shown to compress under pressure and this alters their permeability. As a general rule vegetative bacteria and fungi can be reduced by at least one log cycle by 400 MPa applied for 5 min.

Bacterial endospores are more resistant to hydrostatic pressure, tolerating pressures as high as 1200 MPa. Their susceptibility can be increased considerably by modest increases in temperature, when quite low pressures (100 MPa) can produce spore germination, a process in which the spores lose their resistance to heat and to elevated pressure.

High pressure processing is typically a batch process employing a pressure vessel, the pressure transmission fluid (usually water) and pumps to generate the pressure. Although the capital cost of equipment is quite high, hydrostatic processing has a number of appealing features for the food technologist. It acts instantly and uniformly throughout a food so that the processing time is not related to container size and there are none of the penetration problems associated with heat processing.

With the exceptions noted above, adverse effects on the product are slight; nutritional quality, flavour, appearance and texture resemble the fresh material very closely. To the consumer it is a 'natural' process with none of the negative associations of processes such as irradiation or chemical preservatives.

Initially, commercial application of high-pressure technology was limited mainly to acidic products. The yeasts and moulds normally responsible for spoilage in these products are pressure sensitive and the bacterial spores that survive processing are unable to grow at the low pH. In 1990, the Meidi-Ya company in Japan launched a range of jams treated at 400–500 MPa in pack. These have a chill shelf-life of 60 days and have sensory characteristics quite different from conventional heat-processed jams since more fresh fruit flavour and texture are retained. Refrigeration is necessary to limit residual enzyme activities which give rise to browning and flavour changes. Other products introduced include salad dressings, fruit sauces, and fruit flavoured yoghurts. More recently a number of pressure-treated foods have been introduced in Europe, the United States and elsewhere. These include fruit purees and juices and some more novel products such as guacamole, cooked ham and oysters. Pressure-treated guacamole has been a success in the USA where pressures of around 500 MPa for 2 minutes extend its chill shelf life from 7 to 30 days. Similar treatments are applied to packs of sliced cooked ham and other delicatessen meat products in Italy, Spain, Germany, the USA and Japan to reduce the risk posed by any post-cooking contamination with *Listeria monocytogenes*. In the United States and South Korea, pressure-treated oysters are also available. The process used releases the adductor muscle which holds the oyster shell closed, so it has the dual safety benefits of eliminating any *Vibrio* species present as well as reducing the number of stab injuries incurred during abortive attempts to open the shell.

In the future, the range of products may be increased by coupling moderate pressure with a heat treatment equivalent to pasteurization. In one trial, shelf stable, low acid foods were produced by combining a pressure of just 0.14 MPa with heating at temperatures of 82–103 °C. Other developments such as equipment capable of semi- or fully-continuous operation will also considerably improve commercial feasibility, so that we may see and hear a lot more about pascalization.

4.4   LOW-TEMPERATURE STORAGE – CHILLING AND FREEZING

The rates of most chemical reactions are temperature dependent; as the temperature is lowered so the rate decreases. Since food spoilage is usually a result of chemical reactions mediated by microbial and endogenous enzymes, the useful life of many foods can be increased by storage at low temperatures. Though this has been known since antiquity, one of the earliest recorded experiments was conducted by the English natural philosopher Francis Bacon who in 1626 stopped his coach in Highgate in order to fill a chicken carcass with snow to confirm that it delayed putrefaction. This experiment is less notable for its results, which had no immediate practical consequences, than for its regrettable outcome. As a result of his exertions in the snow, it is claimed Bacon caught a cold which led to his death shortly after.

Using low temperatures to preserve food was only practicable where ice was naturally available. As early as the 11th century BC the Chinese had developed ice houses as a means of storing ice through the summer months, and these became a common feature of large houses in Europe and North America in the 17th and 18th centuries. By the 19th century, the cutting and transporting of natural ice had become a substantial industry in areas blessed with a freezing climate.

Mechanical methods of refrigeration and ice making were first patented in the 1830s. These were based on the cooling produced by the vaporization of refrigerant liquids, originally ether but later liquid ammonia. Much early development work was done in Australia where there was considerable impetus to find a way of transporting the abundant cheap meat available locally to European population centres. At the 1872 Melbourne Exhibition, Joseph Harrison exhibited an 'ice house' which kept beef and mutton carcasses in good condition long enough for some of it to be eaten at a public luncheon the following year. This banquet was to send off a steamship to London carrying 20 tons of frozen mutton and beef packed in tanks cooled by ice and salt. Unfortunately it was an inauspicious start, during passage through the tropics the ice melted and most of the meat had been thrown overboard before the ship reached London. Chilled rather than frozen meat had however already been successfully shipped the shorter distance from North America to Europe and by the end of the century techniques had been refined to the extent that shipping chilled and frozen meat from North and South America and Australia to Europe was a large and profitable enterprise.

Since then, use of chilling and freezing has extended to a much wider range of perishable foods and to such an extent that refrigeration is now arguably the technology of paramount importance to the food industry.

4.4.1 Chill Storage

Chilled foods are those foods stored at temperatures near, but above their freezing point, typically 0–5 °C. This commodity area has shown a massive increase in recent years as traditional chilled products such as fresh meat and fish and dairy products have been joined by a huge variety of new products including complete meals, prepared and delicatessen salads, dairy desserts and many others. Three main factors have contributed to this development:

(1) the food manufacturers' objective of increasing added value to their products;
(2) consumer demand for fresh foods and ease of preparation while at the same time requiring the convenience of only occasional shopping excursions; and
(3) the availability of an efficient cold chain – the organization and infrastructure which allows low temperatures to be maintained throughout the food chain from manufacture/harvest to consumption.

Chill storage can change both the nature of spoilage and the rate at which it occurs. There may be qualitative changes in spoilage characteristics, as low temperatures exert a selective effect preventing the growth of mesophiles and leading to a microflora dominated by psychrotrophs. This can be seen in the case of raw milk which in the days of milk churns and roadside collection had a spoilage microflora comprised largely of mesophilic lactococci which would sour the milk. Nowadays in the UK, milk is chilled almost immediately it leaves the cow so that psychrotrophic Gram-negative rods predominate and produce an entirely different type of spoilage. Low temperatures can also cause physiological changes in micro-organisms that modify or exacerbate spoilage characteristics. Two such examples are the increased production of phenazine and carotenoid pigments in some organisms at low temperatures and the stimulation of extracellular polysaccharide production in *Leuconostoc* spp. and some other lactic acid bacteria. In most cases, such changes probably represent a disturbance of metabolism due to the differing thermal coefficients and activation energies of the numerous chemical reactions that comprise microbial metabolism.

Though psychrotrophs can grow in chilled foods they do so only relatively slowly so that the onset of spoilage is delayed. In this respect temperature changes within the chill temperature range can have pronounced effects. For example, the generation time for one pseudomonad isolated from fish was 6.7 hours at 5 °C compared with 26.6 hours at 0 °C. Where this organism is an important contributor to spoilage, small changes of temperature will have major implications for shelf-life. The keeping time of haddock and cod fillets has been found to double if the storage temperature is decreased from 2.8 °C to −0.3 °C. Mathematical modelling techniques of the sort described in Section 3.4 can be useful in predicting the effect of temperature fluctuations on shelf-life, but, as a general rule, storage temperature should be as low, and as tightly controlled, as possible.

The ability of organisms to grow at low temperatures appears to be particularly associated with the composition and architecture of the plasma membrane (see Section 3.3.2). As the temperature is lowered, the plasma membrane undergoes a phase transition from a liquid crystalline state to a rigid gel in which solute transport is severely limited. The temperature of this transition is lower in psychrotrophs and psychrophiles largely as a result of higher levels of unsaturated and short chain fatty acids in their membrane lipids. If some organisms are allowed to adapt to growth at lower temperatures they increase the proportion of these components in their membranes.

There seems to be no taxonomic restriction on psychrotrophic organisms which can be found in the yeasts, moulds, Gram-negative and Gram-positive bacteria. One feature they share is that in addition to their ability to grow at low temperatures, they are inactivated at moderate temperatures. A number of reasons for this marked heat sensitivity have been put forward including the possibility of excessive membrane fluidity at higher temperatures. Low thermal stability of key enzymes and other functional proteins appears to be an important factor, although thermostable extracellular lipases and proteases produced by psychrotrophic pseudomonads can be a problem in the dairy industry.

Though mesophiles cannot grow at chill temperatures, they are not necessarily killed. Chilling will produce a phenomenon known as cold shock which causes death and injury in a proportion of the population but its effects are not predictable in the same way as heat processing. The extent of cold shock depends on a number of factors such as the organism (Gram-negatives appear more susceptible than Gram-posi-tives), its phase of growth (exponential-phase cells are more susceptible than stationary phase cells), the temperature differential and the rate of cooling (in both cases the larger it is, the greater the damage), and the growth medium (cells grown in complex media are more resistant).

The principal mechanism of cold shock appears to be damage to membranes caused by phase changes in the membrane lipids which create hydrophilic pores through which cytoplasmic contents can leak out. An increase in single-strand breaks in DNA has also been noted as well as the synthesis of specific cold-shock proteins to protect the cell.

Since chilling is not a bacteriocidal process, the use of good microbiological quality raw materials and hygienic handling are key requirements for the production of safe chill foods. Mesophiles that survive cooling, albeit in an injured state, can persist in the food for extended periods and may recover and resume growth should conditions later become favourable. Thus chilling will prevent an increase in the risk from mesophilic pathogens, but will not assure its elimination. There are however pathogens that will continue to grow at some chill temperatures and the key role of chilling in the modern food industry has focused particular attention on these. Risks posed by these organisms, which are given more detailed attention in Chapter 7, may increase with duration of storage but this process is likely to be slow and dependent on the precise storage temperature and composition of the food.

Some foods are not suitable for chill storage as they suffer from cold injury where the low temperature results in tissue breakdown which leads to visual defects and accelerated microbiological deterioration. Tropical fruits are particularly susceptible to this form of damage.

4.4.2   Freezing

Freezing is the most successful technique for long-term preservation of food since nutrient content is largely retained and the product resembles the fresh material more closely than in appertized foods.

Foods begin to freeze somewhere in the range -0.5 to $-3\,°C$, the freezing point being lower than that of pure water due to the solutes present. As water is converted to ice during freezing, the concentration of solutes in the unfrozen water increases, decreasing its freezing point still further so that even at very low temperatures, *e.g.* $-60\,°C$, some water will remain unfrozen. The temperatures used in frozen storage are generally less than $-18\,°C$. At these temperatures no microbial growth is possible, although residual microbial or endogenous enzyme activity such as lipases can persist and eventually spoil a product. This is reduced in the case of fruits and vegetables by blanching before freezing to inactivate endogenous polyphenol oxidases which would otherwise cause the product to discolour during storage. Freezer burn is another non-microbiological quality defect that may arise in frozen foods, where surface discolouration occurs due to sublimation of water from the product and its transfer to colder surfaces in the freezer. This can be prevented by wrapping products in a water-impermeable material or by glazing with a layer of ice.

Low temperature is not the only inhibitory factor operating in frozen foods; they also have a low water activity produced by removal of water in the form of ice. Table 4.11 describes the effect of temperature on water activity. As far as microbiological quality is concerned, this effect is only significant when frozen foods are stored at temperatures where microbial growth is possible (above $-10\,°C$). In this situation, the organisms that grow on a product are not those normally associated with its spoilage at chill temperatures but yeasts and moulds that are both psychrotrophic and tolerant of reduced water activity. Thus meat and poultry stored at -5 to $-10\,°C$ may slowly develop surface defects such as black spots due to the growth of the mould *Cladosporium herbarum*, white spots caused by *Sporotrichum carnis* or the feathery growth of *Thamnidium elegans*.

Table 4.11 *Effect of freezing on the water activity of pure water-ice*

Temperature (°C)	a_w
0	1
−5	0.953
−10	0.907
−15	0.864
−20	0.823
−40	0.68

Micro-organisms are affected by each phase of the freezing process. In cooling down to the temperature at which freezing begins, a proportion of the population will be subject to cold shock discussed in Section 4.4.1 above. At the freezing temperature, further death and injury occur as the cooling curve levels out as latent heat is removed and the product begins to freeze. Initially ice forms mainly extracellularly, intracellular ice formation being favoured by more rapid cooling. This may mechanically damage cells and the high extracellular osmotic pressures generated will dehydrate them. Changes in the ionic strength and pH of the water phase as a result of freezing will also disrupt the structure and function of numerous cell components and macromolecules which depend on these factors for their stability. Cooling down to the storage temperature will prevent any further microbial growth once the temperature has dropped below −10 °C. Finally, during storage there will be an initial decrease in viable numbers followed by slow decline over time. The lower the storage temperature, the slower the death rate.

As with chilling, freezing will not render an unsafe product safe – its microbial lethality is limited and preformed toxins will persist. Frozen chickens are, after all, an important source of *Salmonella*.

Survival rates after freezing will depend on the precise conditions of freezing, the nature of the food material and the composition of its microflora, but have been variously recorded as between 5 and 70%. Bacterial spores are virtually unaffected by freezing, most vegetative Gram-positive bacteria are relatively resistant and Gram-negatives show the greatest sensitivity. While frozen storage does reliably inactivate higher organisms such as pathogenic protozoa and parasitic worms, food materials often act as cryoprotectants for bacteria so that bacterial pathogens may survive for long periods in the frozen state. In one extreme example *Salmonella* has been successfully isolated from ice cream stored at −23 °C for 7 years.

The extent of microbial death is also determined by the rate of cooling. Maximum lethality is seen with slow freezing where, although there is little or no cold shock experienced by the organisms, exposure to high solute concentrations is prolonged. Survival is greater with rapid freezing where exposure to these conditions is minimized. Food freezing processes are not designed however to maximize microbial lethality but to minimize loss of product quality. Formation of large ice crystals and prolonged exposure to high osmotic pressure solutions during slow cooling also damage cells of the food material itself causing greater drip loss and textural deterioration on thawing, so fast freezing in which the product is at storage temperature within half an hour is the method of choice commercially. The rate of freezing in domestic freezers is much slower so, although microbial lethality may be greater, so too is product quality loss.

Thawing of frozen foods is a slower process than freezing. Even with moderate size material the outside of the product will be at the thawing temperature some time before the interior. So with high thawing temperature, mesophiles may be growing on the surface of a product while the interior is still frozen. Slow thawing at lower temperature is generally preferred. It does have some lethal effect as microbial cells experience adverse conditions in the 0 to $-10\,°C$ range for longer, but it will also allow psychrotrophs to grow. Provided the product is not subject to contamination after thawing, the microflora that develops will differ from that on the fresh material due to the selective lethal effect of freezing. Lactic acid bacteria are often responsible for the spoilage of defrosted vegetables whereas they generally comprise only about 1% of the microflora on fresh chilled produce which is predominantly Gram-negative.

Freezing and defrosting may make some foods more susceptible to microbiological attack due to destruction of antimicrobial barriers in the product and condensation, but defrosted foods do not spoil more rapidly than those that have not been frozen. Injunctions against refreezing defrosted products are motivated by the loss of textural and other qualities rather than any microbiological risk that is posed.

4.5 CHEMICAL PRESERVATIVES

The addition of chemicals to food is not a recent innovation but has been practised throughout recorded history. Doubtless too, there has also always been a certain level of misuse but this must have gone largely undetected until modern analytical techniques became available. When chemical analysis and microscopy were first applied to foods in the early 19th century, they revealed the appalling extent of food adulteration then current. Pioneering work had been done by the 18th century chemist Jackson, but publication of the book 'A Treatise on Adulterations of Food, and Culinary Poisons' by Frederick Accum in 1820 marks a watershed. Accum exposed a horrifying range of abuses such as the sale of sulfuric acid as vinegar, the use of copper salts to colour pickles, the use of alum to whiten bread, addition of acorns to coffee, blackthorn leaves to tea, cyanide to give wines a nutty flavour and red lead to colour

Gloucester cheese. These and subsequent investigations, notably those sponsored by the journal *Lancet*, led directly to the introduction of the first British Food and Drugs Act in 1860. Despite the protection of a much stricter regulatory framework, occasional triumphs of human cupidity are still recorded today. Further examples include the use of ethylene glycol in some Austrian wines, the intrepid entrepreneur who sold grated umbrella handles as Parmesan cheese and the grim case of the Spanish toxic cooking oil scandal which killed or maimed hundreds.

Although some would regard all chemical additions to food as synonymous with adulteration, many are recognized as useful and are allowed. Additives may be used to aid processing, to modify a food's texture, flavour, nutritional quality or colour but, here, we are concerned with those which primarily effect keeping quality: preservatives.

Preservatives are defined as 'substances capable of inhibiting, retarding or arresting the growth of micro-organisms or of any deterioration resulting from their presence or of masking the evidence of any such deterioration'. They do not therefore include substances which act by inhibiting a chemical reaction which can limit shelf-life, such as the control of rancidity or oxidative discolouration by antioxidants. Neither does it include a number of food additives which are used primarily for other purposes but have been shown to contribute some antimicrobial activity. These include the antioxidants, butylated hydroxytoluene (BHT) and butylated hydroxyanisole (BHA), and the phosphates used as acidity regulators and emulsifiers in some products.

Preservatives may be microbicidal and kill the target organisms or they may be microbistatic in which case they simply prevent them growing. This is very often a dose-dependent feature; higher levels of an antimicrobial proving lethal while the lower concentrations that are generally permitted in foods tend to be microbistatic. For this reason chemical preservatives are useful only in controlling low levels of contamination and are not a substitute for good hygiene practices.

Recently consumers have shown an inclination to regard preservatives as in some way 'unnatural', even though the use of salts, acid, or smoke to preserve foods goes back to the beginning of civilization. Usage of chemical preservatives is now more restricted and controlled than ever and in many areas it is declining. It is perhaps well to remember though that only the fairly recent advent of technologies such as canning and refrigeration has allowed us any alternative to chemical preservation or drying as a means of extending the food supply.

4.5.1 Organic Acids and Esters

The most important organic acids and esters that are used as food preservatives are listed in Table 4.12 along with their E-numbers (EU

Table 4.12 *Organic acid food preservatives*

Preservative	[a]ADI (mg kg^{-1} body wt).	Typical usage and levels	(mg kg^{-1})
E200 Sorbic acid E201 Sodium salt E202 Potassium salt E203 Calcium salt	25	Salad dressing bakery products fruit desserts	<2000
E210 Benzoic acid E211 Sodium salt E212 Potassium salt E213 Calcium salt	5	Cider, soft drinks, fruit products, bottled sauces	<3000
E260 Acetic acid	No limit	Pickles, sauces chutneys	up to % levels (1% = 10 000 mg kg^{-1})
E270 Lactic acid	No limit	Fermented meats dairy and vegetable products. Sauces and dressings. Drinks.	up to % levels (1% = 10 000 mg kg^{-1})
E280 Propionic acid E281 Sodium salt E282 Calcium salt E283 Potassium salt	10	Bakery goods Cheese spread	1000–5000
Parabens E214 *p*-Hydroxybenzoic acid ethyl ester E215 Sodium salt	10	Bakery goods, pickles, fruit products, sauces	<2000
E216 *p*-Hydroxybenzoic acid *n*-propyl ester E217 Sodium salt		Bakery goods, pickles, fruit products, sauces	<2000
E218 *p*-Hydroxybenzoic acid methyl ester E219 Sodium salt		Bakery goods, pickles, fruit products, sauces	<2000

[a] ADI Acceptable daily intake

codes for food additives used throughout the European Union). Their structures are presented in Figure 4.11.

The antimicrobial effect of organic acids such as *acetic* and *lactic acids* has been discussed in Chapter 3. Both are produced microbiologically, although food-grade acetic acid derived petrochemically is also sometimes used as an alternative to vinegar. They can be an added ingredient in formulated products such as pickles and sauces, or they can be generated *in situ* in the large range of lactic-fermented products described

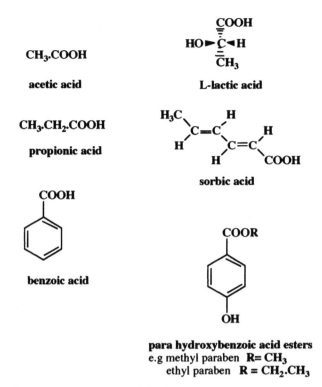

CH₃.COOH

acetic acid

CH₃.CH₂.COOH

propionic acid

COOH

benzoic acid

L-lactic acid

sorbic acid

para hydroxybenzoic acid esters
e.g methyl paraben R= CH₃
ethyl paraben R = CH₂.CH₃

Figure 4.11   *Structures of organic acid food preservatives*

in Chapter 9. They differ from the other acids and esters described here in that they are usually present in amounts sufficient to exert an effect on flavour and on product pH, thus potentiating their own action by increasing the proportion of undissociated acid present.

Benzoic acid occurs naturally in cherry bark, cranberries, greengage plums, tea and anise but is prepared synthetically for food use. Its antimicrobial activity is principally in the undissociated form and since it is a relatively strong acid (pK_a 4.19) it is effective only in acid foods. As a consequence, its practical use is to inhibit the growth of spoilage yeasts and moulds. Activity against bacteria has been reported but they show greater variability in their sensitivity.

Inhibition by benzoic acid appears multifactorial. The ability of the undissociated molecule to interfere with membrane energetics and function appears to be of prime importance since growth inhibition has been shown to parallel closely the inhibition of amino acid uptake in whole cells and membrane vesicles. Some inhibition may also result from benzoic acid once it is inside the cell as a number of key enzyme activities have also been shown to be adversely affected.

Parabens (*para*-hydroxybenzoic acid esters) differ from the other organic acids described here in the respect that they are phenols rather than carboxylic acids. They are much weaker acids with pK_a values of 8.5 and so are predominantly uncharged even at neutral pH. This means that they can be used effectively in non-acidic foods. Their antimicrobial activity increases with the length of the ester group carbon chain, although this also decreases their water solubility and may lead to poor performance in some foods where partition into the fatty phase may occur. Some Gram-negatives are resistant to the higher homologues and this has been ascribed to the cell's outer membrane acting as a barrier.

Parabens appear to act mainly at the cell membrane eliminating the ΔpH component of the protonmotive force and affecting energy transduction and substrate transport. In contrast to other weak acid preservatives, there is little evidence suggesting that parabens interfere directly with specific enzymic activities.

Sorbic acid is an unsaturated fatty acid, 2,4-hexadienoic acid, found naturally in the berries of the mountain ash. It has a pK_a of 4.8 and shows the same pH dependency of activity as other organic acids. It is active against yeasts, moulds and catalase-positive bacteria but, interestingly, is less active against catalase-negative bacteria. This has led to its use as a selective agent in media for clostridia and lactic acid bacteria and as a fungal inhibitor in lactic fermentations.

As with the other weak acids, the membrane is an important target for sorbic acid, although inhibition of a number of key enzymes of intermediary metabolism, such as enolase, lactate dehydrogenase and several Krebs cycle enzymes, has been shown. In contrast to its use as a selective agent for clostridia, some studies have shown that sorbic acid inhibits the germination and outgrowth of *C. botulinum* spores. At one time this attracted some interest in the possibility that sorbic acid could be used as an alternative or adjunct to nitrite in cured meats.

Propionic acid (pK_a 4.9) occurs in a number of plants and is also produced by the activity of propionibacteria in certain cheeses. It is used as a mould inhibitor in cheese and baked products where it also inhibits rope-forming bacilli. Objections to the use of preservatives led, in the late 1980s, to the increased use of acetic acid in the form of vinegar as an alternative to propionate but the complete omission of a rope inhibitor has had serious consequences for the public on at least one occasion (see Section 7.2.5).

4.5.2 Nitrite

The antibacterial action of nitrite was first described in the 1920s though it had long been employed unwittingly in the production of cured meats where it is also responsible for their characteristic colour and flavour. In early curing processes nitrite was produced by the bacterial reduction of nitrate present as an impurity in the crude salt used, but now nitrate, or more commonly nitrite itself, is added as the sodium or potassium salt.

Nitrite is inhibitory to a range of bacteria. Early workers showed that a level of $200\,mg\,kg^{-1}$ at pH 6.0 was sufficient to inhibit strains of *Escherichia, Flavobacterium, Micrococcus, Pseudomonas* and others, although *Salmonella* and *Lactobacillus* species were more resistant. Of most practical importance though is the ability of nitrite to inhibit spore-forming bacteria such as *Clostridium botulinum* which will survive the heat process applied to many cured meats. To achieve this commercially, initial levels of nitrite greater than $100\,mg\,kg^{-1}$ are used. The mechanism of its action is poorly understood partly due to the complexity of the interaction of several factors such as pH, salt content, presence of nitrate or nitrite and the heat process applied to the cured meat. Descriptive mathematical models of these interactions have however been produced which quantify the precise contribution of nitrite to safety (see Section 3.5).

Bacterial inhibition by nitrite increases with decreasing pH, suggesting that nitrous acid (HNO_2, pK_a 3.4) is the active agent. In the case of spores, it appears that nitrite acts by inhibiting the germination and outgrowth of heated spores and by reacting with components in the product to form other inhibitory compounds. The latter effect was first noted in the 1960s by Perigo who observed that when nitrite was heated in certain bacteriological media, the resulting medium proved more inhibitory to clostridia than when filter-sterilized nitrite was added after heating. Clostridia are very sensitive to these 'Perigo factors' which differ from nitrite in displaying activity that is independent of pH. However, they do not seem to be formed in meat and their effect in bacteriological media could be removed if meat was added. The presence of 'Perigo-type factors' has been reported in heated cured meats but these are only produced by severe heating and have minor antibacterial activity.

Studies into the nature of Perigo and Perigo-type factors have looked particularly at the production of Roussin's salts; complex salts of iron, nitrosyl and sulfydryl groups. Although these compounds have not been shown to be present in cured meats in sufficient quantity to cause the inhibition observed, their formation may give an indication of the way nitrite itself interferes with bacterial metabolism. It has been proposed that the biochemical mechanism of inhibition involves nitrite reacting with iron and sulfhydryl groups of key cell constituents. Iron-containing proteins such as ferredoxins are very important in electron transport and energy production in clostridia. For example the phosphoroclastic system is used by clostridia to generate additional ATP by substrate-level phosphorylation. Pyruvate, produced by glycolysis, is oxidized to acetate *via* acetyl-CoA and acetyl phosphate which phosphorylates ADP to

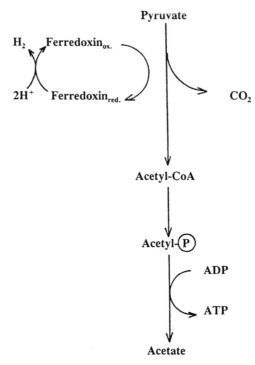

Figure 4.12   *The phosphoroclastic system*

produce ATP (Figure 4.12). Ferredoxin acts as a carrier for the electrons removed in the oxidation step and which are ultimately used to reduce hydrogen ions to hydrogen gas. Support for this hypothesis has come from the observation that nitrite addition leads to an accumulation of pyruvate in *C. sporogenes* and *C. botulinum*.

The use of nitrite and nitrate in food has attracted scrutiny since it was discovered in the 1950s that *N*-nitrosamines, formed by the reaction of nitrite with secondary amines, especially at low pH, can be carcinogenic. Concern developed that they may be present in food or formed in the body as a result of ingestion of nitrate or nitrite with food. Surveys have indicated that cured meats, particularly fried bacon, beer and, in some countries, fish, make the most significant contribution to dietary intakes of nitrosamines, although a US survey made the point that a smoker inhaled about 100 times the amount of volatile nitrosamines per day as were provided by cooked bacon. Dietary intake of nitrite is low, generally less than $2\,\text{mg NaNO}_2\,\text{day}^{-1}$, and comes mainly from cured meats, although it is also present in fish, cheese, cereals, and vegetable products. Nitrate is also of concern since it can be reduced to nitrite by the body's own microflora. Cured meats are not a significant source; vegetables contribute more than 75% of the dietary intake of nitrate, although water can be an important source in some areas.

Awareness of the problem has led to changes in production practices for cured meats such as the use of low levels of nitrite in preference to nitrate and the increased use of ascorbic acid which inhibits the nitrosation reaction. These measures have produced significant reductions in nitrosamine levels.

Mention should also be made here of the other contributions made by nitrite to the quality of cured meats. Reduction of nitrite to nitric oxide produces the characteristic red colour of cured meats. The nitric oxide co-ordinates to the haem ferrous ion in the muscle pigment myoglobin converting it to nitrosomyoglobin (Figure 4.13). When raw cured meats such as bacon are cooked this pigment decomposes to produce nitrosylhaemochrome which has the pink colour also seen in cooked cured hams. Only small quantities of nitrite are required to produce the cured meat colour: theoretically $3 \, \text{mg kg}^{-1}$ is sufficient to convert half the myoglobin present in fresh meat, but because of competing reactions, $25 \, \text{mg kg}^{-1}$ are required to give a stable colour.

Nitrite also contributes to the typical cured meat flavour. Taste panels can distinguish cured meats where nitrite has not been used but the

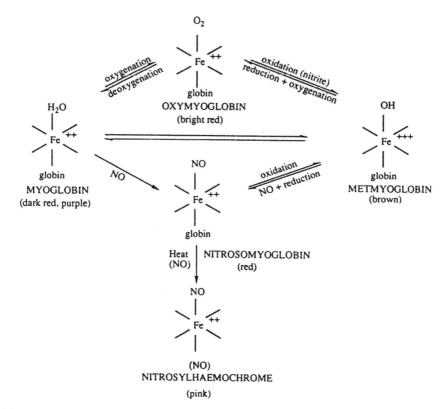

Figure 4.13 *Reactions of nitrite with meat myoglobin*

precise reasons for this are not known. It is thought that nitrite acts as an antioxidant to inhibit lipid degradation in the meat although this may only be part of the story.

4.5.3 Sulfur Dioxide

Sulfur dioxide (SO_2) has long enjoyed a reputation for its disinfecting properties and its earliest use in the food industry was when sulfur candles were burnt to disinfect the vessels used to produce and store wine. Nowadays, it is also used as an antioxidant to inhibit enzymic and non-enzymic browning reactions in some products.

Sulfur dioxide is a colourless gas that readily dissolves in water to establish a pH-dependent equilibrium similar to CO_2 (see Section 3.3.3).

$$SO_{2(gas)} \rightleftharpoons SO_2 + H_2O \rightleftharpoons H_2SO_3 \rightleftharpoons H^+ + HSO_3^- \rightleftharpoons 2H^+ + SO_3^{2-} \quad (4.15)$$

Sulfurous acid (H_2SO_3) is a dibasic acid with pK_a values of 1.86 and 6.91.

The unionized forms of SO_2 which can readily penetrate the cell have the greatest antimicrobial activity. It has been reported that they are between 100 and 1000 times more active than the bisulfite anion. Since the unionized forms predominate at low pH values, it follows that SO_2 is used to best effect in acidic foods. At neutral pH, SO_2 is present as a mixture of the relatively inactive bisulfite (HSO_3^-) and sulfite (SO_3^{2-}) ions, although salts of these anions prove the most convenient way of handling the preservative in the food industry.

SO_2 is a reactive molecule and can disrupt microbial metabolism in a number of ways. As a reducing agent, it can break disulfide linkages in proteins and interfere with redox processes. It can also form addition compounds with pyrimidine bases in nucleic acids, sugars and a host of key metabolic intermediates. One disadvantageous consequence of this reactivity is its ability to destroy the vitamin thiamine in foods and the once widespread practice of using it in meat and meat products has now been prohibited, with the exception of British fresh sausage.

Sulfur dioxide is active against bacteria, yeasts and moulds, although some yeasts and moulds are more resistant. Gram-negative bacteria are most susceptible and in British fresh sausage where sulfite is permitted up to a level of $450\,mg\,kg^{-1}$, the Gram-negative spoilage flora normally associated with chilled meats is replaced by one dominated by Gram-positive bacteria and yeasts. In winemaking the tolerance of the wine yeast *Saccharomyces cerevisiae* to SO_2 levels around $100\,mg\,l^{-1}$ is exploited to control the growth of wild yeasts and acetifying bacteria. Seasonal surpluses of soft fruits are also preserved by the addition of high levels of SO_2 to permit jam production throughout the year.

4.5.4 Natamycin

Natamycin (Figure 4.14), formerly know as pimaricin, is a polyene macrolide antibiotic produced by the bacterium *Streptomyces natalensis.* It is a very effective antifungal agent as it binds irreversibly to the fungal sterol, ergosterol, disrupting the fungal cell membrane leading to a loss of solutes from the cytoplasm and cell lysis. Natamycin is poorly soluble in water and is used as an aqueous suspension for the surface treatment of cheeses and sausages to control yeast and mould growth. It has some advantages over sorbate in this respect since it remains localised on the surface of the product, is not dependent on a low pH for its activity and has no effect on bacteria important in the fermentation and maturation of such products.

4.5.5 'Natural' Food Preservatives

The uncertainty voiced by consumer organisations and pressure groups over the use of food additives including preservatives has already been referred to. One approach to reassuring the consumer has been recourse to methods of preservation that can be described as 'natural'. The whole area though is riddled with inconsistency and contradiction; it can be argued that any form of preservation which prevents or delays the recycling of the elements in plant and animal materials is unnatural.

On the other hand there is nothing more natural than strychnine or botulinum toxin. Smoking of foods might be viewed as a natural method of preservation. Its antimicrobial effect is a result of drying and the activity of woodsmoke components such as phenols and formaldehyde which would probably not be allowed were they to be proposed as chemical preservatives in their own right.

The use of natural food components possessing antimicrobial activity such as essential oils and the lactoperoxidase system in milk (see Section

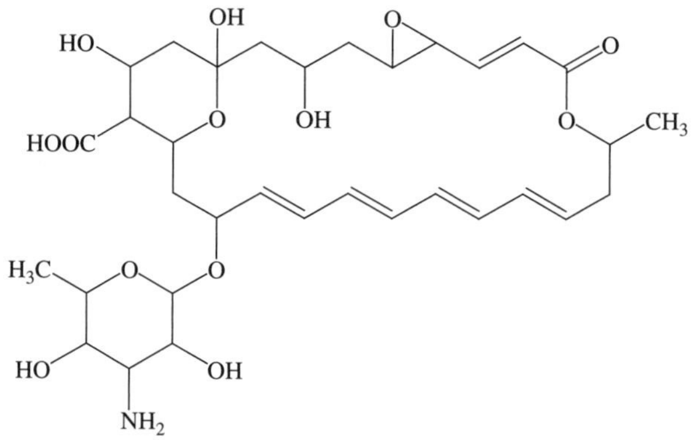

Figure 4.14 *Natamycin*

3.2.4) have attracted some attention in this respect. Attention has also been paid to the bacteriocins produced by food-grade micro-organisms such as the lactic acid bacteria. Nisin (see Section 9.4.1) is an already well established example and its use can be extended by expedients such as inclusion of whey fermented by a nisin-producing strain of *Lactococcus lactis* as an ingredient in formulated products like prepared sauces.

4.6   MODIFICATION OF ATMOSPHERE

At the start of the 19th century it was believed that contact with air caused putrefaction and that food preservation techniques worked by excluding air. We have already seen (Section 4.1) how this misapprehension applied in the early days of canning and it was thought that drying operated in a similar way, expelling air from the interior of food. Some preservation techniques, such as covering a product with melted fat and allowing it to set, did in fact rely on the exclusion of air but it is only in the last 30 years or so that shelf-life extension techniques based on changing the gaseous environment of a food have really come to be widely used.

Modified atmospheres exert their effect principally through the inhibition of fast-growing aerobes that would otherwise quickly spoil perishable products. Obligate and facultative anaerobes such as clostridia and the Enterobacteriaceae are less affected. Thus keeping quality is improved but there is generally little effect on pathogens, if present, and the technique is invariably applied in conjunction with refrigerated storage.

In practice three different procedures are used to modify the atmosphere surrounding a product: vacuum packing, modified-atmosphere packing or gas flushing, and controlled atmospheres. Here we will discuss some of their important characteristics although other aspects will be dealt with under specific commodities in Chapter 5.

An essential feature of all three techniques is that the product is packed in a material which helps exclude atmospheric oxygen and retain moisture. This requires that it should have good barrier properties towards oxygen and water and be easily sealed. The packaging materials used are usually plastic laminates in which the innermost layer is a plastic such as polyethylene which has good heat sealing properties. Mechanical closures on packs are far less effective as they often leave channels through which high rates of gas exchange can occur. Overlying the layer of polythene is usually another layer with much better gas barrier properties. No plastics are completely impermeable to gases, although the extent of gas transmission across a plastic film will depend on the the type of plastic, its temperature, the film thickness and the partial pressure difference across the film. In some cases, it can also be affected by factors such as humidity and the presence of fat. Polyvinylidene chloride, PVDC, is a material commonly used as a gas barrier; the oxygen permeability of a 25 μm thick film is $10\,cm^3\,m^{-2}\,(24h)^{-1}\,atm^{-1}$ compared with values of 8500 and 1840 for low density and high density polythene respectively. Higher rates of transfer occur with CO_2 for which the permeability values are about five times those for oxygen. If a film is required to exclude oxygen transfer completely, then a non-plastic material such as aluminium foil must be included. This is seen for example in the bags used to pack wines. In addition to the sealing- and gas barrier-layers, laminates may also contain an outer layer such as nylon which gives the pack greater resistance to damage.

In *vacuum packing* the product is placed in a bag from which the air is evacuated, causing the bag to collapse around the product before it is sealed. Residual oxygen in the pack is absorbed through chemical reactions with components in the product and any residual respiratory activity in the product and its microflora. To achieve the best results, it is important that the material to be packed has a shape that allows the packaging film to collapse on to the product surface entirely – without pockets and without the product puncturing the film.

Vacuum packing has been used for some years for primal cuts of red meats. At chill temperatures, good quality meat in a vacuum pack will keep up to five times longer than aerobically stored meats. The aerobic microflora normally associated with the spoilage of conventionally stored meats is prevented from growing by the high levels of CO_2 which develop in the pack after sealing (see Section 3.3.3) and the low oxygen tension. The microflora that develops is dominated by lactic acid bacteria which are metabolically less versatile than the Gram-negative aerobes, grow more slowly and reach a lower ultimate population (see Section 5.3).

In recent years vacuum packing has been increasingly used for retail packs of products such as cooked meats, fish and prepared salads. It has been used less often for retail packs of red meats since the meat acquires the purple colour of myoglobin in its unoxygenated form. This does not appeal to consumers even though oxygenation occurs very rapidly on opening a vacuum pack and the meat assumes the more familiar bright red, fresh meat appearance of oxymyoglobin. Cured meats, on the other hand, are often vacuum packed for display since the cured meat pigment nitrosomyoglobin is protected from oxidation by vacuum packing.

The expanding range of chilled foods stored under vacuum and the availability of vacuum packing equipment for small-scale catering and domestic use has prompted concern about increasing the risk from psychrotrophic *Clostridium botulinum*. A number of surveys have been conducted to determine the natural incidence of *C. botulinum* in these products and the concensus is that it is very low. In one recent example, workers failed to isolate *C. botulinum* or detect toxin in more than 500 samples analysed. When they deliberately inoculated these products with *C. botulinum* spores and incubated at the abuse temperature of 10 °C, only in the case of vacuum packed whole trout was toxin produced within the declared shelf-life of the product. Nevertheless, misuse of the technique does have the potential for increasing risk and a Government committee has recommended that all manufacturers of vacuum packing machinery should include instructions alerting the user to the risks from organisms such as *C. botulinum*.

In a variant of vacuum packing, known as *cuisine sous-vide* processing, food is vacuum packed before being given a pasteurization treatment which gives it a longer shelf-life under chill storage. The technique was developed in the 1970s in France and is said to give an improved flavour, aroma and appearance. It is used for the manufacture of chilled ready meals for various branches of the catering industry and *sous-vide* meals are also available in the retail market in some European countries. They have been slow to appear on the UK market due to the lack of appropriate UK regulations and concern over their microbiological safety with respect to psychrotrophic *C. botulinum*. It has been recommended that *sous-vide* products with an intended shelf-life of longer than 10 days at <3 °C should receive a minimum heat process equivalent to 90 °C for 10 minutes; 70 °C for 100 minutes should be sufficient for products with shorter shelf-lives.

For other types of chilled, vacuum or modified atmosphere packed foods, where there are no other controlling factors such as preliminary heat treatment, low pH or high salt, an advisory 10 day rule is sometimes applied. This states that such foods should have a designated shelf life of ≤10 days at storage temperatures ≤ 8 °C. The same rule can also be applied to foods packed in air as anaerobic niches suitable for *C. botulinum* growth can exist in these products too.

In *modified atmosphere packing, MAP*, a bulk or retail pack is flushed through with a gas mixture usually containing some combination of carbon dioxide, oxygen and nitrogen. The composition of the gas atmosphere changes during storage as a result of product and microbial respiration, dissolution of CO_2 into the aqueous phase, and the different rates of gas exchange across the packing membrane. These changes can be reduced by increasing the ratio of pack volume to product mass although this is not often practicable for other reasons.

The initial gas composition is chosen so that the changes which occur do not have a profound effect on product stability. Some examples of MAP gas mixtures used in different products are presented in Table 4.13. Carbon dioxide is included for its inhibitory effect, nitrogen is non-inhibitory but has low water solubility and can therefore prevent pack collapse when high concentrations of CO_2 are used. By displacing oxygen

Table 4.13 *MAP gas mixtures used with foods*

Product	% CO_2	% O_2	% N_2
Fresh meat	30	30	40
	15–40	60–85	–
Cured meat	20–50	0	50–80
Sliced cooked roast beef	75	10	15
Eggs	20	0	80
	0	0	100
Poultry	25–30	0	70–75
	60–75	5–10	>20
	100	0	0
	20–40	60–80	0
Pork	20	80	0
Processed meats	0	0	100
Fish (white)	40	30	30
Fish (oily)	40	0	60
	60	0	40
Cheese (hard)	0–70		30–100
Cheese	0	0	100
Cheese; grated/sliced	30	0	70
Sanwiches	20–100	0–10	0–100
Pasta	0	0	100
	70–80	0	20–30
Bakery	0	0	100
	100	0	0

From *J. Food Protection*, 1991, **54**, 58–70, with permission

it can also delay the development of oxidative rancidity. Oxygen is included in the gas flush mixtures for the retail display of red meats to maintain the bright red appearance of oxymyoglobin. This avoids the acceptability problem associated with vacuum packs of red meats, although the high oxygen concentration (typically 60–80%) helps offset the inhibitory effect of he CO_2 (around 30%) so that the growth of aerobes is slowed rather than suppressed entirely.

In *controlled-atmosphere storage, CAP*, the product environment is maintained constant throughout storage. It is used mainly for bulk storage and transport of foods, particularly fruits and vegetables, such as the hard cabbages used for coleslaw manufacture. CAP is used for shipment of chilled lamb carcasses and primal cuts which are packed in an aluminium foil laminate bag under an atmosphere of 100% CO_2. It is more commonly encountered though with fruits such as apples and pears which are often stored at sub-ambient temperatures in atmospheres containing around 10% CO_2. This has the effect of retarding mould spoilage of the product through a combination of the inhibitory effect of CO_2 on moulds and its ability to act as an antagonist to ethylene, delaying fruit senescence and thus maintaining the fruit's own ability to resist fungal infection.

4.7   CONTROL OF WATER ACTIVITY

The water activity of a product can be reduced by physical removal of liquid water either as vapour in drying, or as a solid during freezing. It is also lowered by the addition of solutes such as salt and sugar. Freezing has already been discussed in this chapter (Section 4.4) and so here we will confine ourselves to drying and solute addition. The primal role of these techniques in food preservation has been alluded to in a number of places. It was the earliest food preservation technique and, until the 19th century, water activity reduction played some part in almost all the known procedures for food preservation.

Nature provided early humans with an object lesson in the preservative value of high solute concentrations in the form of honey produced by bees from the nectar of plants. The role of salt in decreasing a_w accounts for its extreme importance in the ancient economy as evidenced today in the etymology of the word, salary, and of place names such as Salzburg, Nantwich, Moselle and Malaga. It can also be seen in the extraordinary hardship people were prepared to endure (or inflict on others) to ensure its availability; to this day the salt mine remains a by-word for arduous and uncomfortable labour.

Solar drying, while perhaps easy and cheap, is subject to the vagaries of climate. Drying indoors over a fire was one way to avoid this problem and one which had the incidental effect of imparting a smoked flavour to the food as well as the preservative effect of chemical components of the smoke.

Salting and drying in combination have played a central role in the human diet until very recently. One instance of this is the access it gave the population of Europe to the huge catches of cod available off Newfoundland. From the end of the 15th century, salted dried cod was an important item in trans-Atlantic trade and up until the 18th century accounted for 60% of all the fish eaten in Europe. It remains popular today in Portugal and in the Caribbean islands where it was originally imported to feed the slave population. Other traditional dried and salted products persist in the modern diet such as dried hams and hard dry cheeses but the more recent development and application of techniques such as refrigeration, MAP, and heat processing and the preference for 'fresh' foods has meant that their popularity has declined. Nevertheless this should not obscure the important role that low a_w foods still play in our diet in the form of grains, pulses, jams, bakery products, dried pasta, dried milk, instant snacks, desserts, soups, *etc.*

Among the main features of the effect of a_w on the growth and survival of micro-organisms discussed in Section 3.2.5 it was noted that microbial growth does not occur below an a_w of 0.6. This applies to a number of food products (Figure 3.9), but the fact that microbial spoilage is not possible given proper storage conditions, does not mean that they do not pose any microbiological problems. Micro-organisms that were in the product before drying or were introduced during processing can survive for extended periods. This is most important with respect to pathogens if they were present in hazardous numbers before drying or if time and temperature allow them to resume growth in a product that is rehydrated before consumption. There have been a number of instances where the survival of pathogens or their toxins has caused problems in products such as chocolate, pasta, dried milk and eggs. Generally *Salmonella* and *Staphylococcus aureus* have been the principal pathogens involved – there have been about 20 major outbreaks associated with these organisms and dried milk since 1955, but spore formers are particularly associated with some other dried products such as herbs or rice.

Intermediate moisture foods, IMFs, are commonly defined as those foods with an a_w between 0.85 and 0.6. This range, which corresponds roughly to a moisture content of 15–50%, prohibits the growth of Gram-negative bacteria as well as a large number of Gram-positives, yeasts and moulds, giving the products an extended shelf-life at ambient temperature. When spoilage does occur, it is often a result of incorrect storage in a high relative humidity environment. In correctly stored products growth of xerophilic moulds, osmophilic yeasts or halophilic bacteria may occur, depending on the product, and in many IMFs the shelf-life is further protected by the inclusion of antifungal agents such as sulfur dioxide or sorbic acid.

At the a_w of IMFs, pathogens are also prevented from growing. Although *Staph. aureus* is capable of growing down to an a_w of 0.83, it cannot produce toxin and is often effectively inhibited by the combination of a_w with other antimicrobial hurdles.

There are a number of traditional IMFs such as dried fruits, cakes, jams, fish sauce and some fermented meats. Sweetened condensed milk is one interesting example. Milk is homogenized, heated to 80 °C and sugar added before it is concentrated in a multi-effect vacuum evaporator at 50–60 °C. When the product emerges from the concentration stage it is cooled and seeded with lactose crystals to induce crystallization of the lactose. This gives sweetened condensed milk its characteristic gritty texture. Although the product is packed into cans and has an almost indefinite shelf-life, it is not an appertized food. Its stability is a result of its high sugar content (62.5% in the aqueous phase) and low a_w (<0.86). Spoilage may sometimes occur due to growth of osmophilic yeasts or, if the can is under-filled leaving a headspace, species of *Aspergillus* or *Penicillium* may develop on the surface.

Some years ago, our developing understanding of the stability of IMFs led to considerable interest in applying the same principles to the development of new shelf stable foods. Novel humectants such as glycerol, sorbitol and propylene glycol were often used to adjust a_w in these products in addition to the solutes salt and sugar. They were not however well received in the market for human food because of acceptability problems, although a number of successful pet food products were developed. One interesting observation made during this work is that products with the same water activity differ in their keeping quality depending on how they are made. Traditional IMFs are generally made by a process of desorption whereby water is lost from the product during processing but a number of the new IMFs used an adsorption process in which the product is first dried and its moisture content readjusted to give the desired a_w. The hysteresis effect in water sorption isotherms (see Section 3.2.5 and Figure 3.11) means that although products made using the two techniques will have the same initial a_w they will have different moisture contents and so will eventually equilibrate to different a_w values. It was found that products made by desorption and having the higher water content were also more susceptible to microbial spoilage.

Solar drying is still widely practised in hot climates for products such as fruits, fish, coffee and grain. The traditional technique of spreading the product out in the sun with occasional turning often gives only rudimentary or, sometimes, no protection from contamination by birds, rodents, insects and dust. Rapid drying is essential to halt incipient spoilage; this is usually achievable in hot dry climates, though in tropical countries with high humidity drying is usually slower so that products such as fish are often pre-salted to inhibit microbial growth during drying.

There are a number of procedures for mechanical drying which are quicker, more reliable, albeit more expensive than solar drying. The drying regime must be as rapid as possible commensurate with a high-quality product so factors such as reconstitution quality must also be taken into account. With the exception of freeze-drying where the product is frozen and moisture sublimed from the product under vacuum, these techniques employ high temperatures. During drying a proportion of the microbial population will be killed and sub-lethally injured to an extent which depends on the drying technique and the temperature regime used. It is however no substitute for bactericidal treatments such as pasteurization. Although the air temperature employed in a drier may be very high, the temperature experienced by the organisms in the wet product is reduced due to evaporative cooling. As drying proceeds and the product temperature increases, so too does the heat resistance of the organisms due to the low water content. This can be seen for example in the differences between spray dried milk and drum dried milk. In spray drying, the milk is pre-concentrated to about 40–45% solids before being sprayed into a stream of air heated to temperatures up to 260 °C at the top of a tower. The droplets dry very rapidly and fall to the base of the tower where they are collected. In drum drying, the milk is spread on the surface of slowly rotating metal drums which are heated inside by steam to a temperature of about 150 °C. The film dries as the drum rotates and is scraped off as a continuous sheet by a fixed blade close to the surface of the drum. Although it uses a lower temperature, drum drying gives greater lethality since the milk is not subject to the same degree of pre-concentration used with spray-dried milk and the product spends longer at high temperatures in a wet state. Spray drying is however now widely used for milk drying because it produces a whiter product which is easier to reconstitute and has less of a cooked flavour. Milk is pasteurized before drying although there are opportunities for contamination during intervening stages. Most of the organisms which survive drying are thermoduric but Gram-negatives may survive and have on occasion been the cause of food poisoning outbreaks.

The limited lethality of drying processes and the long storage life of dried products means that manufacturers are not exempt from the stringent hygiene requirements of other aspects of food processing. Good quality raw materials and hygienic handling prior to drying are essential. Outbreaks of *Staph. aureus* food poisoning have been caused by dried foods which were stored at growth temperatures for too long prior to drying allowing the production of heat resistant toxin which persisted through to the final product. The dried product must also be protected from moisture by correct packaging and storage in a suitable environment otherwise pockets of relatively high a_w may be created where microbial growth can occur.

4.8   COMPARTMENTALIZATION

Butter is an interesting example of a rather special form of food preservation where microbial growth is limited by compartmentalization within the product.

Essentially there are two types of butter: sweet cream butters, which are often salted, and ripened cream butters. In ripened-cream butters, the cream has been fermented by lactic acid bacteria to produce *inter alia* diacetyl from the fermentation of citrate which gives a characteristically buttery flavour to the product. They have a stronger flavour than sweet-cream butters but are subject to faster chemical deterioration. Sweet-cream butter is most popular in the United States, Ireland, the UK, Australia and New Zealand whereas the ripened cream variey is more popular in continental Europe.

Butter is an emulsion of water droplets in a continuous fat phase in contrast to milk which is an emulsion of fat globules in a continuous

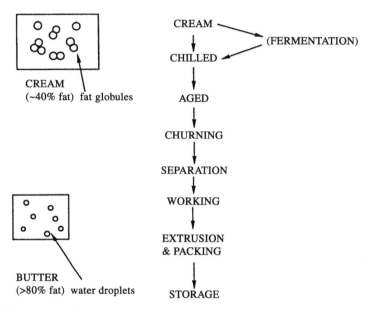

Figure 4.15 *Buttermaking*

water phase (Figure 4.15). It has a higher fat content than milk (80%) and uses pasteurized cream as its starting point. Typically, the cream is pasteurized using an HTST process of 85 °C for 15 s and held at 4–5 °C for a period to allow the fat globules to harden and cluster together. In making a conventional ripened cream butter, the starter culture is added at this stage and the cream incubated at around 20 °C to allow flavour production to take place. A more recent method developed at NIZO, the Dutch Dairy Research Institute, employs a concentrated starter added to sweet-cream butter after manufacture. Phase inversion, the conversion from a fat-in-water emulsion to a water-in-fat emulsion, is achieved by the process of churning. During this process fat globules coalesce, granules of butter separate out, and considerable amounts of water are lost from the product in the form of buttermilk. The buttermilk phase retains most of the micro-organisms from the cream and numbers may show an apparent increase due to the breaking of bacterial clumps.

Traditional farmhouse buttermaking used wooden butter churns and these were originally scaled up for the earliest commercial butter-making. However, the impossibility of effectively cleaning and sanitizing wood has led to its replacement by churns made of stainless steel or aluminium–magnesium alloys. After the butter has formed, the buttermilk is drained off, the butter grains washed with water and, in the case of sweet-cream butter, salt is added usually at a level of 1–2%. The butter is then 'worked' to ensure further removal of moisture and an even distribution of water and salt throughout the fat phase. In properly produced butter the water is distributed as numerous droplets ($> 10^{10} \, g^{-1}$) mostly less than 10 μm in diameter. Since the butter should contain at most around $10^3 \, cfu \, g^{-1}$, most of these droplets will be sterile. In those that do contain micro-organisms, the nutrient supply will be severely limited by the size of the droplet. If the butter is salted, the salt will concentrate in the aqueous phase along with the bacteria which will therefore experience a higher, more inhibitory salt level. For example, bacteria in a butter containing 1% salt and with a moisture content of 16% would experience an effective salt concentration of 6.25%.

Few micro-organisms survive pasteurization so the microbiological quality of butter depends primarily on the hygienic conditions during subsequent processing, particularly the quality of the water used to wash the butter. Good microbiological quality starting materials are essential though, as preformed lipases can survive pasteurization and rapidly spoil the product during storage. Butter spoilage is most often due to the development of chemical rancidity but microbiological problems do also occur in the form of cheesy, putrid or fruity odours or the rancid flavour of butyric acid produced by butterfat hydrolysis. Pseudomonads are the most frequently implicated cause and are thought to be introduced mainly in the wash water. Psychrotrophic yeasts and moulds can also cause lipolytic spoilage and these are best controlled by maintaining low humidity and good air quality in the production environment and by ensuring the good hygienic quality of packaging materials. In this respect aluminium foil wrappers are preferred to oxygen-permeable parchment wrappers as they will help discourage surface mould growth.

Butter is a relatively safe commodity from a microbiological standpoint, although there was an outbreak of listeriosis in Finland in 1998/9 which affected 18 people, four of whom died (see 7.9.5).

Margarine relies on a similar compartmentalization for its microbiological stability, but uses vegetable fat as its continuous phase. Although skim milk is often included in the formulation, it is possible to make the aqueous phase in margarine even more deficient nutritionally than in butter, thus increasing the microbiological stability further. With the move towards low fat spreads containing 40% fat, the efficacy of this system is more likely to breakdown. A higher moisture content means that the preservative effect of salt or lactic acid, which is often included, is diluted and that micro-organisms can grow to a greater extent in the larger aqueous droplets. In these cases the use of preservatives may be required to maintain stability. An approach developed at Unilever's laboratories in the Netherlands is based on a two stage approach where the composition of the aqueous phase is analysed to determine its capacity to support the growth of different spoilage organisms. If there is some potential for microbial growth to occur, this is then calculated by working out which fraction of water droplets will be contaminated and then summing the growth in each of them. These models have been incorporated into an expert system to predict the stability of any proposed product formulation so that microbiological stability can be designed into the product.

CHAPTER 5

Microbiology of Primary Food Commodities

In this chapter we will examine aspects of the microbiology of some specific commodity groups describing the microflora with particular emphasis on spoilage.

5.1 WHAT IS SPOILAGE?

According to the Oxford English Dictionary to spoil is to 'deprive of good or effective qualities'. When a food is spoiled its characteristics are changed so that it is no longer acceptable. Such changes may not always be microbiological in origin; a product may become unacceptable as a result of insect damage, drying out, discolouration, staling or rancidity for instance, but by and large most food spoilage is a result of microbial activity. Microbiological food spoilage can manifest itself in several different ways, some of which often occur in combination. Visible microbial growth may be apparent in the form of surface slime or colonies, degradation of structural components of the food can cause a loss of texture, but the most common manifestation will be chemical products of microbial metabolism, gas, pigments, polysaccharides, off-odours and flavours.

Spoilage is also a subjective quality; what is spoiled for one person may be perfectly acceptable to another. The perception of spoilage is subject to a number of influences, particularly social; foods acceptable in some cultures are unacceptable in others. Some products such as well matured cheeses and game birds that have been hung for extended periods are esteemed by some people and highly objectionable to others. Affluence is another contributory factor – many are not in the position to be able to discard food due to some slight sensory defect. In the 18th and 19th century navy, sailors often preferred to eat in dark corners so that they could not see the weevils and maggots in their food.

A general feature of microbial spoilage is its relatively sudden onset – it does not appear to develop gradually, day by day a little worse, but more often as an unexpected and unpleasant revelation. This is a reflection of the exponential nature of microbial growth (see Section 3.1) and its consequence that microbial metabolism can also proceed at an exponentially increasing rate. If a microbial product associated with spoilage, for example an off odour, has a certain detection threshold, the level will be well below this threshold for most of the product's acceptable shelf-life. Once reached, however, it will be rapidly passed so that in a comparatively short time after, levels will be well in excess of the threshold and the product will be profoundly spoiled. This is represented in Figure 5.1 where growth of a spoilage organism is plotted on a linear scale.

Prediction or early detection of spoilage is not always easy since the mechanisms underlying microbiological spoilage can be quite complex. It is generally far easier to identify the chemical responsible for a particular off-odour than to identify the organism(s) responsible. This and the relative speed of chemical analysis have led to the use of chemical indices of spoilage in some areas, but more often the ultimate judgement as to whether a food is spoiled remains subjective. Total microbial counts are generally a poor indicator of spoilage potential. Many of the organisms enumerated may not grow in the food and many of those that do will not be responsible for spoilage. The value of microbial enumeration techniques can be improved if they are specific to those organisms associated

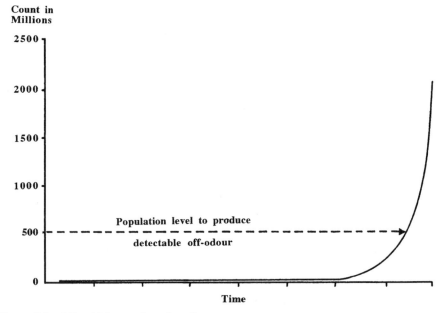

Figure 5.1 *Microbial growth and spoilage*

with spoilage, so called specific spoilage organisms (SSO). This approach has met with some success with aerobically stored fish products and the use of media to detect organisms capable of producing H_2S.

5.2 MILK

5.2.1 Composition

Milk is the fluid, excluding colostrum, secreted by mammals for the nourishment of their young. Colostrum is a much more concentrated liquid containing up to 25% total solids, mainly protein, secreted immediately after parturition. A number of animals are used to produce milk for human consumption, although the cow is by far the most important in commercial terms.

The principal components of milk are water, fat, protein and lactose. The precise composition varies between species so, for example, human milk has lower protein but higher lactose levels than cow's milk (Table 5.1). Generally the protein content of the milk reflects the growth rate of the young animal – the higher the growth rate, the more protein the milk contains.

There can be considerable compositional difference between breeds of a single species – Jersey and Guernsey milks, for instance, are noted for their higher fat content which is reflected in a richer, creamier taste. Even within a single breed variations in composition can arise depending on factors such as the stage of lactation, the stage of milking, the intervals between milking, the time of day, the number of previous lactations and the general nutritional state and health of the cow.

A more detailed analysis of cow's milk is presented in Table 5.2. The lipid content is the most variable feature. It is comprised mainly of C_{14},

Table 5.1   *Typical Milk Composition [% weight(volume)$^{-1}$]*

	Fat	Protein	Lactose	Total solids
Human	3.8	1.0	7.0	12.4
Cow	3.7	3.4	4.8	12.7
Jersey	5.1	3.8	5.0	14.5
Ayrshire	4.0	3.5	4.8	13.0
Short-horn	3.6	4.9	4.9	12.6
Sheep	7.4	5.5	4.8	19.3
Goat	4.5	2.9	4.1	13.2
Water buffalo	7.4	3.8	4.8	17.2
Horse	1.9	2.5	6.2	11.2

Table 5.2 *Composition of fresh cow's milk*

	Concentration g litre^{-1}	
LIPIDS	37	
	of which	% w/w
Triglycerides		95–96
Diglycerides		1.3–1.6
Free fatty acids		0.1–0.5
Total phospholipids		0.8–1.0
PROTEINS	34	
Casein	26	
α_{S1}	11.1	
α_{S2}	1.7	
β	8.2	
γ	1.2	
κ	3.7	
Whey proteins		
α-lactalbumin	0.7	
β-lactoglobulin	3.0	
serum albumin	0.3	
immunoglobulins	0.6	
NON-PROTEIN NITROGEN	1.9	
LACTOSE	48	
CITRIC ACID	1.75	
ASH	7.0	
CALCIUM	1.25	
PHOSPHORUS	0.96	

C_{16}, C_{18}, and $C_{18:1}$ fatty acids and is present in fresh milk mainly in the form of fat globules surrounded by a phospholipid rich layer known as the milk fat globule membrane. Typically these globules have a diameter of about 5 μm and the milk contains about 10^{12} fat globules per litre. If fresh milk is allowed to stand, the fat rises to the surface of the milk to produce a distinct cream line. The tendency for this to happen is reduced if the size of the globules is reduced by passing the milk through a small orifice under pressure; a process known as homogenization.

About 80–85% of the protein in milk is present as caseins. These are milk-specific proteins which are precipitated from milk by decreasing the pH to 4.6. This pH corresponds approximately to their isoelectric point which is relatively low due to the predominance of acidic amino acids and the presence of phosphorylated serine residues in the molecules. There are five main classes of caseins (see Table 5.2); these aggregate together in association with calcium phosphate in milk to form colloidal particles known as micelles. Milk contains around 10^{15} casein micelles l^{-1} with an average diameter of around 0.2 μm. The stability of the micelle is maintained by the presence of κ-casein near or on the surface of the particle. Loss of this stabilizing effect occurs when κ-casein is cleaved by chymosin during cheese production and leads to the micelles sticking together to form a coagulum (see Section 9.6).

The balance of the protein in milk is made up of the whey proteins. These mainly comprise the compact globular proteins β-lactoglobulin and α-lactalbumin but also a number of blood-derived proteins such as serum albumin and immunoglobulins. The latter are present at higher levels in colostrum where they presumably confer some resistance to infection in the newborn calf.

5.2.2 Microflora of Raw Milk

Its high water activity, moderate pH (6.4–6.6) and ample supply of nutrients make milk an excellent medium for microbial growth. This demands high standards of hygiene in its production and processing; a fact recognized in most countries where milk was the first food to be the focus of modern food hygiene legislation.

Milk does possess a number of antimicrobial features (discussed in Section 3.2.4), present either to protect the udder from infection or to protect the newborn calf. Generally these are present at too low a concentration in cow's milk to have a very marked effect on its keeping quality or safety. In some cases the antimicrobial activity is antagonized by other milk constituents such as the effect of citrate and bicarbonate on lactoferrin activity. Stimulation of lactoperoxidase activity through the addition of exogenous hydrogen peroxide has been investigated as a means of preserving raw milk in developing countries where ambient temperatures are high and refrigeration is not often available. In one trial in Africa, use of this technique increased the proportion of samples passing the 10 minute resazurin quality test from 26% to 88%.

Three sources contribute to the micro-organisms found in milk: the udder interior, the teat exterior and its immediate surroundings, and the milking and milk-handling equipment.

Bacteria that get on to the outside of the teat may be able to invade the opening and thence the *udder interior*. Aseptically taken milk from a healthy cow normally contains low numbers of organisms, typically fewer than 10^2–10^3 cfu ml^{-1}, and milk drawn from some quarters may be sterile. The organisms most commonly isolated are micrococci, streptococci and the diptheroid *Corynebacterium bovis*. Counts are frequently higher though due to mastitis, an inflammatory disease of the mammary tissue, which is a major cause of economic loss in the dairy industry. In England and Wales, where it has been estimated to cost the industry around £90 million annually, about 1–2% of cows have a clinical infection at any one time. In the early acute stage of illness the bacterial count in mastitic milk can exceed 10^8 cfu ml^{-1} and macroscopic changes are often visible in the milk. Mastitis is also diagnosed by the presence of high numbers of polymorphonuclear leukocytes which can rise to levels of $10^7\,ml^{-1}$ in infected milk.

In addition to acute mastitis, a substantial proportion of the national dairy herd is subclinically infected. In these cases there may be no obvious signs of infection yet the causative organism can be present in the milk at about $10^5\,cfu\,ml^{-1}$ and will contribute to an increase in the overall count of bulked milk.

Many organisms can cause mastitis, the most important being *Staphylococcus aureus*, *Escherichia coli*, *Streptococcus agalactiae*, *Strep. dysgalactiae*, *Strep. uberis*, *Pseudomonas aeruginosa* and *Corynebacterium pyogenes*. Several of these are potential human pathogens and a number of other human pathogens such as *Salmonella*, *Listeria monocytogenes*, *Mycobacterium bovis* and *Mycobacterium tuberculosis* are also occasionally reported.

Infected cows are treated by injection of antibiotics into the udder. Milk from these cows must be withheld from sale for several days following treatment because antibiotic residues can cause problems in sensitive consumers and inhibit starter culture activity in fermented milks. Attempts to control mastitis by good milking hygiene, use of a disinfectant teat dip after milking and an antibiotic infusion at the end of lactation have helped to reduce streptococcal and staphylococcal infections but have had little success in preventing *E. coli* mastitis.

The *udder exterior and its immediate environment* can be contaminated with organisms from the cow's general environment. This is less of a problem in summer months when cows are allowed to graze in open pasture and is worst when they are housed indoors and under wet conditions. Heavily contaminated teats have been reported to contribute up to $10^5\,cfu\,ml^{-1}$ in the milk. Contamination from bedding and manure can be a source of human pathogens such as *E. coli*, *Campylobacter*, and *Salmonella* and *Bacillus* species may be introduced from soil. Clostridia such as *C. butyricum* and *C. tyrobutyricum* can get into milk from silage fed to cows and their growth can cause the problem known as late blowing in some cheeses.

A number of measures can be taken to minimize milk contamination from the udder exterior and considerable advice on this topic is available to dairy farmers. Some of the recommendations made by the Milk Marketing Board, formerly the principal purchaser of milk in England and Wales, included:

(1) providing enough clean bedding and replacing it as necessary;
(2) removing slurry (faeces and urine) from concrete areas at least twice daily;
(3) preventing muddy areas wherever possible;
(4) shaving udders and trimming tails;

(5) washing teats with warm water containing disinfectant and drying individually with paper towels;
(6) keeping the milking parlour floor clean during milking;
(7) thoroughly cleaning teat cups if they fall off during milking and discarding foremilk.

Although such procedures certainly improve the microbiological quality of milk, economic constraints such as increasing size of individual dairy herds and decreased manning levels in milking parlours encouraged their neglect. The introduction of total bacterial count as a basis for payment in 1982 provided an incentive for their more zealous application and led to a marked decline in bacterial count of milk (see below).

Milk-handling equipment such as teat cups, pipework, milk holders and storage tanks, is the principal source of the micro-organisms found in raw milk. As the overall quality of the milk decreases so the proportion of the microflora derived from this source increases. Milk is a nutritious medium and, if equipment is poorly cleaned, milk residues on surfaces that are frequently left wet will act as a focus for microbial growth which can contaminate subsequent batches of milk. Occasional neglect of cleaning and sanitizing procedures is usually less serious since, although it may contribute large numbers of micro-organisms to the product, these tend to be fast growing bacteria that are heat sensitive and will be killed by pasteurization. They are also sensitive to sanitizing practices used and will be eliminated once effective cleaning is resumed. If cleaning is persistently neglected though, the hydrophobic, mineral-rich deposit known as milkstone can build up on surfaces, particularly heated ones. This will protect organisms from sanitizers and allow slower growing organisms to develop such as micrococci and enterococci. Many of these are thermoduric and may not be removed by pasteurization.

To encourage farmers to apply the available advice on animal husbandry practices, milking procedures, types and design of equipment and cleaning schedules which contribute to good bacteriological quality milk, the Milk Marketing Board (MMB) in England and Wales introduced in 1982 a system of paying farmers based on the total bacterial count (TBC) of their milk. Similar schemes have been introduced in a number of countries but details of the MMB's scheme are presented as Table 5.3. For four months prior to introduction of the scheme, farmers were notified of the TBC count of their milk and in anticipation of its start a dramatic fall in the count was noted (Figure 5.2). Now more than 76% of the milk produced in England and Wales falls into Band A with a mean count of 1.7×10^3 cfu ml^{-1}. The Milk Marketing Board no longer exists as the monopoly purchaser of farm milk in England and Wales, but the bodies that replaced it recognized the value of a payment scheme which includes microbiological quality and have retained similar systems.

Table 5.3   *Milk Marketing Board (England and Wales) total bacterial count payment scheme*

Grade	Count (cfu ml^{-1})	Price adjustment (pence l^{-1})
A	$<2 \times 10^4$	+0.23
B	$>2 \times 10^4$ but $<10^5$	0
C_1	$>10^5$ but no price deduction in previous 6 months	−1.5
C_2	$>10^5$ and Grade C_1 produced in previous 6 months	−6.0
C_3	$>10^5$ and Grade C_2 or C_3 deduction has been applied	−10.0

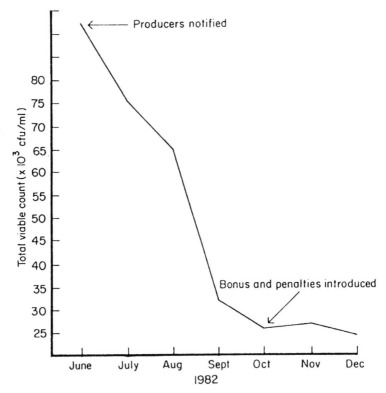

Figure 5.2   *Raw milk counts and the bonus payments scheme. Reproduced from 'Micro-organisms in Agriculture', SAB Symposium Series No. 15*

In most developed countries milk is chilled almost immediately after it issues from the cow and is held at a low temperature thereafter. It is stored in refrigerated holding tanks before being transported by a refrigerated or insulated lorry to the dairy where it is kept in chill storage tanks until use. Throughout this time, its temperature remains below

7 °C and the only organisms capable of growing will be psychrotrophs. There are many psychrotrophic species, but those most commonly found in raw milk include Gram-negative rods of the genera *Pseudomonas, Acinetobacter, Alcaligenes, Flavobacterium*, psychrotrophic coliforms, predominantly *Aerobacter* spp., and Gram-positive *Bacillus* spp.

One consequence of the current extensive use of refrigeration and the change to a microflora dominated by psychrotrophs is that traditional tests for the microbiological quality of milk based on the reduction of a redox dye such as methylene blue or resazurin have become obsolete. Psychrotrophs tend to reduce these dyes poorly and the tests are not very sensitive to low numbers of bacteria.

5.2.3 Heat Treatment of Milk

Proposals for the heat treatment of milk were made as early as 1824, forty years before Pasteur's work on the thermal destruction of micro-organisms in wine and beer. When milk pasteurization was introduced by the dairy industry around 1890, it was as much to retard souring as to prevent the spread of disease. This had become an important commercial requirement since large quantities of milk were now being transported by rail into the large cities rather than being produced locally in cramped and insanitary cowhouses.

Milk has long been recognized as an agent in the spread of human disease and within a few years it was appreciated that pasteurization was also providing protection against milk-borne disease. Nowadays it is safety rather than spoilage considerations which determine the minimum legal requirements for pasteurization.

Originally the main health concerns associated with milk were tuberculosis caused by *Mycobacterium bovis* and *Mycobacterium tuberculosis* (see Section 7.10) and brucellosis caused by *Brucella* spp. (see Section 7.3). In some parts of the world milk is still a significant source of these infections but in the UK and some other countries they have now been effectively eliminated from the national dairy herd by a programme of regular testing and culling of infected animals. Such programmes must be constantly maintained to be effective and there have been occasional problems. Initiatives such as the culling of badgers, thought to be a reservoir of *M. bovis*, have been the subject of some controversy and in 2002 there was an outbreak of brucellosis in a dairy herd in Cornwall, although this was the first recorded in England for ten years. Enteric pathogens such as *Salmonella* and *Campylobacter* are still however prevalent in raw milk and pasteurization remains the most effective measure for their control.

The four types of heat treatment applied to milk are described in Table 5.4. Specification of pasteurization temperatures to the first decimal

Table 5.4 *Heat treatment of milk*

LOW TEMPERATURE HOLDING (LTH)	62.8 °C for 30 min
HIGH TEMPERATURE SHORT TIME (HTST)	71.7 °C for 15 s
ULTRA HIGH TEMPERATURE	135 °C for 1 s
'STERILIZED'	> 100 °C typically 20–40 min

place is not some arcane feature of thermal processing but is simply a result of conversion from the Fahrenheit scale in which they were originally prescribed.

Low temperature holding (LTH) is a batch process that has been superseded in most countries by continuous high temperature/short time (HTST) pasteurization using a plate heat exchanger. Originally the temperatures prescribed for LTH pasteurization were slightly lower. They were increased in 1950 in response to the observation that the rickettsia *Coxiella burnetii*, the causative agent of Q fever, could survive this original process if present in high numbers. Spread of this organism through infected milk is a greater problem in the United States than most of Europe, where transmission appears to be mainly through aerosols in the farm environment. More recent fears that *Listeria* could survive conventional pasteurization treatments appear to be unfounded (see Section 7.9.2).

A simple test, the phosphatase test, is applied to determine whether milk has been properly pasteurized. Milk contains the enzyme alkaline phosphatase which is inactivated by the time/temperature combinations applied during pasteurization. To determine whether a milk sample has been satisfactorily pasteurized and is free from contaminating raw milk, a chromogenic substrate is added. If active phosphatase is still present then it will hydrolyse the substrate producing a colour which can be compared to standards to determine whether the milk is acceptable or not. The same principle is used in the α-amylase test applied to bulk liquid egg (see Section 7.12.5).

The microbiological quality of pasteurized milk is now also governed by EU-based regulations which require pasteurized milk to contain less than 1 coliform ml^{-1}, to have a count at 30 °C of less than 3×10^4, and also that, after 5 days storage at 6 °C, its count at 21 °C should be less than 10^5 cfu ml^{-1}.

UHT milk is a commercially sterile product of the sort described in Section 4.1. It is interesting to note that although UHT milk is a low-acid appertized food, the minimum heat process specified falls well short of the botulinum cook required for equivalent canned foods. It has been claimed that the redox potential in milk is too high to support the growth of *Clostridium botulinum* but more probable explanations for the fact that botulism has never been associated with this product are a low incidence of *C. botulinum* spores in milk and the fact that manufacturers employ heat processes well in excess of the minimum legal requirement.

Sterilized milk described in Table 5.4 is a rather specialized product of diminishing importance. In the UK it is defined as a product which is heat processed in-bottle at temperatures above 100 °C for sufficient time for it to pass the turbidity test. This test is based on the principal that the heat process is sufficient to denature whey proteins. In it, casein is precipitated with ammonium sulfate and filtered off. The filtrate is then heated; if it remains clear the milk is acceptable because the whey proteins have already been denatured and removed with the precipitated casein. Turbidity indicates an inadequate heat process which has left some whey protein undenatured.

Gram-negative psychrotrophs will not survive pasteurization, although some pseudomonads produce extracellular lipases and proteases which are heat resistant. If enough of these bacteria are present ($>10^5\,\mathrm{ml}^{-1}$), sufficient enzyme can be produced pre-pasteurization to cause rancidity and casein degradation in the processed milk.

Raw milk may also contain a number of organisms known as thermodurics that can survive mild pasteurization treatments. These are generally Gram-positives such as the sporeforming bacteria and members of the genera *Microbacterium, Micrococcus, Enterococcus* and *Lactobacillus*, but 1–10% of strains of the Gram-negative *Alcaligenes tolerans* may also survive.

In the main, spoilage of pasteurized milk is due to the growth of psychrotrophic Gram-negative rods such as *Pseudomonas, Alcaligenes,* and *Acinetobacter,* introduced as post-pasteurization contaminants. Product shelf-life will depend on the number of contaminants introduced and the efficiency of the cold chain; a 2 °C decrease in storage temperature will approximately double the shelf life, but pasteurized milk produced under conditions of good manufacturing practice should keep for more than 10 days under refrigeration. Spoilage usually manifests itself as off odours and flavours described as fruity or putrid but visual defects such as clotting due to proteolytic activity can also arise. The souring traditionally associated with milk spoilage and due to the growth of lactic acid bacteria is now rare.

In milk which is subject to very low levels of post-pasteurization contamination, spoilage can result from the growth of thermoduric *Bacillus* spp. This may be associated with flavour defects but the most studied example is the bitty cream phenomenon produced by the lecithinase activity of *Bacillus cereus.* This enzyme hydrolyses the phospholipids associated with the milk fat globule membrane to produce small proteinaceous fat particles which float on the surface of hot drinks and adhere to surfaces of crockery and glasses. Bitty cream is associated mostly with milk that has been subject to temperature abuse, although psychrotrophic *Bacillus* species are becoming increasingly associated with the spoilage of refrigerated milk.

To extend the shelf life of milk beyond 5–15 days requires measures to reduce the number of spores present. UHT processing achieves this but introduces a cooked flavour which is not always appreciated. To avoid this problem, alternative techniques to produce so-called extended shelf life (ESL) milks which retain the taste of pasteurised milk have been explored. These include direct heating by steam injection to temperatures of 85–127 °C for 2 seconds, the removal of spores by centrifugation combined with a pasteurisation treatment, or microfiltration of skim milk and its subsequent recombination with UHT treated cream. Such techniques combined with hygienic handling and an effective cold chain can achieve shelf lives of 21 days.

5.2.4   Milk Products

Fresh milk is the starting point for a number of other food products, some of which are shown in Figure 5.3. A number of these are described in detail elsewhere in this book. Yoghurt, cheese and other fermented products are dealt with in Chapter 9, dried milk in Section 4.7, and the principles behind the production of evaporated milk and UHT milk are described in Section 4.1.

Fat can be concentrated from milk by centrifugal separation to produce a number of different types of cream. These are distinguished

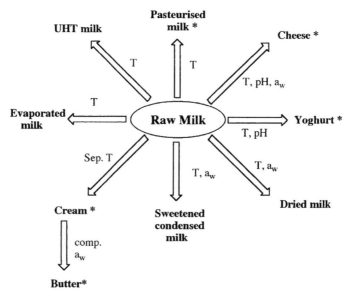

Figure 5.3   *Milk and milk products. T indicates elevated temperature; pH, reduced pH; a$_w$, reduced a$_w$; sep., separation, comp., compartmentalization; and * stored at chill temperatures*

primarily by their fat content, which can vary from around 12% for half cream up to 55% in clotted cream, but also by differences in other aspects of their processing such as the heat treatment they receive and whether they are homogenized, whipped or fermented.

After separation the microbial count of the cream fraction is usually higher than that of the skim milk and, despite the fact that some bacteria are removed as slime from the separator, the combined count of skim milk and cream often exceeds that of the original milk. These observations are a result mainly of physical processes for, although mechanical separators operate at temperatures at which growth can occur (25–30 °C), there is limited time for microbial growth during efficient processing. During separation it is thought that the fat globules rising through the milk act as a moving sieve to which bacteria adhere and become concentrated in the fat layer. The increase in combined count of the two fractions is attributed to the breaking up of bacterial clumps which increases the number of colony forming units, a phenomenon also noticed during the churning of cream to produce butter.

Pasteurization treatments applied to cream are generally in excess of those used with milk because of the protective effect of fat and also because a longer product shelf-life is often necessary. As with milk, the spoilage of cream is due to growth of post-pasteurization contaminants such as pseudomonads and surviving thermodurics such as *B. cereus*. Generally lipolytic activity leading to rancidity is a more important feature of spoilage than proteolysis. Butter, made from cream, is described in Section 4.9 as the principal example of food preservation by compartmentalization.

5.3 MEAT

Originally meat was a term used to describe any solid food, but has now come to be applied almost solely to animal flesh. As such, it has played a significant role in the human diet since the days of hunting and gathering, and animals (sheep) were first domesticated at the beginning of the Neolithic revolution around 8500 BC. Though abjured by some on moral or religious grounds, meat eating remains widely popular today. In the main, this is due to its desirable texture and flavour characteristics, although meat protein does also have a high biological value.

Meat consumption is often something of a status symbol and is generally far greater in wealthy societies. This is because large-scale meat production is a relatively inefficient means of obtaining protein. It requires agriculture to produce a surplus of plant proteins which can be fed to animals: with modern production techniques, it takes two kilos of grain to obtain 1 kilo of chicken, four for 1 kilo of pork and eight for 1 kilo of beef.

Though numerous species are used as a source of meat around the world, ranging from flying foxes to frogs and from kangaroos to crocodiles, the meat animals of principal importance in economic terms are cattle, pigs, sheep, goats and poultry.

5.3.1   Structure and Composition

Edible animal flesh comprises principally the muscular tissues but also includes organs such as the heart, liver, and kidneys. Most microbiological studies on meat have been conducted with muscular tissues and it is on these that the information presented here is based. Though in many respects the microbiology will be broadly similar for other tissues, it should be remembered that differences may arise from particular aspects of their composition and microflora.

Structurally muscle is made up of muscle fibres; long, thin, multinucleate cells bound together in bundles by connective tissue. Each muscle fibre is surrounded by a cell membrane, the sarcolemma, within which are contained the myofibrils, complexes of the two major muscle proteins, myosin and actin, surrounded by the sarcoplasm. The approximate chemical composition of typical adult mammalian muscle after *rigor mortis* is presented in Table 5.5. Its high water activity and abundant nutrients make meat an excellent medium to support microbial growth. Though many of the micro-organisms that grow on meat are proteolytic, they grow initially at the expense of the most readily utilized substrates— the water soluble pool of carbohydrates and non-protein nitrogen. Extensive proteolysis only occurs in the later stages of decomposition when the meat is usually already well spoiled from a sensory point of view.

The carbohydrate content of muscle has a particularly important bearing on its microbiology. Glycogen is a polymer of glucose held in the liver and muscles as an energy store for the body. During life, oxygen is supplied to muscle cells in the animal by the circulatory system and glycogen can be broken down to provide energy by the glycolytic and respiratory pathways to yield carbon dioxide and water. After death the supply of oxygen to the muscles is cut off, the redox potential falls and respiration ceases, but the glycolytic breakdown of glycogen continues leading to an accumulation of lactic acid and a decrease in muscle pH. Provided sufficient glycogen is present, this process will continue until the glycolytic enzymes are inactivated by the low pH developed. In a typical mammalian muscle the pH will drop from an initial value of around 7 to 5.4–5.5 with the accumulation of about 1% lactic acid. Where there is a limited supply of glycogen in the muscle, acidification will continue only until the glycogen runs out and the muscle will have a higher ultimate pH. This can happen if the muscle has been exercised before slaughter

Table 5.5 *Chemical composition of typical adult mammalian muscle after rigor mortis*

		% weight
Water		75.0
Protein		19.0
Myofibrillar	11.5	
Sarcoplasmic	5.5	
Connective	2.0	
Lipid		2.5
Carbohydrate		1.2
Lactic acid	0.9	
Glycogen	0.1	
Glucose and glycolytic intermediates	0.2	
Soluble non-protein nitrogen		1.65
Creatine	0.55	
Inosine monophosphate	0.30	
NAD/NADP	0.30	
Nucleotides	0.10	
Amino acids	0.35	
Carnosine, anserine	0.35	
Inorganic		0.65
Total soluble phosphorus	0.20	
Potassium	0.35	
Sodium	0.05	
Magnesium	0.02	
Other metals	0.23	
Vitamins		

After R.A. Lawrie, 'Meat Science', 3rd edn., Pergamon Press, Oxford, 1979

but can also result from stress or exposure to cold. When the ultimate pH is above 6.2, it gives rise to dark cutting meat, a condition also known as dry, firm, dark (DFD) condition. Because the pH is relatively high, the meat proteins are above their isoelectric point and will retain much of the moisture present. The fibres will be tightly packed together giving the meat a dry, firm texture and impeding oxygen transfer. This, coupled with the higher residual activity of cytochrome enzymes, will mean that the meat has the dark colour of myoglobin rather than the bright red oxymyoglobin colour. The higher pH will also mean that microbial growth is faster so spoilage will occur sooner.

Another meat defect associated with post mortem changes in muscle carbohydrates is known as pale, soft, exudative (PSE) condition. This occurs mainly in pigs and has no microbiological implications but does give rise to lower processing yields, increased cooking losses and reduced juiciness. The PSE condition results when normal non-exercised muscle is stimulated just before slaughter leading to a rapid post mortem fall in pH while the muscle is still relatively warm. This denatures sarcoplasmic proteins, moisture is expelled from the tissues which assume a pale colour due to the open muscle texture and the oxidation of myoglobin to metmyoglobin.

5.3.2 The Microbiology of Primary Processing

The tissues of a healthy animal are protected against infection by a combination of physical barriers and the activity of the immune system. Consequently, internal organs and muscles from a freshly slaughtered carcass should be relatively free from micro-organisms. Microbial numbers detected in aseptically sampled tissues are usually less than $10\,\mathrm{cfu\,kg^{-1}}$, although there is evidence that numbers can increase under conditions of stress and they will of course be higher if the animal is suffering from an infection. Since some animal diseases can be transmitted to humans, meat for human consumption should be produced only from healthy animals. Visual inspection before and after slaughter to identify and exclude unfit meat is the general rule, although it will only detect conditions which give some macroscopic pathological sign. In the UK this and other duties are performed by the Meat Hygiene Service, a body established in 1995 to protect public and animal health and unify all aspects of meat inspection and enforcement.

The most heavily colonized areas of the animal that may contaminate meat are the skin (fleece) and gastrointestinal tract. Numbers and types of organisms carried at these sites will reflect both the animal's indigenous microflora and its environment. The animal hide, for example, will carry a mixed microbial population of micrococci, staphylococci, pseudomonads, yeasts and moulds as well as organisms derived from sources such as soil or faeces. Organisms of faecal origin are more likely to be encountered on hides from intensively reared cattle or from those transported or held in crowded conditions.

The various processing steps in slaughter and butchering are described in Chapter 11 for beef and will be summarized only briefly here. With reasonable standards of hygienic operations, contamination of meat carcasses from processing equipment, knives and process workers is less important than contamination from the animals themselves. The greatest opportunity for this occurs during dressing, the stages during which the head, feet, hides, excess fat, viscera and offal are separated from the bones and muscular tissues.

Skinning can spread contamination from the hide to the freshly exposed surface of the carcass through direct contact and *via* the skinning knife or handling. Washing the animal prior to slaughter can reduce microbial numbers on the hide but control is most effectively exercised by skillful and hygienic removal of the hide. The viscera contain large numbers of micro-organisms, including potential pathogens, and great care must be taken to ensure the carcass is not contaminated with visceral contents either as a result of puncture or leakage from the anus or oesophagus during removal.

After dressing, carcasses are washed to remove visible contamination. This will have only a minor effect on the surface microflora, although bactericidal washing treatments such as hot water (80 °C), chlorinated water (50 mg l^{-1}) or dilute lactic acid (1–2%) have been shown to reduce the surface microflora by amounts varying between about 1 and 3.5 log cycles.

After dressing the carcass is cooled to chill temperatures during which cold shock may cause some reduction in numbers. At chill temperatures, microbial growth among the survivors is restricted to those psychrotrophs present and these can be further inhibited by the partial surface drying that takes place. Surface numbers of bacteria at the end of dressing will typically be of the order of 10^2–10^4 cfu cm^{-2}. Counts are generally higher in sheep carcasses than beef and higher still in pigs which are processed differently, the skin not being removed from the carcass but scalded and dehaired.

Psychrotrophic organisms form only a small percentage of the initial microflora but come to predominate subsequently as the meat is held constantly at chill temperatures. An increase in microbial numbers is seen during cutting and boning, but this is due less to microbial growth, since the operation is usually completed within a few hours at temperatures below 10 °C, than to the spreading of contamination to freshly exposed meat surfaces by equipment such as knives, saws and cutting tables.

The primary processing of poultry differs from red meat in a number of respects that have microbiological implications. First among these is the sheer scale of modern poultry operations where processing plants can have production rates up to 12 000 birds per hour. This leaves little opportunity for effectively sanitizing equipment and exacerbates problems associated with some of the procedures and equipment used which favour the spread of micro-organisms between carcasses.

During transport to the plant contamination can be spread between birds by faeces and feathers and from inadequately cleaned transport cages. Once at the plant, birds are hung by their feet on lines, electrically stunned and killed by cutting the carotid artery. The close proximity of the birds and the flapping wings further contribute to the spread of contamination. This is followed by scalding where the birds are immersed in hot water at about 50 °C to facilitate subsequent removal of the feathers. Each bird contributes large numbers of micro-organisms to the scald water and these will be spread between birds. This can be reduced to some extent by using a counter-current flow of birds and water so that the birds leaving the scalder are in contact with the cleanest water. Higher scald water temperatures will eliminate most vegetative bacteria but cause an unacceptable loosening of the skin cuticle.

After scalding, birds are mechanically defeathered by a system of rotating rubber fingers. A number of studies have demonstrated how these can pass organisms, for example *Salmonella*, from one carcass to others following it and when the fingers become worn or damaged they are liable to microbial colonization. As the poultry carcass is not skinned, skin-associated organisms will not be removed.

The intestinal tract of poultry will contain high numbers of organisms including pathogens such as *Salmonella* and *Campylobacter*. Poultry evisceration therefore poses similar microbiological hazards to those with other animals but the size and structure of the carcass make it a much more difficult operation to execute hygienically. To allow high processing rates, poultry evisceration is usually automated but this too leads to a high incidence of carcass contamination with gut contents. Since the carcasses are not split like those of sheep and cattle, effective washing of the gut cavity after evisceration is more difficult.

Poultry to be frozen is usually chilled in water and this offers a further opportunity for cross contamination. This is controlled by chlorination of the cooling water, use of a counter-current flow as in scalding, and a sufficient flow rate of water to avoid the build up of contamination.

5.3.3 Spoilage of Fresh Meat

Aerobic storage of chilled red meats, either unwrapped or covered with an oxygen permeable film, produces a high redox potential at the meat surface suitable for the growth of psychrotrophic aerobes. Non-fermentative Gram-negative rods grow most rapidly under these conditions and come to dominate the spoilage microflora that develops. Taxonomic description of these organisms has been somewhat unsettled over the years with some being described as *Moraxella* and *Moraxella*-like. Such terms have now been largely abandoned in favour of a concensus that has emerged from numerical taxonomy studies. In this, the principal genera are described as *Pseudomonas, Acinetobacter* and *Psychrobacter* with *Pseudomonas* species such as *P. fragi, P. lundensis* and *P. fluorescens* generally predominating. A dichotomous key describing the differential characteristics of these organisms and some of the names used previously to describe them is presented as Figure 5.4. Other organisms are usually only a minor component of the spoilage microflora, but include psychrotrophic Enterobacteriaceae such as *Serratia liquefaciens* and *Enterobacter agglomerans*, lactic acid bacteria and the Gram-positive *Brochothrix thermosphacta*.

The first indication of spoilage in fresh meat is the production of off odours which become apparent when microbial numbers reach around

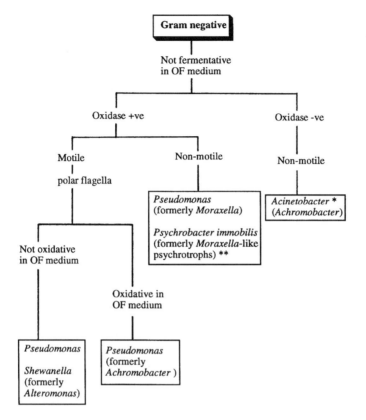

* see Bouvet, P.J.M. & Grimont, P.A.D. (1986). Taxonomy of the genus *Acinetobacter. International Journal of Systematic Bacteriology* **36**, 228-240.
see Juni, E. & Heym, G.A. (1986). *Psychrobacter immobilis* gen. nov., sp. nov. Genospecies composed of Gram-negative, aerobic, oxidase-positive coccobacilli. *International Journal of Systematic Bacteriology* **36, 388-391.

Figure 5.4 *Characteristics of some Gram-negatives associated with meat*

10^7 cfu cm^{-2}. At this point it is believed that the micro-organisms switch from the diminishing levels of glucose in the meat to amino acids as a substrate for growth. In meat with lower levels of residual glucose this stage is reached earlier (10^6 cfu cm^{-2}) and this accounts for the earlier onset of spoilage in high pH meat.

Bacterial metabolism produces a complex mixture of volatile esters, alcohols, ketones and sulfur-containing compounds which collectively comprise the off odours detected. Such mixtures can be analysed by a combination of gas chromatography and mass spectrometry and the origin of different compounds can be established by pure culture studies. These have confirmed the predominant role of pseudomonads in spoilage of aerobically stored chilled meat. Usually the different spoilage taints appear in a sequence reflecting the order in which components of the meat are metabolized. The first indication of spoilage is generally the buttery or cheesy odour associated with production of diacetyl (2,3-butanedione), acetoin (3-hydroxy-2-butanone), 3-methyl-butanol and 2-methylpropanol. These compounds are produced from glucose by members of the Enterobacteriaceae, lactic acid bacteria and *Brochothrix thermosphacta*. Pseudomonads then begin to increase in importance and the meat develops a sweet or fruity odour. This is due to production of a range of esters by *Pseudomonas* and *Moraxella* species degrading glucose and amino acids and by esterification of acids and alcohols produced during the first phase of spoilage. Ester production is particularly associated with *Pseudomonas fragi* which can produce ethyl esters of acetic, butanoic and hexanoic acids from glucose, but other pseudomonads and *Moraxella* species are also capable of producing esters when grown on minced beef. As the glucose becomes exhausted, the meat develops a putrid odour when *Pseudomonas* species and some *Acinetobacter* and *Moraxella* species turn their entire attention to the amino acid pool, producing volatile sulfur compounds such as methane thiol, dimethyl sulfide and dimethyl disulfide.

In the later stages of spoilage an increase in the meat pH is seen as ammonia and a number of amines are elaborated. Some of these have names highly evocative of decay and corruption such as putrescine and cadaverine but in fact do not contribute to off odour. When microbial numbers reach levels of around 10^8 cfu cm^{-2}, a further indication of spoilage becomes apparent in the form of a visible surface slime on the meat.

Vacuum and modified-atmosphere packing of meat (see also Section 4.6) changes the meat microflora and consequently the time-course and character of spoilage. In vacuum packs the accumulation of CO_2 and the absence of oxygen restrict the growth of pseudomonads giving rise to a microflora dominated by Gram-positives, particularly lactic acid bacteria of the genera *Lactobacillus, Carnobacterium* and *Leuconostoc*.

Spoilage of vacuum packed meat is characterized by the development of sour acid odours which are far less objectionable than the odour associated with aerobically stored meat. The micro-organisms reach their maximum population of around 10^7 cfu cm^{-2} after about a week's storage but the souring develops only slowly thereafter. Organic acids may contribute to this odour, although the levels produced are generally well below the levels of endogenous lactate already present. Some work has suggested that methane thiol and dimethyl sulfide may contribute to the sour odour.

The extension of shelf-life produced by vacuum packing is not seen with high pH (>6.0) meat. In this situation *Shewanella putrefaciens*, which cannot grow in normal pH meat, and psychrotrophic Enterobacteriaceae can grow and these produce high levels of hydrogen sulfide giving the meat an objectionable odour.

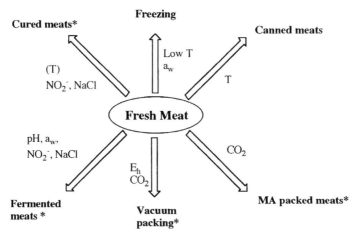

Figure 5.5 *Meat and meat products. T indicates elevated temperature; E_h, low redox potential; pH, reduced pH; a_w, reduced a_w; and * stored at chill temperatures*

In modified-atmospheres containing elevated levels of both CO_2 and O_2 growth of pseudomonads is restricted by the CO_2 while the high levels of O_2 maintain the bright red colour of oxygenated myoglobin in the meat. Here the microflora depends on the type of meat, its storage temperature, and whether it was vacuum packed or aerobically stored previously. In general terms though, the microflora and spoilage tend to follow a similar pattern to that of vacuum packed meat. Hetero-fermentative lactic acid bacteria can be more numerous due to the stimulatory effect of oxygen on their growth and, under some circum-stances, *Brochothrix thermosphacta*, Enterobacteriaceae and pseudo-monads can be more important.

Meat can be processed in a number of different ways which affect its characteristics, shelf-life and microbiology. The variety of these is illustrated by Figure 5.5; they will not be discussed further here but are treated in greater detail under the generic technologies in Chapters 4 and 9.

5.4 FISH

Here we are mainly concerned with what most people think of as fish; principally the free swimming teleosts and elasmobranchs. The same term can also encompass all seafoods including crustaceans with a chitinous exoskeleton such as lobsters, crabs and shrimp, and molluscs such as mussels, cockles, clams and oysters. Microbiologically these share many common features with free swimming fish but some specific aspects are discussed in Section 5.4.3.

Historically the extreme perishability of fish has restricted its con-sumption in a reasonably fresh state to the immediate vicinity of where the catch was landed. This has detracted only slightly from it playing a significant role in human nutrition as, throughout the world, traditional curing techniques based on combinations of salting, drying and smoking were developed which allowed more widespread fish consumption. The importance of dried salted cod in economic and social history has already been alluded to in Section 4.7.

Poor keeping quality is a special feature of fish which sets it apart even from meat and milk. The biochemical and microbiological reasons for this are discussed in Section 5.4.4.

5.4.1 Structure and Composition

Although broadly similar in composition and structure to meat, fish has a number of distinctive features. Unlike meat, there are no visually obvious deposits of fat. Although the lipid content of fish can be up to 25%, it is largely interspersed between the muscle fibres. A further feature which contributes to the good eating quality of fish is the very low content of connective tissue, approximately 3% of total weight compared with around 15% in meat. This, and the lower proportion of body mass contributed by the skeleton, reflect the greater buoyancy in water compared with that in air.

Muscle structure also differs. In land animals it is composed of very long fibres while in fish they form relatively short segments known as myotomes separated by sheets of connective tissue known as myocommata. This gives fish flesh its characteristically flaky texture.

Fish flesh generally contains about 15–20% protein and less than 1% carbohydrate. In non-fatty fish such as the teleosts cod, haddock and whiting, fat levels are only about 0.5%, while in fatty fish such as mackerel and herring, levels can vary between 3 and 25% depending on factors such as the season and maturity.

5.4.2 The Microbiology of Primary Processing

As with meat, the muscle and internal organs of healthy, freshly caught fish are usually sterile but the skin, gills and alimentary tract all carry substantial numbers of bacteria. Reported numbers on the skin have ranged from 10^2–10^7 cfu cm^{-2}, and from 10^3–10^9 cfu g^{-1} in the gills and the gut. These are mainly Gram-negatives of the genera *Pseudomonas, Shewanella, Psychrobacter, Vibrio, Flavobacterium* and *Cytophaga* and some Gram-positives such as coryneforms and micrococci. Since fish are cold blooded, the temperature characteristics of the associated flora will reflect the water temperatures in which the fish live. The microflora of fish from northern temperate waters where the temperatures usually range between -2 and $+12\,°C$ is predominantly psychrotrophic or psychrophilic. Most are psychrotrophs with an optimum growth temperature around 18 °C. Far fewer psychrotrophs are associated with fish from warmer tropical waters and this is why most tropical fish keep far longer in ice than temperate fish.

Bacteria associated with marine fish should be tolerant of the salt levels found in sea water. Though many do grow best at salt levels of 2–3%, the most important organisms are those that are not strictly halophilic but euryhaline, *i.e.* they can grow over a range of salt concentrations. It is these that will survive and continue to grow as the salt levels associated with the fish decline, for example when the surface is washed by melting ice.

After capture at sea, fish are commonly stored in ice or refrigerated sea water until landfall is made. It is important that fresh, clean cooling agent is used as re-use will lead to a rapid build up of psychrotrophic contaminants and accelerated spoilage of the stored fish. Gutting the fish prior to chilling at sea is not a universal practice, particularly with small fish and where the time between harvest and landing is short. It does however remove a major reservoir of microbial contamination at the price of exposing freshly cut surfaces which will be liable to rapid spoilage. Similarly any damage to the fish from nets, hooks, *etc.* that breaches the fish's protective skin will provide a focus for spoilage. Subsequent processing operations such as filleting and mincing which increase the surface area to volume of the product also increase the rate of spoilage.

Fish can be further contaminated by handling on board, at the dock and at markets after landing, particularly where they are exposed for sale and are subject to contamination with human pathogens by birds and flies. Generally though, fish have a far better safety record than mammalian meat. A number of types of foodborne illness are associated with fish (Table 5.6), and these are discussed in detail in Chapters 7 and 8.

5.4.3 Crustaceans and Molluscs

The propensity of crustaceans to spoil rapidly (see Section 5.4.4) can be controlled in the case of crabs and lobsters by keeping them alive until

Table 5.6 *Foodborne illness and pathogens associated with fish*

Vibrio cholerae
Vibrio parahaemolyticus
Vibrio vulnificus
Clostridium botulinum Type E
Enteric viruses
Scombroid fish poisoning
Paralytic shellfish poisoning

immediately before cooking or freezing. This is not possible with shrimp or prawns, which are of far greater overall economic importance but die soon after capture. In addition to their endogenous microflora, shrimp are often contaminated with bacteria from the mud trawled up with them and are therefore subject to rapid microbiological deterioration following capture. Consequently they must be processed either by cooking or by freezing immediately on landing.

Some aspects of the production and processing of frozen cooked peeled prawns can pose public health risks. Increasingly prawns are grown commercially in farms where contamination of the ponds, and thence the product, with pathogenic bacteria can occur *via* bird droppings and fish feed. After cooking, which should be sufficient to eliminate vegetative bacterial contaminants derived from the ponds, the edible tail meat is separated from the chitinous exoskeleton. Peeling machines are used in some operations but large quantities are still peeled by hand, particularly in countries where labour is cheap. The handling involved gives an opportunity for the product to be contaminated with human pathogens after the bactericidal cooking step and prior to freezing.

The flesh of molluscs such as cockles, mussels, oysters and clams differs from that of crustaceans and free swimming fish by containing appreciable ($\approx 3\%$) carbohydrate in the form of glycogen. Though many of the same organisms are involved, spoilage is therefore glycolytic rather than proteolytic, leading to a pH decrease from around 6.5 to below 5.8.

Molluscs are usually transported live to the point of sale or processing where the flesh can often be removed by hand. Although contamination may occur at this stage, the significant public health problems associated with shellfish arise more from their ability to concentrate viruses and bacteria from surrounding waters, the frequent pollution of these waters with sewage and the practice of consuming many shellfish raw or after relatively mild cooking. This is discussed in more detail in Chapter 9.

5.4.4   Spoilage of Fresh Fish

A number of factors contribute to the unique perishability of fish flesh. In the case of fatty fish, spoilage can be non-microbiological; fish lipids contain a high proportion of polyunsaturated fatty acids which are more reactive chemically than the largely saturated fats that occur in mammalian meat. This makes fish far more susceptible to the development of oxidative rancidity.

In most cases though, spoilage is microbiological in origin. Fish flesh naturally contains very low levels of carbohydrate and these are further depleted during the death struggle of the fish. This has two important consequences for spoilage. Firstly it limits the degree of post mortem acidification of the tissues so that the ultimate pH of the muscle is 6.2–6.5

Table 5.7 *Nitrogen-containing extractives in fish*

	Cod	Herring	Dogfish	Lobster
Total nitrogenous extractives gkg^{-1}	12	12	30	55
Free amino acids (mM l^{-1})	7	30	10	300
TMAO (mM l^{-1})	5	3	10	2
Urea (mM l^{-1})	0	0	33	0
Creatine (mM l^{-1})	3	3	2	0
Betaine (mM l^{-1})	0	0	2	1
Anserine	1	0	0	0

Adapted from D.M. Gibson, 'Microbial spoilage of foods in Micro-organisms in Action: Concepts and Applications in Microbial Ecology'. J. M.Lynch and J. E.Hobbie, (eds.) Blackwell, Oxford, 1988

compared with around 5.5 in mammalian muscle. Fish which have a lower pH such as halibut (approx. 5.6) tend to have better keeping qualities. Secondly, the absence of carbohydrate means that bacteria present on the fish will immediately resort to using the soluble pool of readily assimilated nitrogenous materials, producing off-odours and flavours far sooner. This can be less pronounced in fish produced by intensive aquaculture since they are normally fed to satiation which increases glycogen levels in the liver and muscle.

The composition of the non-protein nitrogen fraction differs significantly from that in meat (Table 5.7). Trimethylamine oxide (TMAO) occurs in appreciable quantities in marine fish as part of the osmoregulatory system. TMAO is used as a terminal electron acceptor by non-fermentative bacteria such as *Shewanella putrefaciens* and this allows them to grow under microaerophilic and anaerobic conditions. The product of this reduction is trimethylamine which is an important component in the characteristic odour of fish (Figure 5.6.) TMAO also contributes to a relatively high redox potential in the flesh since the E_h of the TMAO/TMA couple is +19 mV.

Elasmobranchs such as dogfish and shark contain high levels of urea. Bacterial urease activity in the flesh can produce ammonia very rapidly giving the product a pungent odour. Not only does this render the flesh itself uneatable but it can also taint the flesh of other fish stored nearby. It is for this reason that in many areas fishermen will discard all but the fins of shark when they catch them.

Shellfish such as lobster have a particularly large pool of nitrogenous extractives and are even more prone to rapid spoilage; a factor which accounts for the common practice of keeping them alive until immediately prior to consumption.

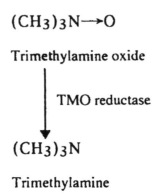

Figure 5.6   *Reduction of trimethylamine oxide*

Fish proteins are less stable than mammalian protein. As with meat, extensive proteolysis does not become apparent until the product is already well spoiled, but limited protein degradation may improve bacterial access to the nutrient pool of extractives.

The speed with which a product spoils is also related to the initial microbial load on the product: the higher the count the sooner spoilage occurs. Since fish from cold waters will have a larger proportion of psychrotrophs among their natural microflora, this can shorten the chill shelf-life appreciably.

Spoilage of chilled fish is due principally to the activity of psychrotrophic Gram-negative rods also encountered in meat spoilage, particularly *Shewanella putrefaciens* and *Pseudomonas* spp. The uniquely objectionable smell of decomposing fish is the result of a cocktail of chemicals, many of which also occur in spoiling meat. Sulfurous notes are provided by hydrogen sulfide, methyl mercaptan and dimethyl sulfide and esters contribute the 'fruity' component of the odour. A number of other amines in addition to TMA are produced by bacterial catabolism of amino acids. Skatole, a particularly unpleasant example produced by the degradation of tryptophan, also contributes to the smell of human faeces. The level of volatile bases in fish flesh has provided an index of spoilage, although this and other chemical indices used are often poor substitutes for the trained nose and eyes.

Figure 5.7 illustrates some of the different products made from fish, most of which are discussed elsewhere in terms of the general processing technologies used. One interesting aspect that relates to some of the discussion above will be discussed here. The combination of a near neutral pH and availability of TMAO as an alternative electron acceptor means that vacuum and modified-atmosphere packing of fish does not produce the same dramatic extension of keeping quality seen with meat. Typically the shelf-life extension of vacuum and modified-atmosphere-packed cod will vary from less than 3 days to about 2 weeks. *Shewanella*

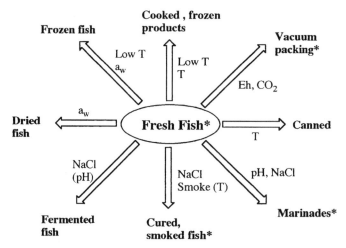

Figure 5.7 *Fish and fish products.* T *indicates elevated temperature;* E_h *low redox potential: pH, reduced pH;* a_w, *reduced* a_w, *and* * *stored an chill temperatures*

putrefaciens can grow under these conditions producing TMA and hydrogen sulfide to spoil the product. Work in Denmark has also demonstrated that CO_2 tolerant marine vibrios like *Photobacterium phosphoreum* may be responsible for a non-sulfurous spoilage of these products in some instances. This is a large (5 μm diameter), almost yeast-like, bacterium that has been isolated from the intestines of several different fish. Because of its size it produces 10–100-fold more TMA per cell than smaller organisms such as *Shewanella* and therefore can cause spoilage at lower populations, typically around 10^7 cfu g^{-1} compared with 10^8 cfu g^{-1} for more conventional bacteria. Whether *Shewanella* or *Photobacterium* is ultimately responsible for spoilage in MAP fish products probably depends on relative numbers present initially and whether any other selective factors operate.

5.5 PLANT PRODUCTS

The plant kingdom provides a considerable part of human food requirements and, depending on the particular plant species, use is made of every part of the plant structure from root, tuber, stem, leaves, fruit and seeds. Plants have evolved many strategies to survive the predation of herbivores and omnivores, including humans, and these strategies include, not only protective mechanisms to protect vegetative parts such as leaves, stems and roots, but also the development of rich, succulent fruits to encourage animals to help in the dispersal of seeds.

As the human race settled down from the nomadic hunter–gatherer state to form increasingly large stable communities so a wider range of plants have been brought into cultivation. Plant products are grown on an ever larger scale and many are stored for significant periods after harvest and may be transported from one part of the world to another.

Microbiological problems may occur at all stages in the production of plant products. During growth in the field there is a wide range of plant pathogens to contend with and, although these are dominated by the fungi, there are a significant number of bacteria and viruses. A further range of micro-organisms may cause post-harvest spoilage although there cannot be an absolutely clear boundary between plant pathogens and spoilage organisms because many plant products are made up of living plant tissue even after harvest.

Plants have evolved many mechanisms to prevent microbial invasion of their tissues. The outer surface is usually protected by a tough, resistant cuticle although the need for gas exchange requires specialized openings in parts of the leaf surface, the stomata and lenticels, which may provide access by some micro-organisms to plant tissue. Plant tissues may contain antimicrobial agents which are frequently phenolic metabolites, indeed the complex polyphenolic polymer known as lignin is especially resistant to microbial degradation. Many plants produce a special group of antimicrobial agents, the phytoalexins, in response to the initiation of microbial invasion. The low pH of the tissues of many fruits provides considerable protection against bacteria and the spoilage of these commodities is almost entirely by fungi. In contrast, many vegetables have somewhat higher pH and may be susceptible to bacterial spoilage (Table 5.8).

Another physical factor influencing the pattern of spoilage is the availability of water. Cereals, pulses, nuts and oilseeds are usually dried post harvest and the low water activity should restrict the spoilage flora to xerophilic and xerotolerant fungi. These three groups of plant products, *i.e.* fruits, vegetables and cereals *etc.*, are sufficiently distinct that they will now be considered separately.

Table 5.8 *pH values of some fruits and vegetables*

Fruits	pH	Vegetables	pH
Apples	2.9–3.3	Asparagus	5.4–5.8
Apricots	3.3–4.4	Broccoli	5.2–6.5
Bananas	4.5–5.2	Cabbage	5.2–6.3
Cherries	3.2–4.7	Carrots	4.9–6.3
Grapefruit	3.0	Cauliflower	6.0–6.7
Grapes	3.4–4.5	Celery	5.6–6.0
Limes	2.0–2.4	Lettuce	6.0–6.4
Melons	6.2–6.7	Parsnip	5.3
Oranges	3.3–4.3	Rhubarb	3.1–3.4
Pears	3.4–4.7	Runner beans	4.6
Plums	2.8–4.6	Spinach	5.1–6.8
Raspberries	2.9–3.5	Sweet potato	5.3–5.6
Tomatoes	3.4–4.9	Turnips	5.2–5.6

5.5.1 Cereals

The cereals, which all belong to the Gramineae or the grass family, are one of the most important sources of carbohydrates in the human diet. Some of the more important cereal crops are listed in Table 5.9. Wheat, rice and maize are by far the most important cereal crops on the basis of world-wide tonnage and well over three hundred million tons of each are produced annually. However, each cereal species is adapted to grow in a particular range of climatic conditions although plant breeding pro-grammes have extended the ranges of several of them. The common wheat is the major cereal of temperate parts of the world, and is grown extensively in both northern and southern hemispheres, whereas the durum wheat is grown extensively in the Mediterranean region and in the warmer, drier parts of Asia, North and South America. Rye can be grown in colder parts of the world and is an important crop in central Europe and Russia. Maize, which originated from the New World, is now grown extensively throughout the tropics and subtropical regions of both northern and southern hemispheres. Similarly, although rice was originally of Asian origin, it is also grown in many parts of the world. Although sorghum and some of the millets are not very important in terms of world tonnage, they are especially well adapted to growing in warm, dry climates and may be locally the most important cereals in such regions as those bordering the southern edge of the Sahara desert.

The microbiology of cereals, during growth, harvest and storage is dominated by the moulds and it is convenient to consider two groups of fungi. The field fungi are well adapted to the sometimes rapidly changing conditions on the surfaces of senescing plant material in the field. Although they require relatively high water activities for optimum growth, genera such as *Cladosporium, Alternaria* and *Epicoccum* are able to survive the rapid changes that can occur from the desiccation of a hot sunny day to the cool damp conditions of the night. The genus *Fusarium* includes species which have both pathogenic and saprophytic activities.

Table 5.9 *Some of the more important cereal crops*

Botanical name	Common name
Triticum aestivum	Common wheat (bread wheat)
Triticum durum	Durum wheat (pasta wheat)
Hordeum spp.	Barley
Avena sativa	Oats
Secale cereale	Rye
Zea mays	Maize (American corn)
Oryza sativa	Rice
Sorghum vulgare	Sorghum
Panicum miliaceum	Millet
Pennisetum typhoideum	Bulrush millet

Thus *F. culmorum* and *F. graminearum* can cause both stem rot and head blight of wheat and barley in the field and these field infections may lead to more extensive post harvest spoilage of these commodities if they are stored at too high a water activity. By contrast the so-called storage fungi seem to be well adapted to the more constant conditions of cereals in storage, and generally grow at lower water activities (Table 5.10). The most important genera of the storage fungi are *Penicillium* and *Aspergillus*, although species of *Fusarium* may also be involved in spoilage when grain is stored under moist conditions.

Water activity and temperature are the most important environmental factors influencing the mould spoilage of cereals, and the possible production of mycotoxins, and Table 5.11 shows the relationship between water content and water activity for barley, oats and sorghum while Figure 3.10 shows the same information as an isotherm for wheat. Although xerophilic moulds such as *Eurotium* spp. and *Aspergillus*

Table 5.10 *Minimum water activity requirements of some common field and storage fungi*

Species	Minimum a_w
Field Fungi	
Fusarium culmorum	0.89
Fusarium graminearum	0.89
Alternaria alternata	0.88
Cladosporium herbarum	0.85
Storage fungi	
Penicillium aurantiogriseum	0.82
Penicillium brevicompactum	0.80
Aspergillus flavus	0.78
Aspergillus candidus	0.75
Eurotium amstelodami	0.71
Wallemia sebi	0.69

Table 5.11 *Equilibrium relative humidity, water activity and moisture content (as % wet weight) of cereals at 25 °C*

Equilibrium relative humidity (%)	Water activity a_w	Water potential (MPa)	Water content (% wet weight)		
			Barley	Oats	Sorghum
15	0.15	−261	6.0	5.7	6.4
30	0.30	−166	8.4	8.0	8.6
45	0.45	−110	10.0	9.6	10.5
60	0.60	−70	12.1	11.8	12.0
75	0.75	−39	14.4	13.8	15.2
90	0.90	−14.5	19.5	18.5	18.8
100	1.00	0.0	26.8	24.1	21.9

restrictus may grow very slowly at the lower limit of their water activity range (0.71 corresponding to about 14% water content in wheat at 25 °C) once they start growing and metabolizing they will produce water of respiration and the local water activity will steadily rise allowing more rapid growth. Indeed it could increase sufficiently to allow mesophilic mould spores to germinate and grow; the process being, in a sense, autocatalytic. There is a sequence of observable consequences of the process of mould growth on cereals starting with a decrease in germinability of the grain. This is followed by discolouration, the production of mould metabolites including mycotoxins, demonstrable increase in temperature (self-heating), the production of musty odours, caking and a rapid increase in water activity leading finally to the complete decay with the growth of a wide range of microorganisms.

5.5.2 Preservation of High-moisture Cereals

Although not directly relevant to human foods, the availability of high-moisture cereals, such as barley, provides a highly nutritious winter feed for cattle. Long-term storage of such material can be achieved by a lactic acid fermentation comparable to the making of silage, or by the careful addition of fatty acids such as propionic acid. If this process is not carried out carefully then it may be possible to have sufficient propionic acid to inhibit the normal spoilage moulds associated with cereals in a temperate climate, but not enough to inhibit *Aspergillus flavus*. It has been shown that, even though partially inhibited in its growth, this mould can produce aflatoxin B_1 at enhanced levels under these conditions. If such material is fed to dairy cattle there is the possibility of aflatoxin M_1 being secreted in the milk and it then becomes a problem in human foods and not just a problem of animal feeds (see Section 8.4.2).

5.5.3 Pulses, Nuts and Oilseeds

The pulses are members of the huge legume family of plants, the Fabaceae also known as the Papilionaceae and Leguminosae, which form a major source of vegetable proteins and include such important crops as peas, beans, soya, groundnuts and lentils. Although many species of peas and beans are familiar to us as fresh vegetables, millions of tons of the mature seeds of soya beans and groundnuts are harvested for longer term storage every year and may be susceptible to mould spoilage if not stored under appropriate conditions. Several of the leguminous seeds, such as groundnuts and soya beans, are also valuable sources of vegetable oils but there are plants from many other diverse families which are now used to provide food quality vegetable oils, rapeseed from the crucifers, sunflower seed from the daisy family, oilpalm and olives to mention just a few. Edible nuts may also come from a botanically wide range of tree species and many of them are rich in oil and give similar microbiological problems as oilseeds.

Seeds rich in oil, such as groundnuts, have a much lower water content at a particular water activity than cereals, thus groundnuts with a 7.2% water content have a water activity of about 0.65–0.7 at 25 °C. Apart from the problem of mycotoxin formation in moulded oilseeds, several mould species have strong lipolytic activity leading to the contamination of the extracted oils with free fatty acids which may in turn undergo oxidation to form products contributing to rancidity. The most important lipolytic moulds are species of *Aspergillus*, such as *A. niger* and *A. tamarii, Penicillium* and *Paecilomyces*, while at higher water activities species of *Rhizopus* may also be important. Figure 5.8 shows the influence of moisture content, and damage on the formation of free fatty

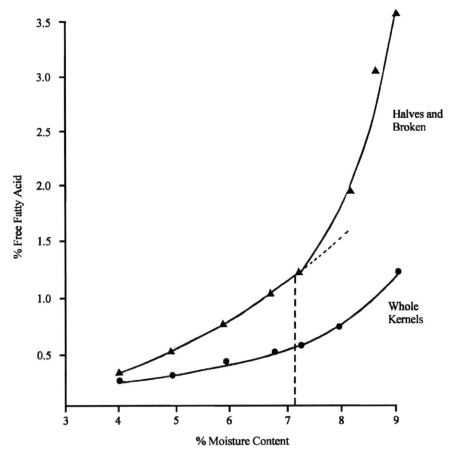

Figure 5.8   *The influence of moisture content on free fatty acid formation in groundnuts stored for four months. 7.2% moisture content corresponds to a$_w$ 0.65–0.70 at 25 °C so fungal growth may occur*

acids (FFA) in groundnuts stored for 4 months. It can be seen that in whole nuts there is a steady increase in FFA formation with an increase in moisture content and this is due to the plants own lipolytic enzymes. However, damaged groundnuts show a more rapid rise at low moisture contents, presumably due to increased contact of enzymes and substrate as a result of damage, but they also show an especially rapid rise in FFA at moisture contents greater than 7.2% corresponding to the active growth of lipolytic moulds.

If cereals, pulses, oilseeds and tree nuts are harvested with as little damage as possible and dried to an appropriate water content it should be possible to store them for considerable periods of time so long as they are not exposed to excessive temperature abuse during storage. The problems which may arise when large storage facilities, such as silos, are not carefully designed to avoid temperature differentials arising within the stored commodity were discussed in Section 3.3.1. The migration of water in these circumstances can result in the germination of fungal spores and the growth of mycelium creating a localized region of fungal activity releasing further water of respiration into the region. In this way, despite the commodity initially going into store at what was judged to be a safe water content, it may nevertheless go mouldy over a period of time.

It should be noted that, although a commodity may be dry enough to avoid direct microbiological spoilage, it may not be secure against the ravages of pests such as insects and rodents and their activity may lead to secondary invasion and mould spoilage.

5.5.4 Fruits and Fruit Products

Despite the high water activity of most fruits, the low pH leads to their spoilage being dominated by fungi, both yeasts and moulds but especially the latter. The degree of specificity shown by many species of moulds, active in the spoilage of harvested fruits in the market place or the domestic fruit bowl, reflects their possible role as pathogens or endophytes of the plant before harvest. Thus *Penicillium italicum* and *P. digitatum* show considerable specificity for citrus fruits, being the blue mould and green mould respectively of oranges, lemons and other citrus fruits. *Penicillium expansum* causes a soft rot of apples and, although the rot itself is typically soft and pale brown, the emergence of a ring of tightly packed conidiophores bearing enormous numbers of blue conidiospores, has led to this species being referred to as the blue mould of apples. This particular species has a special significance because of its ability to produce the mycotoxin patulin which has been detected as a contaminant in unfermented apple juices but not in cider (see Section 8.4.3).

Other common diseases of apples and pears include the black spot or scab, caused by the ascomycete *Venturia inaequalis* (anamorph *Spilocaea pomi = Fusicladium dendriticum*), and a brown rot caused by another ascomycete, *Monilinia fructigena* (= *Sclerotinia fructigena*, anamorph *Monilia fructigena*). Apple scab spoils the appearance of fruit, and would certainly reduce its commercial value, but does not cause extensive rotting of the tissue. The brown rot, however, can lead to extensive damage of fruit both on the tree and in storage. The typical brown rot is usually associated with rings of brown powdery pustules of the imperfect, or anamorph, stage, however fruit which is infected, but apparently healthy when it goes into store, can be reduced to a shiny black mummified structure in which much of the fruit tissue has been replaced by fungal material and the whole apple has become a functional sclerotium, or overwintering resting body, of the fungus. Although rarely seen in the United Kingdom, it is this structure which may germinate in the spring to produce the stalked apothecia of the perfect, or teleomorph, stage.

An especially widespread mould on both fruits and vegetables is the grey mould *Botrytis cinerea*, which is the imperfect stage of another ascomycete, *Botryotinia fuckeliana* (= *Sclerotinia fuckeliana*). Its role in the spoilage of strawberries was described in Section 2.5. Infection of grapes on the vine by this same mould can lead to drying out of the grape and an increase in sugar concentration and wines made from such contaminated fruit are considered to be very special. Under these circumstances the fungus has been referred to as *La Pourriture Noble –* the noble rot!

To avoid excessive mould spoilage of harvested fruit during storage and transport it is necessary to harvest at the right stage of maturity and avoid damage and bruising. Mouldy fruit should be removed and destroyed and good hygiene of containers and packaging equipment is essential to prevent a build-up of mould propagules. The development of international trade in many fruit species has led to the use of some biocides (Figure 5.9) to prevent mould spoilage. Benomyl has proved useful where it can be applied to the surface of fruits, such as citrus and bananas, in which the skins would normally be discarded (this, of course, is not the case for citrus used for marmalade and other preserves). In some parts of the world moulds like *Penicillium digitatum* have developed increased resistance to benomyl. Biphenyl is quite an effective protectant when incorporated into the wrapping tissues of fruit such as oranges when they are individually wrapped. Captan has been used as a spray for strawberries in the field to control *Botrytis* but its use must be stopped well before harvest.

Reduced temperature and increased carbon dioxide concentration may also be useful in controlling mould spoilage during storage and transport but many fruits are themselves sensitive to low temperatures

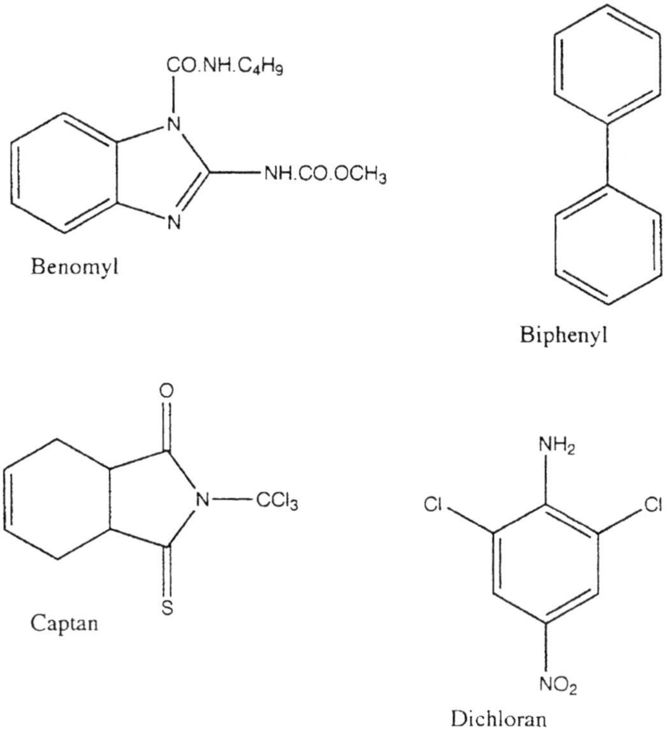

Figure 5.9 *Examples of antifungal biocides which may be used to protect fruits from mould spoilage*

and enhanced CO_2 levels and appropriate conditions need to be established for each commodity.

Canned fruits are normally given a relatively low heat treatment because of their low pH and although most mould propagules would be killed the ascospores of some members of the Eurotiales are sufficiently heat resistant to survive. Species of *Byssochlamys* are the best known but the increasing use of more exotic fruits is providing cases where spoilage of canned fruits have been due to such organisms as *Neosartorya fischeri* (anamorph *Aspergillus fischerianus*) and *Talaromyces flavus* var. *macrosporus* (anamorph *Penicillium* sp.).

5.5.5 Vegetables and Vegetable Products

The higher pH values of the tissues of many vegetables makes them more susceptible to bacterial invasion than fruits although there are also a number of important spoilage fungi of stored vegetables. The bacteria involved are usually pectinolytic species of the Gram-negative genera *Pectobacterium, Pseudomonas* and *Xanthomonas*, although pectinolytic strains of *Clostridium* can also be important in the spoilage of potatoes

Table 5.12 *Some micro-organisms involved in the spoilage of fresh vegetables*

Micro-organism	Vegetable	Symptom
BACTERIA		
Corynebacterium sepedonicum	Potato	ring rot of tubers
Ralstonia solanacearum	Potato	soft rot
Pectobacterium carotovorum var. *atrosepticum*	Potato	soft rot
Streptomyces scabies	Potato	scab
Xanthomonas campestris	Brassicas	black rot
FUNGI		
Botrytis cinerea	Many	grey mould
Botrytis allii	Onions	neck rot
Mycocentrospora acerina	Carrots	liquorice rot
Trichothecium roseum	Tomato Cucurbits	pink rot
Fusarium coeruleum	Potato	dry rot
Aspergillus alliaceus	Onion Garlic	black rot

under some circumstances, and the non-sporing Gram-positive organism *Corynebacterium sepedonicum* causes a ring rot of potatoes. Table 5.12 lists a range of micro-organisms which may cause spoilage of fresh vegetables.

The role of plant pathogens in subsequent spoilage post-harvest may be complex, thus *Phytophthora infestans* causes a severe field disease of the potato plant, frequently causing death of the plant, but it may also remain dormant within the tubers and either cause a rot of the tubers during storage, or a new cycle of disease in the next season's crop. However, the most frequent agents of spoilage are not the plant pathogens themselves but opportunistic micro-organisms which gain access to plant tissue through wounds, cracks, insect damage or even the lesions caused by the plant pathogens. All freshly harvested vegetables have a natural surface flora, including low numbers of pectinolytic bacteria, and it is becoming increasingly evident that healthy tissue of the intact plant may also contain very low numbers of viable micro-organisms (endophytic). The onset and rate of spoilage will depend on the interactions between the physiological changes occurring in the tissues after harvest and changes in microbial activity. Harvesting itself will produce physiological stress, principally as a result of water loss and wilting, and cut surfaces may release nutrients for microbial growth. This stress may also allow growth of the otherwise quiescent endophytic flora.

The most frequently observed form of spoilage is a softening of the tissue due to the pectinolytic activity of micro-organisms. Pectin, the methyl ester of α-1,4-poly-D-galacturonic acid, and other pectic substances are major components of the middle lamella between the cells making up plant tissue and once it is broken down the tissue loses its integrity and individual plant cells are more easily invaded and killed. Pectic substances may be quite complex and include unesterified pectic acid as well as having side chains of L-rhamnose, L-arabinose, D-galactose, D-glucose and D-xylose. Several distinct enzymes are involved in the degradation of pectin and their role is illustrated in Figure 5.10.

As described in the case of fruits, the prevention of spoilage during storage and transport of vegetables must involve a range of measures. The control of the relative humidity and the composition of the atmosphere in which vegetables are stored is important but there is a limit to the reduction of relative humidity because at values below 90–95%, loss of water from vegetable tissues will lead to wilting. It is essential to avoid the presence of free water on the surfaces of vegetables and temperature

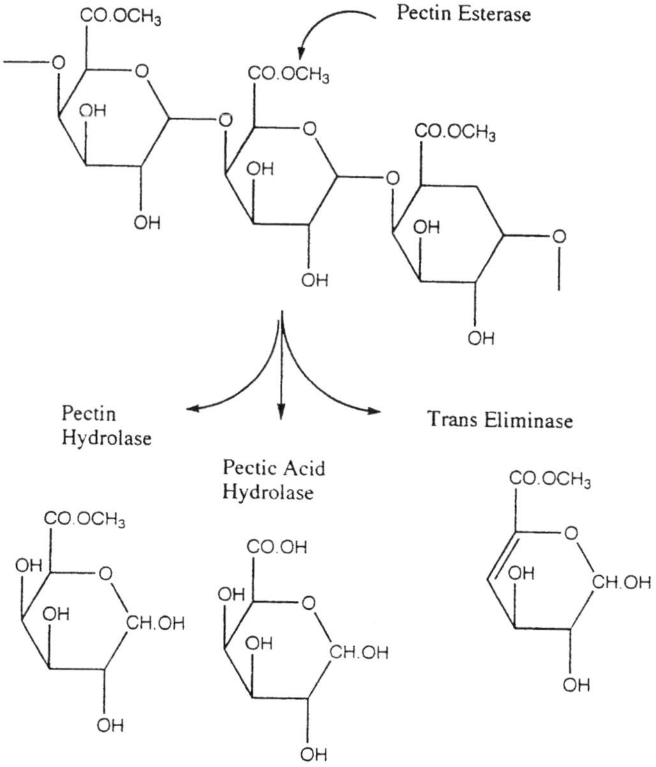

Figure 5.10 *Enzymic activities leading to the degradation of pectin*

control may be just as important to prevent condensation. The presence of a film of water on the surface will allow access of motile bacteria such as *Erwinia Pectobacterium* and pseudomonads to cracks, wounds and natural openings such as stomata. A combination of constant low temperature, controlled relative humidity, and a gas phase with reduced oxygen (*ca.* 2–3%) and enhanced CO_2 (*ca.* 2–5%) has made it possible to store the large hard cabbages used in coleslaw production for many months making the continuous production of this commodity virtually independent of the seasons.

Vegetables should not normally be a cause of public health concern but the transmission of enteric pathogens such as *Salmonella*, VTEC and *Shigella* is possible by direct contamination from farmworkers and the faeces of birds and animals, the use of manure or sewage sludge as fertilizer, or the use of contaminated irrigation water. Celery, watercress, lettuce, endive, cabbage and beansprouts have all been associated with *Salmonella* infections, including typhoid and paratyphoid fevers, and an outbreak of shigellosis has been traced to commercial shredded lettuce. Since salad vegetables are not usually cooked before consumption, it is important to follow good agricultural practices to avoid their contamination during production. Contamination can be reduced by washing produce in clean water but even chlorinated water will normally give only a 2–3 log reduction in microbial numbers as some surface bacteria are lodged in hydrophobic folds or pores and thus evade treatment.

Not all pathogens are necessarily transmitted to vegetables by direct or indirect faecal contamination. Organisms such as *Clostridium botulinum* have a natural reservoir in the soil and any products contaminated with soil can be assumed to be contaminated with spores of this organism, possibly in very low numbers. This would not normally present a problem unless processing or storage conditions were sufficiently selective to allow subsequent spore germination, growth and production of toxin. In the past, this has been seen mainly as a problem associated with underprocessed canned vegetables, but now it must be taken into consideration in the context of sealed, vacuum or modified-atmosphere packs of prepared salads. Those salads containing partly cooked ingredients, where spores may have been activated and potential competitors reduced in numbers, could pose particular problems. In 1987 a case of botulism caused by *Clostridium botulinum* type A was associated with a pre-packed rice and vegetable salad eaten as part of an airline meal. Similar risks may occur in foil-wrapped or vacuum packed cooked potatoes or film-wrapped mushrooms and in all these cases adequate refrigeration appears to be the most effective safety factor.

Another group of pathogens naturally associated with the environment includes the psychrotrophic species *Listeria monocytogenes* which is commonly associated with plant material, soil, animals, sewage and a wide range of other environmental sources. Raw celery, tomatoes and lettuce were implicated on epidemiological grounds as a possible cause of listeriosis which occurred in several hospitals in Boston, USA in 1979, although direct microbiological evidence was missing. An outbreak of listeriosis in Canada in 1981 was associated with coleslaw (see Section 7.9.5). Strains of *L. monocytogenes* can certainly grow on shredded cabbage and salad vegetables such as lettuce at temperatures as low as 5 °C and modified-atmospheres seem to have no effect on this organism. In the UK, routine surveillance of foods by the Public Health Laboratories revealed that, out of 567 samples of processed vegetables and salads examined, 87 (15%) were found to contain *Listeria* spp. while 72 (13%) contained *L. monocytogenes* specifically.

Two other psychrotrophic organisms which are readily isolated from the environment are *Yersinia enterocolitica* and *Aeromonas hydrophila*. Both may be expected to be associated with vegetables and could grow to levels capable of causing illness if care is not taken during the growth, harvesting, storage and treatment of these commodities.

CHAPTER 6

Food Microbiology and Public Health

6.1 FOOD HAZARDS

Although food is indispensible to the maintenance of life, it can also be responsible for ill health. A simple insufficiency will lead to marasmus (protein-energy deficiency) while over-reliance on staples low in protein, such as cassava, produces the condition known as kwashiorkor. A diet may provide adequate protein and energy but be lacking in specific minerals or vitamins giving rise to characteristic deficiency syndromes such as goitre (iodine deficiency), pellagra (nicotinic acid), beriberi (thiamine) and scurvy (ascorbic acid).

Foods are complex mixtures of chemicals and often contain compounds that are potentially harmful as well as those that are beneficial (Figure 6.1). Several vitamins are toxic if consumed in excessive amounts and many food plants produce toxic secondary metabolites to discourage their attack by pests.

Potatoes contain the toxic alkaloid solanine. Normally this is more concentrated in aerial parts of the plants and the peel which are not eaten, but high levels are also found in green potatoes and potato sprouts which should be avoided.

Cassava contains cyanogenic glycosides which produce hydrogen cyanide on hydrolysis. Similar compounds are also present in apple seeds, almonds, lima beans, yams and bamboo shoots. The body's detoxification pathway converts cyanide to thiocyanate which can interfere with iodine metabolism giving rise to goitre and cretinism. Traditional methods of preparing cassava eliminate the acute toxicity problem from hydrogen cyanide, but the increased incidence of goitre and cretinism in some areas where cassava is a staple may be a reflection of chronic exposure.

Legumes or pulses contain a number of anti-nutritional factors such as phytate, trypsin inhibitors and lectins (haemagglutinins). Many of these

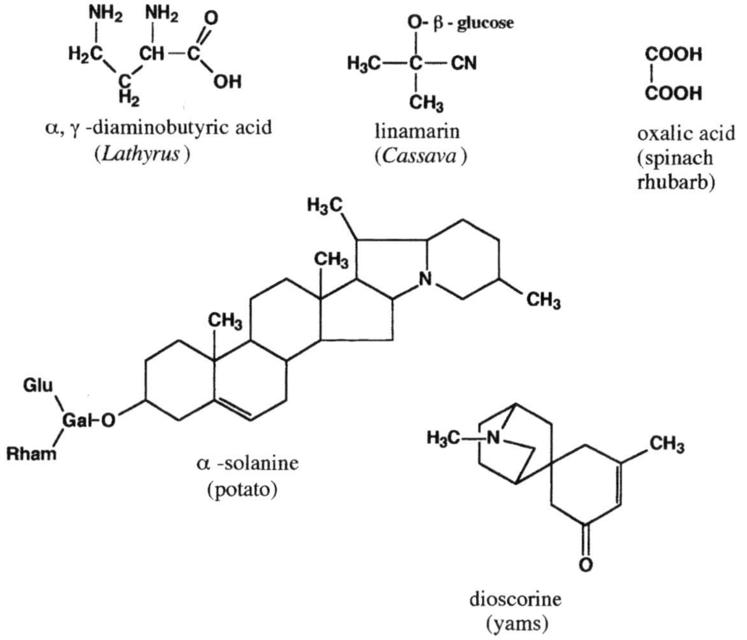

Figure 6.1 *Some natural food toxicants*

are destroyed or removed by normal preparation procedures such as soaking and cooking. Even so, red kidney beans are still responsible for occasional outbreaks of food poisoning when they have been insufficiently cooked to destroy the lectins they contain. Lathyrism is a more serious condition associated with a toxin in the pulse *Lathyrus sativa* which can be a major food item in North African and Asian communities during times of famine. In favism, an enzyme deficiency predisposes certain individuals to illness caused by glycosides in the broad bean, *Vicia faba*.

To some extent toxic or antinutritional characteristics can be bred out of cultivars intended for human consumption, although the problem cannot be eliminated completely. For affluent consumers in the developed world, particularly toxic foods can be avoided since alternatives are normally available in plenty. This is not always the case in poorer countries where these diet-related conditions are far more common.

It is however important to keep a sense of proportion in this. Eating inevitably exposes us to natural chemicals whose long-term effects on health are not known at present. Provided such foods form part of a balanced diet and are correctly prepared, the risks involved are generally acceptable, particularly when compared to the certain outcome should immoderate fear of food lead to complete abstinence.

In addition to the hazards posed by natural toxins that are an intrinsic feature of their composition, foods may also act as the vehicle by which

Table 6.1 *Possible causes of foodborne illness*

Chemical
  Intrinsic, natural toxins, *e.g.* red kidney bean poisoning, toxic mushrooms
  Extrinsic contamination
Algae, *e.g.* paralytic shellfish poisoning
Bacteria (infection and intoxication)
Fungi (mycotoxins)
Parasites
Protozoa
Viruses

an exogenous harmful agent may be ingested. This may be a pesticide, some other chemical contaminant added by design or accident, a micro-organism or its toxin. Various causes of foodborne illness are summarized in Table 6.1.

Here we are concerned primarily with microbiological hazards (Table 6.2). These are considered in some detail subsequently, but to justify this attention, we must first provide some assessment of their importance.

6.2 SIGNIFICANCE OF FOODBORNE DISEASE

Foodborne disease has been defined by the World Health Organization (WHO) as:

'Any disease of an infectious or toxic nature caused by, or thought to be caused by, the consumption of food or water.'

This definition includes all food and waterborne illness and is not confined to those primarily associated with the gastro-intestinal tract and exhibiting symptoms such as diarrhoea and/or vomiting. It therefore encompasses illnesses which present with other symptoms such as paralytic shellfish poisoning, botulism and listeriosis as well as those caused by toxic chemicals, but excludes illness due to allergies and food intolerances. The essential message of this section can be summarized by the conclusions of a WHO Expert Committee which pointed out that foodborne diseases, most of which are of microbial origin, are perhaps the most widespread problem in the contemporary world and an important cause of reduced economic productivity.

A number of assessments of the relative significance of hazards associated with food have concluded that micro-organisms are of paramount importance. A study conducted in the United States found that, although the attention given to different food hazards by the media, pressure groups and regulatory authorities might differ, as far as the food industry was concerned microbial hazards were the highest priority. Similarly, it has been estimated that the risk of becoming ill as a result of microbial contamination of food was 100 000 times greater than the risk from pesticide contamination.

Table 6.2 Some microbiological agents of foodborne illness

Agents	Important reservoir/carrier	Transmission[a]			Multiplication in food	Examples of some incriminated foods
		water	food	person to person		
BACTERIA:						
Aeromonas	Water	+	+	–	+	Cooked rice cooked meats
Bacillus cereus	Soil	–	+	–	+	Vegetables, starchy puddings
Brucella species	Cattle, goats, sheep	–	+	–	+	Raw milk, dairy products
Campylobacter jejuni	Chickens, dogs, cats, cattle, pigs, wild birds	+	+	+	–[b]	Raw milk, poultry
Clostridium botulinum	Soil, mammals, birds, fish	–	+	–	+	Fish, meat, vegetables (home preserved)
Clostridium perfringens	Soil, animals, man	–	+	–	+	Cooked meat and poultry, gravy, beans
Escherichia coli						
Enterotoxigenic	Man	+	+	+	+	Salads, raw vegetables
Enteropathogenic	Man	+	+	+	+	Milk
Enteroinvasive	Man	+	+	0	+	Cheese
Entero-haemorrhagic	Cattle, poultry, sheep	+	+	+	+	Undercooked meat, raw milk, cheese
Listeria monocytogenes		+	+	+	+	Soft cheeses, milk, coleslaw, pate
Mycobacterium bovis	Cattle	–	+	–	–	Raw milk
Salmonella Typhi	Man	+	+	±	+	Dairy produce, meat products, shellfish, vegetable salads
Salmonella (non-Typhi)	Man and animals	±	+	±	+	Meat, poultry, eggs, dairy produce, chocolate
Shigella	Man	+	+	+	+	Potato/egg salads
Staphylococcus aureus (enterotoxins)	Man	–	+	–	+	Ham, poultry and egg salads, cream-filled bakery produce, ice-cream, cheese
Vibrio cholerae O1	Man, marine life?	+	+	±	+	Salad, shellfish
Vibrio cholerae, non-O1	Man and animals, marine life?	+	+	±	+	Shellfish

Table 6.2 (*continued*)

Agents	Important reservoir/carrier	Transmission[a]			Multiplication in food	Examples of some incriminated foods
		water	food	person to person		
Vibrio para-haemolyticus	Seawater, marine life	–	+	–	+	Raw fish, crabs, and other shellfish
Yersinia enterocolitica	Water, wild animals pigs, dogs, poultry	+	+	–	+	Milk, pork and poultry
VIRUSES:						
Hepatitis A virus	Man	+	+	+	–	Shellfish, raw fruit and vegetables
Norovirus	Man	+	+	0	–	Shellfish
Rotavirus	Man	+	0	+	–	0
PROTOZOA:						
Cryptosporidium parvum	Man, animals	+	+	+	–	Raw milk, raw sausage (nonfermented)
Entamoeba histolytica	Man	+	+	+	–	Raw vegetables and fruits
Giardia lamblia	Man, animals	+	±	+	–	0
HELMINTHS:						
Ascaris lumbricoides	Man	+	+	–	–	Soil-contaminated food
Taenia saginata and *T. solium*	Cattle, swine	–	+	–	–	Undercooked meat
Trichinella spiralis	Swine, carnivora	–	+	–	–	Undercooked meat
Trichuris trichiura	Man	0	+	–	–	Soil-contaminated food

[a] Almost all acute enteric infections show increased transmission during the summer and/or wet months, except infections due to rotavirus and *Yersinia enterocolitica*, which show increased transmission in cooler months

[b] Under certain circumstances some multiplication has been observed. The epidemiological significance of this observation is not clear

+ = Yes

± = Rare

– = No

0 = No information

Adapted from WHO 1992

Foodborne diseases range from relatively mild, self-limiting gastrointestinal upsets through to life-threatening conditions such as botulism. Some foodborne infections can develop severe complications such as the haemolytic uraemic syndrome associated with about 10% of *E. coli* O157:H7 infections or the neurological disorder Guillain-Barré syndrome that follows about 0.1% of *Campylobacter* infections. Some groups of people are particularly susceptible to the more serious consequences of food borne disease. These include the elderly, infants and those who are immunocompromised as a result of illness or chemotherapy.

For otherwise healthy, well-nourished people in the developed world, most food poisoning is an unpleasant episode from which recovery is normally complete after a few days. For society as a whole though, it is increasingly being recognized as a largely avoidable economic burden. Costs are incurred in the public sector from the diversion of resources into the treatment of patients and the investigation of the source of infection. To the individual the costs may not always be calculable in strictly financial terms but could include loss of income, as well as costs of medication and treatment. Studies conducted by the Communicable Disease Surveillance Centre (CDSC) in London have even identified as a cost the 'trousseau effect', where an individual who is hospitalized incurs additional expense as a result of having to purchase items such as new night-attire for the occasion. On the larger scale, absence from work will also constitute a cost to the national economy.

A number of attempts have been made to quantify these costs and, while the errors must be large, they do at least give an idea of the magnitudes involved. Thus a study in the United States estimated that the total annual cost to the US economy of bacterial food poisoning is approaching US$ 7 billion. Substantially lower, but still considerable, costs of almost £1 million have been associated with 1482 salmonella cases in the UK in the year 1988/9 (Table 6.3). The Infectious Intestinal Disease (IID) Study conducted in England (see later) estimated the

Table 6.3 *Costs associated with 1482 Salmonella cases in 198819*

	Cost (£)	*Proportion (%)*
Investigation of cases[a]	157 162	16
Treatment of cases[b]	235 660	24
Costs to individuals and families[c]	95 962	9
Loss of production[d]	507 555	51
Total	996 339	100

[a] Local authority and laboratory costs
[b] GP and hospital services
[c] Treatment-related and incidental costs
[d] Absence from work related to illness and caring for sick individuals
Source: P.N. Sockett, PhD thesis; P.N. Sockett and J.R. Roberts (1991)

average cost of a case of IID, whatever its cause, to be £79 at 1993-1995 prices.

For the food industry, the costs can be huge and it is not unusual for the company producing a product implicated in an outbreak of food poisoning to go bankrupt as a result. Companies not directly involved in an outbreak can also suffer. There is often a general decline in demand for a product prompted by public concern that the same problem could occur with similar products from other manufacturers. There was, for instance, a marked downturn in all yoghurt sales after the hazelnut yoghurt botulism outbreak in England in 1989.

Increased vigilance by companies to ensure that the same process failures responsible for an outbreak do not occur elsewhere, also has its attendant costs. For instance, it was estimated that the costs of checking the integrity of spray-drier cladding by dried-milk manufacturers following a salmonella outbreak caused by dried milk were of the order of hundreds of thousands of pounds.

Food retailers can also be affected as a result of a decline in sales, particularly if a suspect product is associated with one particular store.

In the less developed world the consequences of foodborne illness are even more serious. Diarrhoeal disease is a major cause of morbidity and mortality in poor countries, particularly among children. It has been estimated that some 1,500 million children under 5 suffer from diarrhoea each year and that over 3 million die as a result. Diarrhoea can occur repeatedly in the same individual leading to malnutrition which in turn predisposes them to more severe diarrhoeal episodes and other serious infections. This can produce a downward spiral of increasingly poor health which can seriously impair a child's mental and physical development and can lead to its premature death. (Figure 6.2).

Weaning is a particularly hazardous time for the infant. The anti-infective properties of maternal breast milk are lost or diluted and are replaced by foods which often have a low nutrient density. At the same time, the immature immune system is exposed to new sources of infection in the environment. Poor hygienic practices in the preparation of

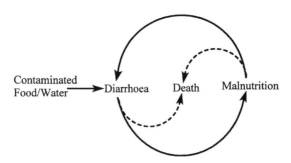

Figure 6.2 *The malnutrition and diarrhoea cycle*

weaning foods and the use of contaminated water are often implicated in weaning diarrhoea and it has been estimated that 15–70% of all diarrhoea episodes in young children are food associated.

6.3 INCIDENCE OF FOODBORNE ILLNESS

Statistics covering foodborne illnesses are notoriously unreliable. Simply quantifying the problem of those diseases initiated by infection through the gastrointestinal tract is difficult enough, but to determine in what proportion food acted as the vehicle is harder still.

Many countries have no system for collecting and reporting data on gastrointestinal infections and even where these exist the reported data is acknowledged to represent only a fraction of the true number of cases. Studies have suggested that the ratio of actual to reported cases can be between 25:1 and 100:1. One should also be circumspect about using published national statistics for comparative purposes since apparent differences can often simply reflect differences in the efficiency of the reporting system. In the United States, reporting of foodborne illness outbreaks to the Center for Infectious Diseases, is not compulsory so that some States report rates 200 times those of other States. In the early 1980s reported outbreaks of foodborne disease for the United States were roughly twice those reported by Canada which has a population only one tenth the size. It seems unlikely that Canadians are markedly more susceptible to foodborne illness or more careless about food hygiene than their neighbours; more probably the disparity reflects a higher level of under-reporting in the United States. Some support for this appears if the statistics for all gastrointestinal disease are compared. These are a much closer reflection of the relative population sizes since these figures are officially notiable in the USA.

Such statistical problems are not unique to North America. The WHO Surveillance Programme for Control of Foodborne Infections and Intoxications in Europe which reports data from more than 30 countries has noted the different national systems of notification and reporting. These include:

(i) notification of cases of foodborne disease without any specification of the causative agent or other epidemiologically important details;
(ii) reporting only laboratory-confirmed cases of foodborne disease collated by a central agency;
(iii) reporting cases of gastrointestinal infection which, in some cases, are regarded as being foodborne regardless of whether the involvement of food has been established;
(iv) reporting only cases of salmonellosis.

As a result, the reported incidence of foodborne illnesses can vary widely from country to country. In the countries which make up the Organisation for Economic Co-operation and Development (OECD), for example, the reported incidence of non-typhoid *Salmonella* infections ranges from 3.9 to 476.2 cases per 100,000 inhabitants and from 0.1 to 271.5 for *Campylobacter*. Some of this variation might be due to differences in food consumption habits but the WHO assumes that the overall burden of foodborne illness is probably of the same order of magnitude in most OECD countries, therefore these differences must in large part reflect the efficiency of various national data collection systems.

Most cases of foodborne illness are described as sporadic; single cases which are not apparently related to any others. Sometimes two or more cases are shown to be linked to a common factor in which case they constitute an outbreak. Outbreaks can be confined to a single family or be more generalized, particularly when commercially processed foods are involved.

In England and Wales, information on sporadic cases of foodborne disease comes from a number of different sources. The Health Protection Agency publishes statistics on clinical cases of food poisoning which comprise notifications by medical practitioners and those cases identified during the course of outbreak investigations but not formally notified by a doctor.

Although notification is statutory, *i.e.* required by law, these data are acknowledged to be incomplete as a result of significant under-reporting. Diagnosis is often made purely on the basis of symptoms, without recourse to any microbiological investigation which could establish both the causative agent and the food vehicle. Similarly, it is probably significant that the league table of the most commonly reported causes of food poisoning in England, Wales and Scotland (Table 6.4) also partially reflects the relative severity of symptoms (with the notable exception of *C. botulinum*). It is reasonable to assume that the more ill you feel the more likely you are to seek medical attention and the more likely your case is to figure in official statistics. The situation can be represented as a pyramid, where the large base reflects the true incidence of food poisoning which is reduced to a small apex of official statistics by the various factors that contribute to under-reporting (Figure 6.3).

The Infectious Intestinal Disease (IID) Study which collected data in England in the period 1993–1996 aimed to estimate some of these uncertainties. Based primarily around 70 representative doctor's practices, volunteers were recruited to notify the doctor each week whether or not they had had symptoms of gastrointestinal illness during that week and, in cases where they had been ill, to submit a faecal specimen to the laboratory. Surveys were also made on the number of people visiting the doctor complaining of IID and the proportion from which faecal specimens were taken, the long-term medical sequelae of IID and the

Table 6.4   *Outbreaks of gastrointestinal illness in England and Wales (2003, 2004) and the USA (1996, 1997)*

| Agent | Cases (outbreaks) | | | |
| | England and Wales[b] | | USA[c] | |
	2003	2004	1996	1997
Salmonella	14963[a]	13125[a]	12450 (69)	1731 (60)
Clostridium perfringens	23 (2)	486 (6)	1011 (10)	255 (6)
Bacillus spp.	55 (5)	0 (0)	22 (1)	438 (4)
Staph. aureus	0 (0)	31 (2)	178 (7)	393 (9)
E. coli O157	675[a]	699[a]	325 (11)	300 (8)
Shigella sonnei	633[a]	815[a]	109 (6)	315 (10)
Clostridium Botulinum	1	2	4 (2)	2 (1)
Campylobacter	46181[a]	44294[a]	101 (5)	104 (2)

[a] Numbers of reported isolations (includes sporadic cases)
[b] Source: Health Protection Agency
[c] Source: MMWR March 17 2000 **49** No. SS-1

numbers of cases presenting to hospitals rather than to their local doctor. The results, published in the British Medical Journal (*Br. Med. J.*, 1999, **318**, 1046–1050), were broadly similar to those found in an earlier Dutch study and indicated that infectious intestinal disease occurs in 1 in 5 people each year amounting to an estimated 9.4 million cases. Of those people affected, only 1 in 6 went to the doctor. The proportion of cases that were not recorded by official statistics was large and varied widely by organism. For example, the degree of under reporting for salmonella was relatively low with 1 in 3.2 cases being reported. It was worse with campylobacter where the ratio was 1 in 7.6 cases. Under reporting of viral IID was much more severe with national surveillance picking up only 1 in 35 cases of diarrhoea caused by rotavirus and 1 in 1562 caused by small round structured viruses (Norovirus). There were also many cases for which no causative organism was identified.

An additional source of statistics is the voluntary, nonstatutory reporting system from public health and hospital laboratories of isolations of gastrointestinal pathogens. The statistics generated in this way include cases where food was not the vehicle but the pathogen was acquired by some other means such as person-to-person spread or from domestic pets.

Information on outbreaks is collected by the HPA from microbiologists and environmental health officials around the country. Sometimes the existence of an outbreak is impossible to ignore if it involves a large number of people or a readily defined commercial or institutional context, for example a large public reception, diners at the same restaurant or passengers in the same airliner. Sometimes the existence of an outbreak may emerge from follow-up investigations on sporadic cases. This is often possible where highly discriminating typing schemes are available that enable the pathogen strain causing an outbreak to be

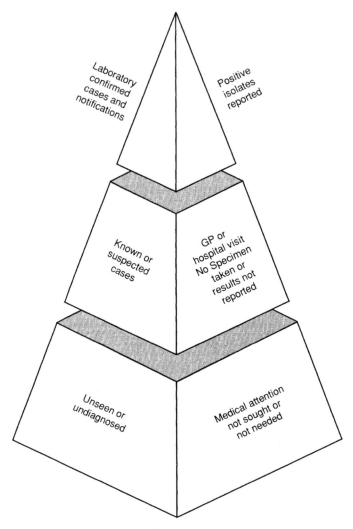

Figure 6.3 *The food poisoning pyramid*
(P. Sockett, PhD thesis)

distinguished from strains responsible for the statistical background 'noise' of sporadic cases. Such schemes have also enabled international outbreaks of salmonellosis to be identified through the Enter-net surveillance network which was started by the European Union but now includes countries such as Australia, Canada, New Zealand, Japan and Mexico. Even so, it is probable that many outbreaks remain undetected, submerged in the numbers for sporadic cases.

Annual reports of statistics on food poisoning and isolations of *Salmonella* have been published for England and Wales since 1949 and showed no discernible trend until the 1980s when a steady increase was apparent (Figure 6.4). A similar but smaller increase was noted in Scotland. Nearly

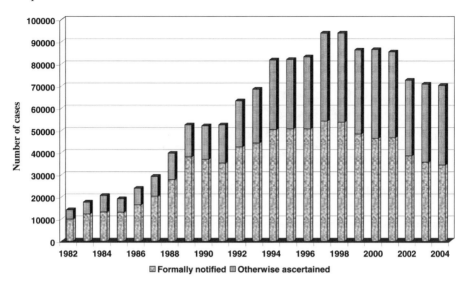

Figure 6.4 *Food poisoning in England and Wales 1982–2004*
Source OPCS

all European countries have reported an increase in foodborne illness starting in the mid-1980s, and in the United States over the period 1983–87 more cases, although fewer outbreaks, were reported. In Austria, the rate of foodborne salmonellosis has increased from 19 cases per 100 000 inhabitants in 1985 to 62 in 1989 and 89 in 1996.

Since 1997 numbers of salmonella isolations have declined throughout Europe although cases of campylobacteriosis have continued to climb. Higher numbers of foodborne disease cases may reflect improved reporting and data collection procedures, better methods of isolation and identification and heightened awareness of the problem but they are also held to reflect a real underlying increase in the incidence.

6.4 RISK FACTORS ASSOCIATED WITH FOODBORNE ILLNESS

Outbreaks of food poisoning involve a number of people and a common source and are consequently more intensively investigated than the more numerous sporadic cases that occur. Valuable information is derived from these investigations about the most common contributory factors and faults in food hygiene that lead to outbreaks of foodborne illness. Specific examples will be given in the following chapter when bacterial pathogens are considered individually, but analysis of this information does allow a number of generalizations to be made.

The foods that are most frequently incriminated in foodborne disease in Europe and North America are those of animal origin: meat, poultry,

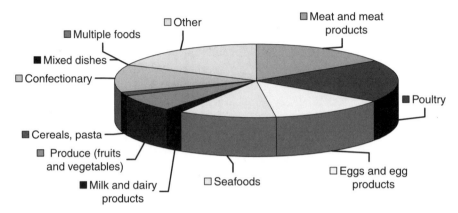

Figure 6.5 *Food implicated in foodborne disease outbreaks in the UK*

milk, eggs, and products derived from them. This is particularly true of illness caused by *Salmonella* and *Clostridium perfringens*. Data on the association of particular foods with foodborne disease outbreaks in England and Wales are presented in Figure 6.5. The same general picture is true of most industrialized countries although the relative importance of some animal products does differ. For example, in Spain between 1985 and 1989 eggs and egg products such as mayonnaise were incriminated in 62% of outbreaks for which a cause was established. In the Netherlands in 1991 and 1992 Chinese food was the most common vehicle associated with outbreaks, ahead of both poultry and eggs and other meats.

Fish and shellfish are less commonly implicated but can be an important vehicle in some countries, often reflecting local dietary habits. In Japan in the year 2000 seafoods were responsible for 25% of outbreaks where a causative food was identified.

Outbreaks can result from the distribution of a contaminated food product or from situations where meals are being produced for large numbers of people. Evidence from numerous countries has shown that mass-catering is by far the most frequent cause of outbreaks, whether it comes under the guise of restaurants, hotels, canteens, hospitals or special events such as wedding receptions. There are a number of reasons why this should be, but inadequacies of management, staff training and facilities are often identified.

Analyses of the specific failures in food hygiene that have contributed to outbreaks have been conducted on a number of occasions and results of two of these, from the United States and from England and Wales are presented in Table 6.5. Comparing the two is not entirely straight-forward since, in most outbreaks more than one contributory factor has been identified so that the columns do not add up neatly to 100%. Also, the surveys differ in the categories used and even where they are nominally the same they may still not be equivalent in all respects. Even

Table 6.5 *Factors contributing to outbreaks of food poisoning*

Factor	England and Wales[a]	USA[b]
Preparation too far in advance	57	29
Storage at ambient temperature	38	63
Inadequate cooling	32	
Contaminated processed food	17	n.i.
Undercooking	15	5
Contaminated canned food	7	n.i.
Inadequate thawing	6	n.i.
Cross contamination	6	15
Food consumed raw	6	n.i.
Improper warm handling	5	27
Infected food handlers	4	26
Use of left overs	4	7
Extra large quantities prepared	3	n.i.

[a] 1320 outbreaks between 1970 and 1982 from Roberts 1985
[b] Outbreaks occurring between 1973 and 1976 from Bryan 1978
n.i. category not included in analysis
Figures are expressed as percentages. Since several factors may contribute to a single outbreak columns do not total 100%

so, inspection of Table 6.5 reveals two major contributory factors; temperature and time. Failure to cool foods and hold them at temperatures inimical to microbial growth, or to heat them sufficiently to kill micro-organisms, coupled with prolonged storage giving micro-organisms time to multiply to dangerous levels. An interesting difference between the two sets of data is the lower incidence of infected food handlers contributing to illness in England and Wales.

6.5 THE CHANGING SCENE AND EMERGING PATHOGENS

So far we have tried to present a brief overview of the current situation with regard to foodborne disease. It is important to remember however that what we see now is just a snapshot of a dynamic situation. Major food hazards such as *Listeria monocytogenes* and *Campylobacter* were simply not recognised as such 40 years ago and other "new" pathogens such as Verotoxin-producing *E. coli* and *Enterobacter sakazakii* have emerged even more recently. This has led to some organisms being designated emerging pathogens. These are not necessarily entirely new species (though they can be) but can also be old favourites in new guises. In their broadest definition emerging pathogens are organisms causing illnesses that have only recently appeared or been recognised in a population, or organisms that are well recognised but are rapidly increasing in incidence or geographical range.

A number of factors contribute to this evolving pattern of foodborne disease:

Changes in farming practices – there are constant economic pressures to increase agricultural productivity and this can impact on food safety as when, for example, intensified animal production contributes to the spread of zoonotic pathogens or pressure on land use results in contamination of field crops from manure.

Increased international trade in foods – countries increasingly source their food on a global basis and this can pose problems with control of foodborne hazards. High standards and efficient control methods in one country can be undermined by importation from a country where such standards do not apply. There have been a number of international outbreaks of foodborne disease cause by imported foods and these are likely to increase with the level of international trade.

Changes in food processing – there is increasing reliance on refrigeration and the cold chain as a way of extending the shelf life of fresh foods and this has contributed to the emergence of psychrotrophic pathogens such as *Listeria monocytogenes* as important concerns.

Increased international movement of people – this can take the form of refugees from wars, social conflict or economic hardship as well as movement of those from more prosperous regions for leisure or business purposes. Both offer new opportunities for the acquisition or transmission of foodborne and other diseases. In Sweden, 90% of *Salmonella* cases are estimated to be imported cases.

Changing character of the population – the very young, the old, the very sick and the immunocompromised are all more at risk from foodborne diseases. As a result of improvements in nutrition and healthcare, the proportion of the population in some of these groups is increasing.

Lifestyle changes – increased affluence, urbanisation and other social changes can lead to increased consumption of exotic or unusual foods or meals prepared away from the home. This can result in changes in the incidence and nature of foodborne illness.

Microbial evolution – micro-organisms are constantly changing their characteristics as a result of evolutionary processes. If these changes affect the virulence or pathogenicity of an organism then a new hazard can emerge. Such changes can occur completely independently of human activity but the latter can sometimes provide selective pressures enabling new strains to thrive.

6.6   THE SITE OF FOODBORNE ILLNESS. THE ALIMENTARY TRACT: ITS FUNCTION AND MICROFLORA

In most of the cases of foodborne illness we consider, the pathogenic (disease producing) effect occurs in the alimentary tract giving rise to symptoms such as diarrhoea and vomiting. Since these are essentially a dysfunction of the gut, a useful starting point would be to outline its normal operation and the role micro-organisms play in this process.

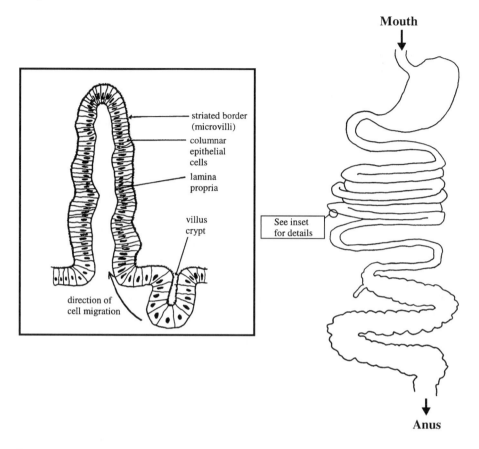

Figure 6.6 *The gastrointestinal tract. Inset: expanded view of inner surface of small intestine*

The alimentary or gastrointestinal tract is not an internal organ of the body but a tube passing through it from the mouth to the anus (Figure 6.6). Its principal functions are the digestion and absorption of food and the excretion of waste. Unlike most of the body's other external surfaces, it is not lined with a dry protective skin and so, although it possesses some protective features, it offers a more congenial environment for micro-organisms and an easier route by which they can penetrate the body.

In the mouth, food is mixed with saliva and broken down mechanically to increase the surface area available for attack by digestive enzymes. Saliva is an alkaline fluid containing starch-degrading (amylase) enzyme and the antimicrobial factors immunoglobulin (IgA), lysozyme, lactoferrin and lactoperoxidase. It provides lubrication to assist chewing and swallowing and performs a cleansing function, rinsing the teeth and mouth to remove debris. On average, an adult secretes and swallows about 1.5 l of saliva each day.

The variety of foods consumed and the range of micro-environments in the mouth result in a diverse and continually changing microflora. On the teeth, bacteria are associated with the formation of dental plaque – an organic film in which bacteria are embedded in a matrix derived from salivary glycoproteins and microbial polysaccharides. The microbial composition of plaque varies with its age but filamentous *Fusobacterium* species and streptococci are common components. Plaque offers a protective environment for bacteria and its development is often a prelude to conditions such as dental caries and periodontal disease.

Swallowed food descends *via* the oesophagus into the stomach; a bulge in the alimentary tract which serves as a balance tank from which food is gradually released into the small intestine for further digestion.

In the stomach, food is blended with gastric juice, an acidic fluid containing hydrochloric acid. Stomach pH can range from 0.8–5.0 (typically 2.0–3.0) and has a marked effect on ingested micro-organisms, killing most. Normally only acid-tolerant vegetative cells and spores survive and the microbial count in the stomach is low, although lactobacilli are frequently found in association with the stomach wall. Gastric acidity generally provides very effective protection for subsequent sections of the intestine but is not, as we shall see, an invulnerable defence. Bacteria can evade prolonged exposure to the acid by being sheltered in food particles or as a result of accelerated passage through the stomach as occurs, for instance, when the stomach is full. Alternatively, acidity may be neutralized by the food or absent as a result of illness.

The digestive functions of the stomach are not confined to those of a mechanical churn with antimicrobial features. Proteases, such as pepsin, and lipase which can operate at low pH partially digest the stomach contents. The gastric mucosa also secretes a protein responsible for efficient absorption of vitamin B_{12}. Little absorption of nutrients occurs in the stomach, with the notable exception of ethanol, but some material transfer is often necessary to adjust the osmotic pressure of the stomach contents to ensure they are isotonic with body fluids.

From the stomach, small quantities of the partially digested mixture of food and gastric juice, known as chyme, are released periodically into the small intestine. In this muscular tube over 6 metres long most of the digestion and absorption of food occur. Its internal lining is extensively folded and the folds covered with finger-like projections or villi which are themselves covered in microvilli. This gives the inner surface the appearance and texture of velvet and maximizes the area available for absorption (Figure 6.6).

In the first section of the small intestine, the duodenum, large-scale digestion is initiated by mixing the chyme with digestive juice from the pancreas and bile from the gallbladder which neutralize the chyme's acidity. The pancreatic juice also supplies a battery of digestive enzymes, and surfactant bile salts emulsify fats to facilitate their degradation and the absorption of fat soluble vitamins. Further digestive enzymes that break down disaccharides and peptides are secreted by glands in the mucous lining of the duodenum called, with evocations of a Gothic horror, the crypts of Lieberkühn.

The duodenum is a relatively short section of the small intestine, accounting for only about 2% of its overall length. Food is swept along by waves of muscle contraction, known as peristalsis, from the duodenum into the jejunum and thence into the ileum. During this passage, nutrients such as amino acids, sugars, fats, vitamins, minerals and water are absorbed into capillaries in the villi from where they are transported around the body. Absorption is sometimes a result of passive diffusion, but more often involves the movement of nutrients against a concentration gradient; an active process entailing the expenditure of energy.

The gut is home to a huge population of bacteria comprised of more than 400 different species with total numbers estimated at around 10^{14}, far more than the number of cells in the human body. The microbial population increases down the length of the small intestine: counts of 10^2–10^3 ml^{-1} in the duodenum increase to around 10^3–10^4 in the jejunum, 10^5 in the upper ileum and 10^6 in the lower ileum. This corresponds with a decreasing flux of material through the small intestine as water is absorbed along its length. In the higher reaches of the duodenum, the flow rate is such that its flushing effect frequently exceeds the rate at which micro-organisms can multiply so that only those with the ability to adhere to the intestinal epithelium can persist for any length of time. As the flow rate decreases further along the small intestine, so the microbial population increases, despite the presence of antimicrobial factors such as lysozyme, secretory immunoglobulin, IgA, and bile.

In the healthy individual, the microflora of the small intestine is mainly comprised of lactobacilli and streptococci, although, as we shall see, other bacteria have the ability to colonize the epithelium and cause illness as a consequence.

Extensive microbial growth takes place in the colon or large intestine where material can remain for long periods before expulsion as faeces. During this time active absorption of water and salts helps to maintain the body's fluid balance and to dry faecal matter. Bacterial cells account for 25–30% of faeces, amounting to 10^{10}–10^{11} cfu g^{-1}, the remainder is composed of indigestible components of food, epithelial cells shed from the gut, minerals, and bile.

Obligate anaerobes such as *Bacteroides* and *Bifidobacterium* make up 99% of the flora of the large intestine and faeces. Members of the Enterobacteriaceae, most commonly *Escherichia coli*, are normally present at around 10^6 g^{-1}, enterococci around 10^5 g^{-1}, *Lactobacillus*,

Clostridium and *Fusobacterium*, 10^3–10^{5-1}g, plus numerous other organisms, such as yeasts, staphylococci and pseudomonads, at lower levels.

The interaction between the gut microflora and its host appears to have both positive and negative aspects and is the subject of much current research and conjecture. Addition of antibiotics to feed has been shown to stimulate the growth of certain animals, suggesting that some gut organisms have a deleterious effect on growth.

A normal gut microflora confers some protection against infection. One example of this effect is the inflammatory disease pseudomembranous colitis caused by *Clostridium difficile*. Normally this organism is present in the gut in very low numbers, but if the balance of the flora is altered by antibiotic therapy, it can colonize the colon releasing toxins. Similarly, the infective doses of some other enteric pathogens have been shown to be lower in the absence of the normal gut flora.

It appears that protection is not simply a result of the normal flora occupying all available niches, since enterotoxigenic *E. coli* adheres to sites that are normally vacant. Some direct antagonism through the production of organic acids and bacteriocins probably plays a part, but stimulation of the host immune system and its capacity to resist infection also appear to be factors.

In monogastric animals such as humans, gut micro-organisms do not play the same central role in host nutrition as they do in ruminants. Some facultative anaerobes found in the gut, such as *E. coli* and *Klebsiella mobilis* (previously known as *K. aerogenes* and *Enterobacter aerogenes*) are known to produce a variety of vitamins *in vitro* and studies using animals reared in a germ-free environment and lacking any indigenous microflora have shown that *in vivo* vitamin production by micro-organisms can be important on certain diets. In humans, however, the evidence is less convincing. Some have questioned the efficiency of absorption of vitamins produced in the large intestine pointing to the fact that vegans have developed vitamin B_{12} deficiency despite its production in the gut and excretion in the faeces. It appears that an adequate balanced diet will probably meet all the body's requirements in this respect and that, short of coprophagy, which is practised by some herbivores such as rabbits, access to vitamins produced *in situ* is limited.

6.7   THE PATHOGENESIS OF DIARRHOEAL DISEASE

Several foodborne illnesses, such as typhoid fever, botulism and listeriosis, involve body sites remote from the alimentary tract which serves simply as the route by which the pathogen or toxin gains entry to the body. The pathogenesis of these conditions will be discussed under the individual organisms concerned.

Table 6.6 *Clinical classification of infections diarrhoea*[a]

Type	Symptoms		Typical causative organisms
Acute watery	diarrhoea		Loose or watery stools without visible blood. Duration generally less than 7 days
Vibrio	*cholerae*, Enterotoxigenic *E. coli.* Small round structured viruses		
Acute bloody	diarrhoea		Loose or watery stools with visible blood. Duration generally less than 7 days
Shigella,	*Campylobacter jejuni* Enteroinvasive *E. coli.* Small round structured viruses		
Persistent	diarrhoea		Loose or watery stools with or without visible blood with a duration of 14 days or more
	Multifactorial: enteric infection, malnutrition, impaired immunity, lactose intolerance		

[a] Definition: passage of loose or watery stools three or more times in a 24 hour period

A more common conception of foodborne illness, often described as food poisoning, is where symptoms, like the causative agent, are confined to the gut and its immediate vicinity. The patient presents an acute gastroenteritis characterized by diarrhoea and vomiting. Individual pathogens will be described in some detail subsequently, but for now we will consider some common features of the mechanisms involved.

Diarrhoea is the excessive evacuation of too-fluid faeces (see Table 6.6). Any process which seriously interferes with the gut's capacity to absorb most of the 8–10 l of fluid it receives each day, or increases secretion into the intestinal lumen, will produce this condition. Consequently, the aetiology of diarrhoea can be quite complex and a number of different mechanisms have been identified.

The ability to cause illness is generally the result of a combination of properties that enable a micro-organism to damage its host. These are called virulence factors and include not only those factors most directly responsible for damage such as toxins, but others such as the ability to evade host defence mechanisms e.g. stomach acidity and bile in the gut, the ability to adhere to mucosal surfaces, secrete proteins and, where necessary, invade host cells. Many of these properties are encoded on relatively large, discrete segments of DNA known as pathogenicity islands. *Salmonella* for example possesses at least 5 pathogenicity islands. These are incorporated into the bacterial chromosome but have a $G + C$ content markedly different from the surrounding chromosome. This and other characteristics suggest that they have been acquired by horizontal transfer from a foreign source. Acquisition of relatively large segments of DNA can promote genetic variability and play and important role in microbial evolution. When these encode virulence factors, they allow sudden rapid changes in pathogenicity an the emergence of new hazards.

Toxins are frequently the direct cause of diarrhoea. As their nomenclature often causes students some confusion, one or two preliminary definitions are probably in order. *Exotoxin* is the term used to describe toxins that are released extracellularly by the living organism. These include:

enterotoxins which act on the intestinal mucosa generally causing diarrhoea;
cytotoxins which kill host cells;
neurotoxins which interfere with normal nervous transmission.
Endotoxins are pyrogenic (fever producing) lipopolysaccharides released from the outer membrane of the Gram-negative cell envelope by bacterial lysis.

Bacterial food poisoning can be divided into three principal types.

(1) *Ingestion of pre-formed toxin.* Toxins may be produced in and ingested with the food as in *Staphylococcus aureus* food poisoning and the *Bacillus cereus* emetic syndrome. Botulism is similar in this respect though in this case gastrointestinal symptoms are of minor importance. The absence of person to person spread and a relatively short incubation period between ingestion of food and the onset of symptoms are usual characteristics of this type of food poisoning.

(2) *Non-invasive infection.* In a non-invasive infection, viable bacteria ingested with food, colonize the intestinal lumen. This is principally associated with the small intestine where competition from the endogenous microflora is less intense. To prevent their removal by the flushing action of the high flow rates in this section of the gut, the pathogen generally attaches to and colonizes the epithelial surface. It does this by producing adhesins, molecules often associated with fimbriae on the bacterial cell surface, which recognize and attach to specific receptor sites on the microvilli. Loss of the ability to adhere to the gut wall will dramatically reduce a pathogen's virulence – its ability to cause illness.

Once attached, the pathogen produces a protein enterotoxin which acts locally in the gut changing the flow of electrolytes and water across the mucosa from one of absorption to secretion. Several enterotoxins act by stimulating enterocytes (the cells lining the intestinal epithelium) to over-produce cyclic nucleotides.

Most extensively studied in this respect is the cholera toxin produced by *Vibrio cholerae*. The toxin (MW 84 000) comprises five B subunits and a single A subunit. The B subunits bind to specific ganglioside (an acidic glycolipid) receptors on the enterocyte surface. This creates a hydrophilic channel in the cell membrane through which the A unit can pass. Once inside the cell, a portion of the A unit acts enzymically to transfer an ADP-ribosyl group derived from cellular NAD to a protein regulating the activity of the enzyme adenylate cyclase. As a result, the enzyme is locked into its active state leading to an accumulation of cyclic adenosine monophosphate (cAMP) which inhibits absorption of Na^+ and Cl^- ions while stimulating the secretion of Cl^-, HCO_3^- and Na^+ ions. To maintain an osmotic balance, the transfer of electrolytes is accompanied by a massive outflow of water into the intestinal lumen. This far exceeds the absorptive capacity of the large intestine and results in a profuse watery diarrhoea (Figure 6.7).

A number of other enterotoxins have been shown to act in the same way as the cholera toxin including the heat labile toxin (LT) produced by some types of enterotoxigenic *E. coli*. Other toxins such as the heat stable toxin of *E. coli* are similar in the respect that they stimulate the production of a cyclic nucleotide in enterocytes. In this case it is cyclic guanosine monophosphate (cGMP) which differs slightly from cAMP in its activity but also produces diarrhoea as a result of electrolyte imbalances.

A different enterotoxin is produced by *Clostridium perfringens* as it sporulates in the gut. The toxin binds to receptors on the surface of cells of the intestinal epithelium, producing morphological changes in the

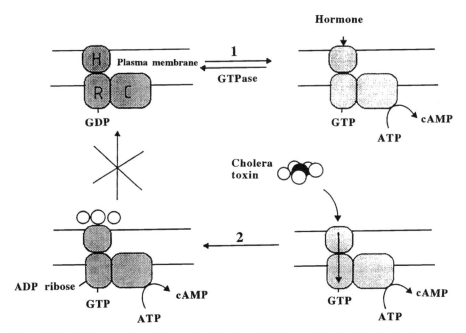

Figure 6.7 *Cholera toxin and its mode of action. (1) Reversible physiological activation of adenylate cyclase (C) through hormone binding to receptor (H) and (2) cholera toxin binding and translocaton of A subunit leads to ADP ribosylation of regulatory subunit (R) and irreversible activation*

membrane which affect absorption/secretion processes thus precipitating diarrhoea. It does not increase intestinal cAMP levels.

A traditional method of analysing for the presence of enterotoxins is based upon their _in vivo_ action. The ileum of a rabbit under anaesthesia is tied off to produce a number of segments or loops which serve as test chambers. These are injected with cultures, culture filtrates or samples under test. If an enterotoxin is present it produces, after about 24 h, an accumulation of fluid in the loop which becomes distended. A number of alternative assays, based on the effects of enterotoxins on cells in tissue culture are also used. These have the advantages of being more econom-ical, more humane and easier to quantify than the ligated ileal loop assay but are less directly related to the clinical action of the toxin.

(3) _Invasive infection._ Other diarrhoea-causing pathogens invade the cells of the intestinal epithelium but do not normally spread much beyond the immediate vicinity of the gut. Some, such as _Salmonella_ preferentially invade the ileum to produce a profuse watery diarrhoea. Bacterial cells invade and pass through the epithelial cells to multiply in the _lamina propria_, a layer of connective tissue underlying the enterocytes. The precise mechanism of fluid secretion into the intestinal lumen is not known and is probably multifactorial. A heat-labile enterotoxin which stimulates adenylate cyclase activity has been identified in some salmo-nellas as well as a cytotoxin. It has also been suggested that the local acute inflammation caused by the infection and responsible for the fever and chills that are often a feature of salmonellosis, causes an increase in levels of prostaglandins, known activators of adenylate cyclase.

Other enteroinvasive pathogens like _Shigella_ and enteroinvasive _E. coli_ invade the colonic mucosa and produce a dysenteric syndrome characterized by inflammation, abscesses and ulceration of the colon and the passage of bloody, mucus-and pus-containing stools. Bacterial cells adhere to the enterocytes _via_ outer membrane protein adhesins. They are then engulfed by the enterocytes in response to a phagocytic signal produced by the bacterium and multiply within the cytoplasm invading adjacent cells and the underlying connective tissue. The strong inflam-matory response to this process causes abscesses and ulcerations of the colon.

Invasiveness can be diagnosed by examination of the fluid accumu-lated and the mucosal surface in rabbit ileal loops. A less definitive test for invasiveness is the Sereny test which measures the ability of an organism to cause keratoconjunctivitis in the eye of guinea pigs or rabbits.

Some shigellas also produce a protein exotoxin, known as Shiga toxin, which has a range of biological activities. It inhibits protein synthesis by inactivating the 60S ribosomal subunit and is a powerful cytotoxin. It has neurotoxic activity causing paralysis and death in experimental animals and is an enterotoxin capable of causing fluid accumulation in ligated rabbit ileal loops. As an enterotoxin, it appears unrelated to cholera toxin since it does not stimulate adenylate cyclase or cross-react with antibodies to cholera toxin. Its role in the pathogenesis of shigellosis is unclear since strains incapable of producing Shiga toxin remain pathogenic. Enteroinvasive *E. coli* causes a similar syndrome but does not produce Shiga toxin.

Some authors have linked the enterotoxin activity of Shiga toxin with the watery diarrhoea which often precedes dysentery. Interestingly, a similar sequence of watery diarrhoea with supervening bloody diarrhoea is seen with enterohaemorrhagic *E. coli*. This organism, which both colonizes the epithelial surface in the colon and multiplies in the *lamina propria*, produces a number of Shiga-like toxins, sometimes known as verotoxins because of their activity against Vero cells in culture.

Common features have been identified among the various diarrhoea-causing toxins and a number of bacterial exotoxins important in other diseases such as diphtheria. Each consists of five, linked B units which are able to bind to the target cell and facilitate transport of the active A unit into the cell.

Having discussed some of the general features of foodborne diseases, in the next chapter we will look more closely at some of the bacterial agents responsible for them.

CHAPTER 7

Bacterial Agents of Foodborne Illness

7.1 AEROMONAS HYDROPHILA

7.1.1 Introduction

Currently, *Aeromonas* (principally *A. hydrophila*, but also *A. caviae* and *A. sobria*) has the status of a foodborne pathogen of emerging importance. Like *Listeria monocytogenes*, *Plesiomonas*, and *Yersinia enterocolitica*, it has attracted attention primarily because of its ability to grow at chill temperatures, prompting the concern that any threat it might pose will increase with the increasing use of chilled foods (Table 7.1). Present uncertainty over its significance however, is reflected in much of the information available which does not, as yet, present a coherent picture. It was first isolated from drinking water by Zimmerman in 1890 and the following year from frog's blood by Sanarelli. They called their isolates *Bacillus punctata* and *Bacillus hydrophilus* respectively and it was not until the 1930s that the genus *Aeromonas* was first described. Although the taxonomy is still not settled, more recent studies have led to the recognition of two major groups within the genus: the Salmonicida group, which contains the non-motile *Aeromonas salmonicida* and several

Table 7.1 *Reported lag times and growth rates of psychrotrophic pathogens at chill temperatures*

	Lag time (days)			Generation time (h^{-1})	
Temperature (°C)	0–1	2–3	5	0–1	4–5
Aeromonas hydrophila	>22	6–10	3–4	>49	9–14
Listeria monocytogenes	3–33	2–8	1–3	62–131	13–25
Yersinia enterocolitica	3	2.4	–	25	20

sub-species, and the Hydrophila–Punctata group containing a number of motile species, including *A. hydrophila A. sobria*, and *A. caviae*.

A. salmonicida is not a human pathogen but causes diseases of freshwater fish which can be an important economic problem in fish farming. Members of the Hydrophila group can cause extra-intestinal infections, commonly in the immunosuppressed or as a result of swimming accidents where the skin is punctured. The first report of gastroenteritis due to *Aeromonas* came from Jamaica in 1958, but evidence of its ability to cause gastroenteritis in otherwise healthy individuals is patchy.

Epidemiological investigations in several countries have reported higher rates of isolation of aeromonads from patients with diarrhoea than from control groups, although this does not necessarily indicate a causal relationship. In one study, the incidence of *A. hydrophila* in American travellers to Thailand with diarrhoea was significantly higher than in unaffected individuals. Interestingly, isolation rates in the Thai population were similar for both groups of patients suggesting that *Aeromonas* may be a cause of 'travellers diarrhoea' in these regions.

Good supporting evidence from sources other than epidemiological studies has proved difficult to obtain. In a feeding trial involving 50 volunteers with doses as high as 5×10^{10}, only two cases of diarrhoea resulted, although a laboratory accident has been reported where approximately 10^9 cells were ingested by a worker mouth pipetting who later suffered acute diarrhoea.

7.1.2 The Organism and its Characteristics

Aeromonads are Gram-negative, catalase-positive, oxidase-positive rods which ferment glucose. They are generally motile by a single polar flagellum.

A. hydrophila is neither salt ($< 5\%$) nor acid (min. pH ≈ 6.0) tolerant and grows optimally at around 28 °C. Its most significant feature with regard to any threat it may pose in foods is its ability to grow down to chill temperatures, reportedly as low as -0.1 °C in some strains. Its principal reservoir is the aquatic environment such as freshwater lakes and streams and wastewater systems. The numbers present will depend on factors such as the nutrient level and temperature but can be as high as 10^8 cfu ml^{-1} in a relatively nutrient rich environment such as sewage. Although it is not resistant to chlorine, it is found in potable water, where it can multiply on the low level of nutrients available in piped water systems. It has also been isolated from a wide range of fresh foods and is a transient component of the gut flora of humans and other animals.

7.1.3   Pathogenesis and Clinical Features

Gastroenteritis associated with *Aeromonas* occurs most commonly in children under five years old. It is normally mild and self-limiting mostly characterized by profuse watery diarrhoea, although dysenteric stools may sometimes be a feature. Vomiting is not usually reported.

Aeromonas spp., particularly *A. hydrophila* and *A. sobria*, produce a range of potential virulence factors including a number of distinct cytotoxic and cytotonic enterotoxins. Most clinical strains of *A. hydrophila* and *A. sobria* produce aerolysin, a heat-labile, β-haemolytic, cytotoxic enterotoxin with a molecular mass of 52 kDa. Three cytotonic enterotoxins have also been described which act like cholera toxin, stimulating accumulation of high levels of cAMP within epithelial cells. Only one of these shows any marked structural similarity to cholera toxin as measured by cross reactivity with cholera toxin antibodies.

7.1.4   Isolation and Identification

In some instances enrichment media such as alkaline peptone water are used, but where high numbers are present direct plating is usually sufficient. Species of the Hydrophila group grow on a wide range of enteric media but may often be misidentified as 'coliforms' since many strains can ferment lactose. Most cannot ferment xylose and this is a useful distinguishing feature used in several media. As well as bile salts, ampicillin is used as a selective agent in media such as starch ampicillin agar, blood ampicillin agar and some commercial formulations. Colonies which give the characteristic appearance of *Aeromonas* on the medium concerned and are oxidase-positive are then confirmed with biochemical tests.

7.1.5   Association with Foods

Apart from their possible role in gastroenteritis, food and water are also probably the source of the severe extra-intestinal *Aeromonas* infections associated with immunocompromised individuals.

Aeromonads of the Hydrophila group have been isolated from a wide range of fresh foods including fish, meat, poultry, raw milk, and salad vegetables as well as water. The ability of some strains to grow at very low temperatures can lead to the development of high numbers under chill conditions and they can be an important part of the spoilage flora of chilled meats.

They are unlikely to survive even mild cooking procedures but may be introduced as post-process contaminants from uncooked produce or contaminated water.

7.2   *BACILLUS CEREUS* AND OTHER *BACILLUS* SPECIES

7.2.1   Introduction

An early report associating food poisoning with *Bacillus* spp. was made in 1906 when Lubenau described an outbreak in a sanatorium where 300 inmates and staff developed symptoms of profuse diarrhoea, stomach cramps and vomiting. A spore forming bacillus was isolated from meatballs from the incriminated meal. Although Lubenau named the organism *Bacillus peptonificans*, the properties he described resemble those of *B. cereus*. Subsequently, aerobic spore formers were implicated in a number of outbreaks in Europe and between 1936 and 1943 they were suspected of causing 117 of 367 cases investigated by the Stockholm Board of Health.

Bacillus cereus was not conclusively established as a cause of food poisoning until 1950, after the taxonomy of the genus had been clarified. Hauge described four outbreaks in Norway involving 600 people. The food vehicle was a vanilla sauce which had been prepared a day in advance and stored at room temperature before serving. Samples of the sauce later tested contained from 2.5×10^7 to 1.1×10^8 *B. cereus* ml^{-1}. This classic report and many of the early ones from Europe described an illness in which diarrhoea was the predominant symptom. It is now known that *B. cereus* is responsible for two distinct types of foodborne illness: a relatively late-onset, 'diarrhoeal syndrome' and a rapid-onset, 'emetic syndrome', first described in 1971 in the UK.

Since 1975 a number of other *Bacillus* species have been associated with foodborne illness. In these episodes, tests have failed to find known pathogens but food remnants and/or clinical specimens have yielded high numbers of *Bacillus* spp. Far less common than outbreaks featuring *B. cereus*, they usually involve very closely related species such as *B. licheniformis* and *B. pumilis* or *B. subtilis*. *B. thuringiensis* has also been reported as causing an outbreak in Canada.

Overall, the reported number of cases of foodborne illness due to *Bacillus* spp. in the UK is much lower than those for *Salmonella*. Since 1992 there have been up to 10 outbreaks each year involving a total of 67 cases. Such statistics though, are likely to underestimate the true level far more than those for *Salmonella* since the data come only from outbreaks and there is no estimate of sporadic cases. In some Northern European countries however the organism appears to have far greater importance. It accounted for 33% of total bacterial food poisoning cases in Norway between 1988 and 1993, 47% in Iceland (1985–1992), 22% in Finland (1992) and 8.5% in the Netherlands (1991). In Denmark, England and Wales, Japan, the USA and Canada the figure ranges between 0.7 and 5.0%.

7.2.2   The Organism and its Characteristics

Members of the genus *Bacillus* are Gram-positive, aerobic, spore-forming rods, though they do, on occasion, display a Gram-negative or variable reaction. Their taxonomy is quite complex and has been subject to considerable revision in recent years. The genus still contains about 80 species, those causing food poisoning being *Bacillus cereus*, a number of species very closely related to *B. cereus* and *Bacillus subtilis*.

   Bacillus cereus is facultatively anaerobic with large vegetative cells, typically 1.0 µm by 3.0–5.0 µm in chains. It grows over a temperature range from 8 to 55 °C, optimally around 28–35 °C, and does not have any marked tolerance for low pH (min. 5.0–6.0, depending on the acidulant) or water activity (min. ~ 0.95).

   Spores are central, ellipsoidal in shape and do not cause swelling in the sporangium. As a spore former, *B. cereus* is widely distributed in the environment and can be isolated from soil, water and vegetation. This ubiquity means that it is also a common component of the transient gut flora in humans. The spores show a variable heat resistance; recorded D values at 95 °C in phosphate buffer range between around 1 min up to 36 min. Resistance appears to vary with serovar.

   In the UK, a serotyping scheme based on the flagellar (H) antigen has been devised, based on a set of 29 agglutinating antisera raised against outbreak and non-outbreak strains isolated from foods. In about 90% of outbreaks it is possible to serotype the causative organism, although only about half of environmental isolates are typable. There does not appear to be a strong association between the two different types of *B. cereus* food poisoning and particular serotypes. Some have been associated with both types of syndrome, although in a study of 200 outbreaks of the emetic syndrome from around the world, serotype 1, which possesses markedly greater heat resistance than other serotypes, was isolated from implicated foods, faeces or vomitus in 63.5% of cases.

7.2.3   Pathogenesis and Clinical Features

Symptoms of the diarrhoeal syndrome resemble those of *Clostridium perfringens* food poisoning. The onset of illness is about 8–16 h after consumption of the food, lasts for between 12 and 24 h, and is characterized by abdominal pain, profuse watery diarrhoea and rectal tenesmus. Nausea and vomiting are less frequent.

   The emetic syndrome resembles the illness caused by *Staphylococcus aureus*. It has a shorter incubation period than the diarrhoeal syndrome, typically 0.5–5 h, and nausea and vomiting, lasting between 6 and 24 h, are the dominant feature.

Table 7.2 *Chatacteristics of the two types of disease caused by* Bacillus cereus

	Diarrhoeal syndrome	*Emetic syndrome*
Infective dose	10^5–10^7 (total)	10^5–10^8 (cells g^{-1})
Toxin produced	In the small intestine of the host	Preformed in foods
Type of toxin	Protein(s) 3 components MW37, 38, 46 kDa	Cyclic peptide MW 1.2 kDa
Heat stability	Inactivated 56 °C, 5 min	Stable 126 °C, 90 min
pH stability	Unstable <4 and >11	stable 2–11
Incubation period	8–16 h (occasionally >24 h)	0.5–5 h
Duration of illness	12–24 h (occasionally several days)	6–24 h
Symptoms	Abdominal pain, watery diarrhoea and occasionally nausea	Nausea, vomiting and malaise sometimes followed by diarrhoea, due to additional enterotoxin production?
Foods most frequently implicated	Meat products, soups, vegetables, puddings/sauces and milk/milk products	Fried and cooked rice, pasta, pastry and noodles

Adapted from Granum and Lund, *FEMS Microbiol. Lett.*, 1997, **157**, 223 – 228

Both syndromes are caused by distinct enterotoxins (Table 7.2). A number of toxins have been associated with the diarrhoeal syndrome but illness appears to be associated primarily with production in the gut of two three-component enterotoxins: a haemolytic enterotoxin HBL consisting of three proteins B, L_1 and L_2 and a non-haemolytic enterotoxin NHE. Some strains produce both HBL and NHE though others contain the genes for only one. The toxins, which are sensitive to heat and proteolytic enzymes such as trypsin and pepsin, are produced in the late exponential/early stationary phase of growth. Like *C. perfringens* toxin, they exert their effect by binding to epithelial cells and disrupting the epithelial membrane, though the precise mechanisms of action are thought to be different. Though the toxins can be produced in food, their sensitivity to low pH and proteolysis, and the relatively long incubation period associated with illness indicate that toxin production in the gut is primarily responsible for the observed symptoms.

The emetic toxin, cereulide, is a 1.2 kDa cyclic peptide that is acid and heat resistant. Closely related to the potassium ionophore valinomycin, cereulide is a dodecadepsipeptide consisting of three repeats of a unit containing 2 aminoacids and 2 oxyacids (D-*O*-Leu-D-Ala-L-*O*-Val-L-Val)$_3$ (Figure 7.1). The toxin is produced in the food in the late exponential to stationary phase of growth and is thought to act by binding to and stimulating the vagus nerve.

Pathogenic features of the illness caused by the other *Bacillus* spp. are not known. The short incubation periods recorded in outbreaks (from

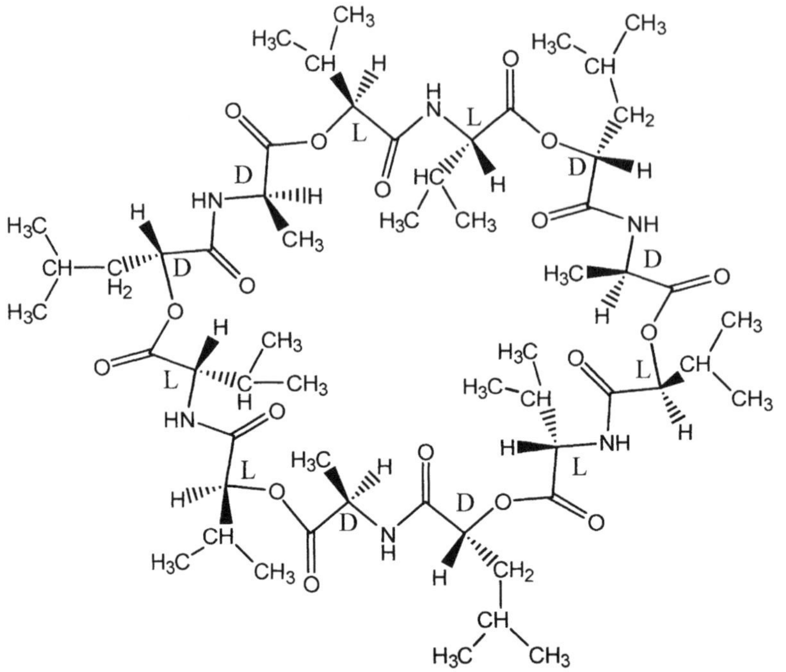

Figure 7.1   *Cereulide, the emetic toxin of* B. cereus

<1 h up to 11 h) suggest an intoxication, though no toxin has been isolated and described as yet.

7.2.4   Isolation and Identification

In an outbreak of *Bacillus cereus* food poisoning, implicated foods will contain large numbers ($>10^5 \, g^{-1}$) of organisms so enrichment techniques are not needed. The same is true of faecal or vomitus specimens and a non-selective medium such as blood agar (sometimes with the addition of polymyxin as a selective agent to suppress Gram-negatives) is commonly used. *B. cereus* can be identified after 24 h incubation at 37 °C by its characteristic colonial morphology of large (3–7 mm diameter), flat or slightly raised, grey-green colonies with a characterisitc granular or ground-glass texture and a surrounding zone of α or β haemolysis. To confirm the identity of a blood agar isolate or to isolate smaller numbers of *B. cereus* from foods, a more selective diagnostic agar is necessary. Several of these have been proposed which have a number of common features. Polymyxin/pyruvate/eggyolk/mannitol/bromothymol blue agar (PEMBA) is one widely used example. It includes polymyxin as a selective agent and where yeasts and moulds are likely to be a problem actidione may also be included. On PEMBA, *B. cereus* produces typical crenated colonies which retain the turquoise-blue of the pH indicator (bromothymol blue) due to their inability to ferment mannitol, they are surrounded by a zone of egg-yolk precipitation caused by lecithinase activity. Pyruvate in the medium improves the egg-yolk precipitation reaction and a low level of peptone enhances sporulation. Colonies of *B. cereus* can be confirmed by a microscopic procedure combining a spore stain with an intracellular lipid stain. Spores appear green in a cell with red vegetative cytoplasm and containing black lipid globules. Biochem-ical confirmation can be based on an isolate's ability to produce acid from glucose but not from mannitol, xylose and arabinose.

Commercial kits are available which claim to detect the diarrhoeal enterotoxin though they have limited use. One detects the L_2 unit of HBL, though some outbreak strains do not produce this toxin. The other detects a protein present in NHE but not in the HBL complex.

7.2.5 Association with Foods

The ability to produce spores resistant to factors such as drying and heat means that the food-poisoning bacilli are widely distributed in foods. In most circumstances however they are only a small part of the total flora and are not present in numbers sufficient to cause illness.

Heat processing will select for spore formers and a number of surveys have reported a higher incidence of *B. cereus* in pasteurized and other heat-processed milks (typically 35–48% of samples positive) compared with raw milk ($\sim$9% positive). In most of these cases the numbers detected were low ($< 10^3 \, ml^{-1}$), but when pasteurized milk or cream are stored at inadequate chill temperatures *B. cereus* can grow and cause the type of spoilage known as 'sweet curdling' or 'bitty cream'. Despite this, milk and dairy products are rarely associated with illness caused by *B. cereus*, although dried milk has been implicated in outbreaks when used as an ingredient in vanilla slices and macaroni cheese. A possible explanation is that though liquid milk is an excellent growth medium for the organism, toxin production is not favoured. One study in Sweden has linked this with the low aeration in static packs of milk.

The ability of spores to resist desiccation allows their survival on dried products such as cereals and flours. In the Norwegian outbreaks de-scribed above (Section 7.2.1), the cornflour used to thicken the vanilla sauce was implicated. Moderate heating during preparation would not inactivate the spores and subsequent extended storage of the high-a_w sauce at ambient temperature was conducive to spore germination and outgrowth.

The emetic syndrome is particularly associated with starchy products such as rice and pasta dishes. In the UK, its association with cooked rice has been sufficiently marked for it to earn the soubriquet 'Chinese

Restaurant Syndrome'. The typical scenario is where rice is prepared in bulk, in advance. Spores, commonly those of the more heat-resistant serotype 1, survive precooking to germinate, grow and produce the emetic toxin in the product during storage. This would be prevented by chilling to below 8 °C, but the rate of cooling in the centre of a bulk of cooked rice, even if transferred to chill storage, can be slow enough for growth and toxin production to occur. Reheating the rice prior to serving will not inactivate the emetic toxin and render the product safe.

A wider range of foods have been implicated with the diarrhoeal syndrome including meat products, soups, vegetables, puddings and sauces. Dried herbs and spices used in food preparation can be an important source of *B. cereus* and this has often been cited as a reason for a relatively high incidence of *B. cereus* food poisoning in Hungary, where between 1960 and 1968 it was the third most common cause of food poisoning accounting for 15.2% of persons affected. More recent figures suggest that its relative importance has declined somewhat but whether this is due to changes in culinary practices, improvements in hygiene, decreased contamination of spices or a statistical artefact is not known.

Meat pies and pasties are common vehicles for the other food-poisoning bacilli along with a range of processed meats and meat and rice dishes. Baked goods such as bread and crumpets have been involved in a number of *B. subtilis* outbreaks. Although *B. subtilis* is responsible for the defect known as ropey bread where spores surviving baking degrade the loaf's internal structure and produce a sticky slime, this does not always prevent people from eating it. In 1988, a bakery in the Isle of Man omitted propionate from their bread in order to claim for it the virtue of being free from artificial preservatives and thereby more healthy. As a result, nine people developed nausea, vomiting, diarrhoea, headache and chills 10 min after consuming ropey bread containing more than 10^8 organisms gm^{-1}.

7.3   BRUCELLA

7.3.1   Introduction

The genus *Brucella* is named after Sir David Bruce who in 1887 recognized it as the causative organism of undulant fever (brucellosis, Malta fever, Mediterranean fever). Each of the four species that are human pathogens is associated with a particular animal host, *B. abortus* (cattle), *B. melitensis* (sheep and goats), *B. suis* (pigs), and *B. canis* (dogs) (Table 7.3). Brucellosis is principally contracted from close association with infected animals and is an occupational disease of farmers, herdsmen, veterinarians and slaughterhouse workers. It can also be contracted by consumption of milk or milk products from an infected animal, although the risk is lower. The

Table 7.3 *Differential characteristics of* Brucella *species*

	B. abortus	B. canis	B. melitensis	B. suis
5% CO_2 required	+	−	−	−
H_2S produced	+	−	−	+
Urease	weak	strong	weak	strong
Growth on dye medium:				
Basic fuchsin $(1:10^5)$	+	−	+	−
Thionin $(1:10^5)$	−	+	+	+

illness has been effectively eliminated from the United States, Scandinavia, the UK and other countries by campaigns to eradicate the organism in the national dairy herds through a programme of testing, immunization of young calves and compulsory slaughter of infected cattle.

7.3.2   The Organism and its Characteristics

Brucella are Gram-negative, catalase-positive, oxidase-positive, short oval rods $(0.3\,\mu m \times 0.4\,\mu m)$ which are non-motile and usually occur singly, in pairs, or, rarely, in short chains. It grows optimally around 37 °C and is killed by heating at 63 °C for 7–10 min. When shed in the milk of an infected animal it can survive for many days provided the acidity remains low ($<0.5\%$ as lactate).

7.3.3   Pathogenesis and Clinical Features

Brucellosis is a protracted and debilitating illness characterized by an incubation period of from one to six weeks followed by a chronic, relapsing fever with accompanying lassitude, sweats, headache, constipation, anorexia, pains in the limbs and back, and weight loss. After the temperature has returned to normal for a few days, another bout of fever may ensue and such episodes recur a number of times over several months. Treatment is commonly with a mixture of tetracycline and streptomycin.

It is a facultative parasite and can live intracellularly or in extracellular body fluids. During the febrile stage, caused by circulating endotoxin, the organism may be isolated from the bloodstream but in the majority of laboratory-confirmed cases diagnosis is based on serological tests rather than cultural techniques.

7.3.4   Isolation and Identification

Brucella are quite fastidious organisms and do not grow in conventional laboratory media. Liver infusions or calf serum are normally added. The organism grows slowly and cultures are normally incubated for three weeks before they are considered negative. In view of this, testing foods for the organism is not practically feasible or useful. Cattle are tested for the presence of antibodies to the organism in the 'Ring Test'. Stained antigen is mixed with the test milk, if antibodies to *Brucella* are present (indicative of infection) then they will cause the antigen to clump and rise with the milkfat on standing to form an intense blue-violet ring at the top of the milk.

7.3.5   Association with Foods

Although brucellosis has sometimes been associated with the consumption of inadequately cooked meat from an infected animal, raw milk or cream are the principal food vehicles. *Brucella* is readily killed by normal milk pasteurization conditions so there is no risk from pasteurized milk or products made from it. Cheeses made from unpasteurized milk can sometimes pose a problem since the organism can survive the cheese-making processes and subsequent storage in the product.

7.4   *CAMPYLOBACTER*

7.4.1   Introduction

Campylobacter has been known as a veterinary problem since the early years of the 20th Century when the original isolate, known then as *Vibrio fetus*, was associated with infectious abortion in sheep and cattle. In 1931 the species *Vibrio jejuni* was described as the cause of winter dysentery in calves and in 1946 a similar organism was isolated from blood cultures of patients in a milk-borne outbreak of acute diarrhoea. Later King, working with human blood isolates, distinguished two groups on the basis of their optimum growth temperature. One group corresponded to *Vibrio fetus* and the second, 'thermophilic', group, which grew best at 42 °C, came from patients with preceding diarrhoea.

Both groups differ from the cholera and halophilic vibrios biochemically, serologically and in their mol% $G + C$ ratio, and were reclassified into the new genus *Campylobacter* in 1963.

In the 1970s, with the development of suitable selective media, it was established that *Campylobacter jejuni*, and to a lesser extent *Campylobacter coli*, are a major cause of diarrhoeal illness, rivalling and even surpassing *Salmonella* in importance in many countries. *Campylobacter laridis*, *C. concisus* and *C. hyointestinalis* have also been isolated occasionally from patients with diarrhoea and *C. pylori*, now reclassified as *Helicobacter pylori*, has been associated with gastritis and stomach and duodenal ulcers.

The *Campylobacter*-like genus *Arcobacter* is frequently associated with abortion and enteritis in cattle and pigs. Two species, *Arcobacter butzleri* and *A. cryaerophilus*, have also been implicated in human infections causing diarrhoea, bacteraemia and other extra-enteric infections.

7.4.2 The Organism and its Characteristics

Campylobacters are non-sporeforming, oxidase-positive, Gram-negative rods. Cells are pleomorphic and may be 0.5–8 µm in length and 0.2–0.5 µm in width. Log-phase cells have a characteristic slender, curved or spiral shape and one or more polar or amphitrichous flagella which confer a rapid, darting motility and may be an important feature in pathogenesis (Figure 7.2). As cultures age, spiral or curved bacilli are replaced by round forms.

Campylobacters cannot ferment or oxidize sugars and are oxygen-sensitive microaerophiles, growing best in an atmosphere containing 5–10% carbon dioxide and 3–5% oxygen.

All *Campylobacter* species grow at 37 °C; *C. jejuni* and *C. coli* have optima at 42–45 °C but cannot survive cooking or pasteurization temperatures (D_{55} 2.5–6.6 min). They do not grow below 30 °C and survive poorly at room temperature. Although their viability declines during chill or frozen storage, they may nevertheless persist under these conditions for prolonged periods; survival has been recorded in milk and water at 4 °C after several weeks storage and in frozen poultry after several

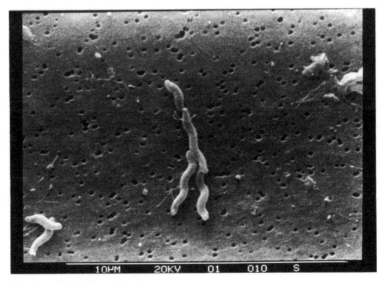

Figure 7.2 Campylobacter jejuni
(Photo S. Boucher)

months. They are also particularly sensitive to other adverse conditions such as drying and reduced pH.

The principal environmental reservoir of pathogenic campylobacters is the alimentary tract of wild and domesticated animals and birds and it is a commonly found commensal of rodents, dogs, cats, dairy cattle, sheep, pigs, poultry and wild birds. The high optimum growth temperature of *C. jejuni* and *C. coli* could be an adaptation to the higher body temperature of birds and reflect their importance as a primary reservoir of the organism. Asymptomatic human carriage also occurs.

Though they would not appear to survive particularly well outside an animal host, campylobacters can be commonly isolated from surface water. Survival is enhanced by low temperatures and studies conducted in Norway have shown that strains of *C. jejuni, C. coli* and *C. laridis* remained viable in unchlorinated tap water at 4 °C for 15 days (10 days at 12 °C) and 10–15 days in polluted river water at the same temperature (6–12 days at 12 °C).

Under adverse environmental conditions campylobacters have been reported to adopt a 'viable non-culturable' state where the organism cannot be isolated by cultural methods but nevertheless remains infective. Evidence for this is conflicting, one study has shown that viable non-culturable *C. jejuni* can revert to a culturable state by passage through an animal host but others have failed to observe this effect.

Studies in the United States and Europe have isolated *Arcobacter* species from pork and poultry meat, though the rates of isolation vary widely (0.5 to 97%). Physiologically, *Arcobacter* species differ significantly from *Campylobacter* and *Helicobacter* in being both aerotolerant and capable of growth at 15 °C, attributes which could give them a considerable advantage when it comes to foodborne transmission. However the evidence for an association between *Arcobacter* and diarrhoeal disease in humans remains circumstantial at present.

7.4.3 Pathogenesis and Clinical Features

Enteropathogenic campylobacters cause an acute enterocolitis which, in the absence of microbiological evidence, is not easily distinguished from illness caused by other pathogens. The incubation period is from 1 to 11 days, most commonly 3–5 days, with malaise, fever, severe abdominal pain and diarrhoea as the main symptoms. The diarrhoea produces stools containing 10^6–10^9 cells g^{-1}, which are often foul-smelling and can vary from being profuse and watery to bloody and dysenteric. Gastrointestinal symptoms are sometimes preceded by a prodromal stage of fever, headache and malaise which lasts about a day. The diarrhoea is self-limiting and persists for up to a week, although mild relapses often occur. Excretion of the organism continues for up to 2–3 weeks. Vomiting is a less common feature.

Complications are rare although reactive arthritis can develop and *Campylobacter* has been shown to cause the serious neurological disease, Guillain–Barre syndrome.

As with other pathogens the infective dose will depend upon a number of factors including the virulence of the strain, the vehicle with which it is ingested and the susceptibility of the individual. Young adults (15–24 years old) and young children (1–4 years) appear particularly susceptible. In an outbreak at a boys' school in England caused by contamination of a water-holding tank with bird droppings, the infective dose was estimated as 500 organisms and in a separate study, a similar dose in milk caused illness in a volunteer. Motility, chemotaxis and the corkscrew morphology of the cells are all important factors in the virulence of *Campylobacter*, enabling it to penetrate the viscous mucus which covers the epithelial surface of the gut. Studies with *C. jejuni* have demonstrated a chemotactic response toward the sugar L-fucose, a number of amino acids, and intestinal mucus from mice and pigs. Although *Campylobacter* does not normally possess fimbriae it probably possesses other adhesins that enable it to adhere to epithelial cells once the mucosal barrier has been penetrated. The production of peritrichous 'pili-like' appendages when cells are grown in the presence of bile has been observed. A nonpiliated mutant did not show any reduced adherence or invasion in a cell line although it did produce significantly reduced disease symptoms in an animal model.

There is considerable uncertainty as to the precise mechanisms by which *Campylobacter* causes illness. It has been shown to be invasive in cell cultures and a number of toxins with cholera-like or cytotoxic activity have been described. Their role in pathogenesis however has been brought into question following the complete sequencing of the *Campylobacter* genome which has failed to identify sequences associated with known toxins. One exception is the so-called cytolethal distending toxin which has also been reported in a small number of *E. coli* and *Shigella* strains. There is evidence that some of the pathogenic features of the illness are the result of the body's inflammatory response to invasion by the organism, a factor also implicated in illness caused by *Salmonella*.

7.4.4 Isolation and Identification

Although most of the isolation procedures and media used were designed for *C. jejuni*, they are also suitable for *C. coli* and *C. laridis*.

Pathogenic campylobacters have a reputation for being difficult to grow but in fact their nutritional requirements are not particularly complex and they can be grown on a number of peptone-based media including nutrient broth. Where problems can sometimes arise is in their sensitivitiy to oxygen and its reactive derivatives. Although pathogenic campylobacters possess catalase and superoxide dismutase, the accumulation of peroxides and superoxide in media during storage or incubation can inhibit growth. For this reason an incubation atmosphere of 5–6% oxygen with about 10% carbon dioxide and media containing oxygen scavenging compounds such as blood, pyruvate, ferrous salts, charcoal and metabisulfite are commonly used.

To isolate the relatively low numbers of cells that may be present in foods, several selective enrichment media are used which include cocktails of antibiotics such as polymyxin B, trimethoprim and others as selective agents. In many cases cells isolated from food or other environmental sources have been sub-lethally injured as a result of stresses such as freezing, drying or heating and, as a result, are more sensitive to antibiotics and toxic oxygen derivatives. This can mean that they will not grow on the usual selective media unless allowed a period for recovery and repair in which case a resuscitation stage of 4 h at 37 °C in a non-selective environment is recommended.

After selective enrichment for 24 and 48 h under microaerobic conditions at 42–43 °C, samples are streaked on to selective plating media. These normally contain a nutrient-rich basal medium supplemented with oxygen scavengers such as blood and/or FBP (a mixture of ferrous sulfate, sodium metabisulfite, and sodium pyruvate), and a cocktail of antibiotics similar to those used for selective enrichment. It is important to store pre-prepared media under nitrogen, at 4 °C and away from light to reduce the build-up of toxic oxides.

Colonies are non-haemolytic and have a rather unimpressive flat, watery appearance with an irregular edge and a grey or light-brown coloration. Suspect colonies are examined microscopically for motility and morphology and subjected to a range of tests after purification. *C. jejuni*, *C. coli*, and *C. laridis* are catalase and oxidase positive, reduce nitrate to nitrite, grow at 42 °C but not at 25 °C microaerobically and cannot grow aerobically at 37 °C. *C. laridis* is resistant to nalidixic acid while *C. jejuni* and *C. coli* are not. *C. jejuni* and *C. coli* can be distinguished by the ability of the former to hydrolyse hippurate. Various typing schemes have been proposed for epidemiological investigations. Biotyping based on biochemical tests and the more discriminating serotyping schemes of Lior and Penner have been used most frequently but not routinely. Molecular methods such as ribotyping and flagellin gene typing have been used but pulsed-field gel electophoresis has become the sub-typing method of choice due to its sensitivity and discriminatory power.

7.4.5   Association with Foods

The incidence of *Campylobacter* infection is characterized by large numbers of sporadic cases rather than single source outbreaks. Infection can be acquired by a number of routes. Direct transmission person-to-person or from contact with infected animals, particularly young pets such as kittens or puppies, has been reported, as have occasional water-borne outbreaks. However, food is thought to be the principal vehicle.

As a common inhabitant of the gastrointestinal tract of warm-blooded animals, *Campylobacter* inevitably finds its way on to meat when carcasses are contaminated with intestinal contents during slaughter and evisceration. Numbers are reduced significantly as a result of chilling in the abattoir; the incidence of *Campylobacter*-positive beef carcasses in Australia was found to decrease from 12.3% to 2.9% on chilling and a similar survey of pig carcasses in the UK found a decrease from 59% down to 2%. This is primarily a result of the sensitivity of *Campylobacter* to the dehydration that takes place on chilling. Subsequent butchering of red-meat carcasses will spread the surviving organisms to freshly cut, moist surfaces where viability will decline more slowly.

Poultry carcasses which cool more rapidly due to their size suffer less surface drying when air-chilled and this, probably coupled with the surface texture of poultry skin, enhances survival. Surveys in Australia, the UK and the USA have found 45%, 72% and 80% respectively of chilled poultry carcasses at the abattoir to contain *Campylobacter*.

The incidence of campylobacters on retail meats in several countries has been found to vary from 0–8.1% for red meats and from 23.1–84% for chicken. Adequate cooking will assure safety of meats but serious under-cooking or cross-contamination from raw to cooked product in the kitchen are thought to be major routes of infection.

Despite its frequent occurrence in poultry, eggs do not appear to be an important source of *Campylobacter*. Studies of eggs from flocks colonized with *C. jejuni* have found the organism on around 1% of egg shells or the inner shell and membranes. Prolonged survival on the dry egg surface is unlikely and egg albumin has been shown to be strongly bactericidal.

Milk can contain *Campylobacter* as a result of faecal contamination on the farm or possibly *Campylobacter* mastitis. The bacterium cannot survive correct pasteurization procedures and the majority of outbreaks, many quite large, have involved unpasteurized milk. More than 2500 children aged 2–7 years in England were infected by consumption of free school milk which is thought to have by-passed pasteurization. In Switzerland, a raw-milk drink was associated with an outbreak at a fun-run which affected more than 500 participants. It is not on record whether any personal best times were achieved that day.

Post-pasteurization contamination may always re-introduce the organism to milk. For example, pecking of doorstep delivered pasteurized milk by birds of the crow family has been strongly implicated in a number of cases of *Campylobacter* enteritis in the UK. Dairy products other than fresh milk do not pose a threat due to the low resistance of *Campylobacter* to conditions of reduced pH or a_w.

Other foods recognized as potential sources of *Campylobacter* infection include shellfish and mushrooms. *C. jejuni* and *C. coli* were detected in 14% of oyster flesh tested, although 2 days depuration was sufficient to cleanse oysters artificially contaminated with 800 cfu of campylobacters g^{-1}. An outbreak in the USA was ascribed to raw clams.

7.5 *CLOSTRIDIUM BOTULINUM*

7.5.1 Introduction

Because of its severity and distinctive symptoms, botulism is the form of bacterial food poisoning for which we have the earliest reliable reports.

In 1793 in Wildbad, Wurttemburg, 13 people fell ill and 6 later died after eating Blunzen, a type of sausage made by packing blood and other ingredients into a pig's stomach. The sausage had been boiled and then smoked, after which it was considered stable at room temperature for several weeks and suitable for consumption without reheating.

Several further incidents of *Wurstvergiftung*, or sausage poisoning, were recorded in the years that followed, usually associated with sausages that contained animal components other than muscle tissue. This prompted a local district medical officer, Justinius Kerner, to undertake a study of the disease which became known as botulism (Latin: *botulus* = sausage). Kerner noted several important features including the facts that heating was an essential precondition for the development of toxicity in sausages and that small sausages or those containing air pockets were less likely to become toxic.

It was not until 1896 that the micro-organism responsible was isolated and described by van Ermengem, Professor of Bacteriology at the University of Ghent and former pupil of Robert Koch. This was a result of his investigation into an outbreak of botulism where 34 members of a music club in Belgium ate raw, unsmoked ham. Several noted that the ham had a slightly 'off' flavour akin to rancid butter but was otherwise unremarkable. About a day later, 23 of the group fell ill and 3 died within a week.

Van Ermengem established that botulism resulted from the consumption of food containing a heat-labile toxin produced by an obligately anaerobic, spore-forming bacillus which he called *Bacillus botulinus*. He further demonstrated that toxin would not be produced in the presence of sufficient salt, that it was resistant to mild chemical agents and was not uniformly active against all animal species.

Although much of the early evidence suggested that botulism was confined to meat products, it was later found to occur wherever foods and their processing offer conditions suitable for survival and growth of the causative organism. It was identified with icthyism, a paralytic illness associated with the consumption of raw, salted fish, known in Russia since 1880, and in 1904 an outbreak of botulism in Darmstadt, Germany was caused by canned white beans.

7.5.2 The Organism and its Characteristics

Van Ermengem's original designation was superseded in 1923 when the organism responsible for botulism was reclassified as *Clostridium botulinum*. The cells are Gram-positive, motile with peritrichous flagella, obligately anaerobic, straight or slightly curved rods 2–10 μm long, and form central or subterminal oval spores.

Strains of *C. botulinum* display sufficient variety of physiological and biochemical characteristics to be inconsistent with their inclusion in a single species. In this instance however, taxonomic rectitude has been sacrificed to avoid any possibility of confusion over nomenclature with potentially fatal consequences. The most important common feature of the species is the production of pharmacologically similar neurotoxins responsible for botulism. Eight serologically distinct toxins are recognized (A, B, C_1, C_2, D, E, F, and G, though C_2 is not a neurotoxin), a single strain of *C. botulinum* will usually only produce one type, although there are exceptions. In 1985, certain strains of *C. barati* and *C. butyricum* responsible for cases of infant botulism (Section 7.5.3) were found to produce similar neurotoxins, although they have not been implicated in any foodborne cases of botulism.

Physiological diversity within the species *C. botulinum* is recognized by its division into four groups (Table 7.4) and molecular studies based on DNA homology and ribosomal RNA sequences have confirmed this grouping. Group I strains are culturally indistinguishable from the non-toxigenic species *Clostridium sporogenes* which can sometimes serve as a useful and safe model in laboratory studies. They are strongly proteolytic and will often betray their presence in food by partial disintegration of the product and a slight rancid or cheesy odour. Unfortunately despite these warning signs the potency of the toxin is such that the amount ingested on sampling the food has often proved sufficient to cause illness. Group I strains are not psychrotrophic and are therefore of little concern in adequately refrigerated products. They do, however, produce the most heat-resistant spores and can pose a problem when foods that depend upon a heating step for their stability and safety are underprocessed.

In contrast, Group II strains represent a greater potential hazard in chilled foods. They are non-proteolytic with native protein, can grow and produce toxin down to about 3 °C and produce spores with a low resistance to heat. They also tend to be more susceptible to inhibition

Table 7.4 *The physiological subdivision of* Clostridium botulinum

Group	Toxin types	Proteolytic	Lipolytic	Saccharolytic	Psychrotrophic (min. growth temp.)	Inhibition by salt (a_w)	Heat resistance	Pathogenicity
I	A, B or F	+	+	+	− (10–12 °C)	10% (0.94)	+ (D_{121} 0.1–0.25 min)	Humans
II	B, E or F	−	+	+	+ (3–5 °C)	5% (0.975)	− (D_{80} 0.6–3.3 min)	Humans
III	C_1, C_2 or D	−	+	+	− 15 °C	3%	±	Usually animals and birds
IV	G	+	−	−	12 °C	>3%	No data	Humans

by salt (Table 7.4). The rate of growth and toxin production at the lower temperature limit is slow and will be reduced still further by any other factors adverse to growth. Experimental studies have indicated that storage periods of 1–3 months are necessary for toxin production at 3.3 °C, although this period can be markedly reduced at higher temperatures still within the chill range. Vacuum-packed herrings inoculated with 100 spores per pack became toxic after 15 days storage at 5 °C.

Most cases of botulism in humans are due to toxin types A, B or E and the incidence of other types in human illness is extremely rare. Group III strains producing toxin types C and D are usually associated with illness in animals and birds. Type G toxin has been incriminated in human illness largely as a result of its isolation at autopsy from people who had died suddenly and unexpectedly. Since botulism was not necessarily the cause of death in these cases, and there have been no reports of the presence of type G in foods, its role in foodborne illness is questionable. Group IV strains which produce type G toxin and some non-toxigenic clostridia have been re-designated as a new species, *Clostridium argentinense.*

Although it is found occasionally, growing in the alimentary tract of birds and mammals, *C. botulinum* is essentially a soil saprophyte. It occurs widely, although the geographical distribution is not uniform. Surveys conducted in the United States found type A to be the most common in the Western States, rare in the Mississippi Valley but less so along the Eastern Seaboard where type B was predominant. This distribution was reflected in outbreaks of botulism in the United States in the period 1950–1979; when 85% of those west of the Mississippi were due to type A toxin and 63% of those to the east were due to type B. In European soils type B tends to be more common than type A.

Aquatic muds provide a moist, anaerobic, nutrient-rich environment in which clostridia can flourish, so isolation of *C. botulinum* from these sources is more frequent than from soils. The psychrotrophic type E has been particularly associated with this environment in regions such as western North America, Japan and the Baltic sea coasts. As a consequence, type E is often responsible for outbreaks of botulism where fish is the vehicle.

The minimum pH at which *C. botulinum* will grow depends very much on factors such as temperature, water activity and the acid used to adjust the pH. The consensus has long been that a pH around 4.7 represents an absolute minimum and this fact has had important practical implications for the canning industry (see Chapter 4). Non-proteolytic strains have a lower acid tolerance and are generally inhibited at pH 5.0–5.2. Reports have appeared of growth and toxin production at pH values as low as 4.0 in protective, high-protein containing media but this does not reflect the situation in acid canned foods which are generally low in protein. In cases where botulism has occurred in foods where acidity is an important protective hurdle, such as canned fruits, it has been as a result of other organisms, yeasts or moulds, growing in the product and increasing the pH.

The maximum pH for growth is 8.5–8.9 and the toxin is unstable at alkaline pH values. This is generally an unimportant feature of the organism's physiology since nearly all foods are slightly acidic. It may be significant however in some North American fermented fish products occasionally associated with botulism where the usual increase in pH on fermentation would be a protective factor.

7.5.3 Pathogenesis and Clinical Features

Three types of botulism are recognised: foodborne botulism, infant or infectious botulism and wound botulism. Only in the first type is food invariably involved.

Foodborne botulism is an example of bacterial food poisoning in its strictest sense: it results from the ingestion of an exotoxin produced by *Clostridium botulinum* growing in the food. The botulinum toxins are neurotoxins; unlike enterotoxins, which act locally in the gut, they affect primarily the cholinergic nerves of the peripheral nervous system.

Experiments in animals have shown that toxin ingested with food and surviving inactivation is absorbed in the upper part of the small intestine and reaches the bloodstream *via* the lymphatics. It binds to the nerve ending at the nerve–muscle junction, blocking release of the acetylcholine responsible for transmission of stimuli, thus producing a flaccid paralysis.

Initial symptoms of botulism occur anything from 8 h to 8 days, most commonly 12–48 h, after consumption of the toxin-containing food. Symptoms include vomiting, constipation, urine retention, double vision, difficulty in swallowing (dysphagia), dry mouth and difficulty in speaking (dysphonia). The patient remains conscious until, in fatal cases, shortly before the end when the progressive weakness results in respiratory or heart failure. This usually occurs 1–7 days after the onset of symptoms. Surviving patients may take as long as 8 months to recover fully.

The clinician can do little to mitigate the effect of toxin already adsorbed at the neuromuscular junction, although neuromuscular block-ade antagonists such as 4-aminopyridine have produced transient improvements. Survival is therefore critically dependent on early diag-nosis and treatment, principally by alkaline stomach washing to remove any remaining toxic food, intravenous administration of specific or polyvalent anti-toxins to neutralize circulating toxin, and mechanical respiratory support where necessary.

The mortality rate is usually high (20–50%), but will depend on a variety of factors such as the type of toxin (type A usually produces a higher mortality than B or E), the amount ingested, the type of food and the speed of treatment.

The botulinum toxins are the most toxic substances known, with a lethal dose for an adult human in the order of 10^{-8} g. They are high molecular mass (150 kDa) proteins and can be inactivated by heating at $80°C$ for 10 min. In culture, they are produced during logarithmic growth as complexes and released into the surrounding medium on cell lysis. In the smallest of these complexes, the M complex, neurotoxin is accompanied by a similar-sized protein with no apparent biological activity, while in the larger L complex, an additional haemagglutinin component is also present. It appears that the neurotoxin is synthesized as a single chain protoxin which is activated by proteolytic cleavage to produce a molecule consisting of light (M_r 50 kDa) and heavy (M_r 100 kDa) chains linked by a disulfide bridge (Figure 7.3). Where the organism does not itself produce appropriate proteolytic enzymes, protoxin can be activated by the gut enzyme trypsin. More extensive proteolysis will lead to toxin inactivation so that, although the structure of the natural complex affords some protection, the lethal oral dose of toxin A in mice is 10^4–10^5 times that observed when administered intraperitoneally. The heavy chain is responsible for specific binding to neuronal cells and cell penetration by the light chain. The light chain is a zinc endopeptidase which is activated by reduction of the interchain disulfide bond.

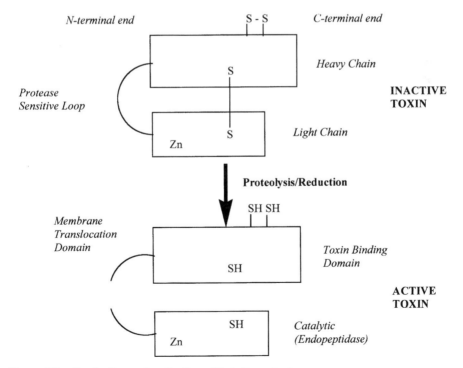

Figure 7.3 *Production and activation of botulinum toxin*

The endopeptidase then cleaves components of the docking and fusing complex of the synaptic vesicle, the vesicle that contains the neurotransmitter acetylcholine. The particular protein attacked and the specific peptide linkage hydrolysed varies with the type of toxin. Toxin types B, D, F, and G each hydrolyse a different peptide bond on vesicle associated membrane protein, also known as synaptobrevin. Types A and E attack different linkages on the synaptosome-associated protein, SNAP-25, and type C_1 degrades both SNAP-25 and syntaxin.

It has been shown, at least for types C and D, that the genetic information coding for toxin production is associated with a temperate bacteriophage. This persists in the bacterial cell as a prophage; its DNA incorporated and replicating with the bacterial chromosome without causing lysis. This lysogenic state occurs widely among bacteria in nature, usually without changing the micro-organism's characteristics, but sometimes, as here, it is associated with the production of toxins. Another example is the production of diphtheria toxin by *Corynebacterium diphtheriae*.

Infant botulism differs from the classical syndrome in that it results from colonization of the infant's gut with *C. botulinum* and production of toxin *in situ*. It was first described (in 1976) and is most frequently reported in the United States, although cases have occurred in Australia, Canada, Europe and South America. Up to 2005 there had been 6 confirmed cases of infant botulism in the UK, mostly involving type B toxin producers.

It occurs mostly in infants aged 2 weeks to 6 months, particularly around the time that non-milk feeds are introduced. At this stage the infant's gut microflora is not fully developed and is less able to outcompete and exclude *C. botulinum*. Since it only requires the ingestion of viable spores, environmental sources other than food may be involved and those foods that do act as vehicles need not be capable of supporting growth of the organism. Honey has been associated with several cases of infant botulism in the USA and some surveys have found viable spores of *C. botulinum* in 10% of the samples examined. Consequently it is thought inadvisable to feed honey to children less than a year old.

The illness is characterized by neuromuscular symptoms related to those of classical botulism and diagnosis can be confirmed by the isolation of the organism and its toxin from the faeces. Although implicated in a small proportion (4%) of cases of sudden infant death syndrome in the United States, the mortality rate is low in treated cases.

Wound botulism is caused by a subcutaneous infection with *C. botulinum*. This can result from trauma, but in recent years has been more commonly associated with intravenous drug use. Accidental overdoses of botulinum toxin during its cosmetic use to remove facial wrinkles have also caused occasional cases.

7.5.4 Isolation and Identification

In view of the metabolic diversity within the species selective media are of limited use in the isolation of *C. botulinum* and identification is based on the ability of typical colonies to produce toxin in culture.

C. botulinum will often constitute only a small proportion of the total microflora so enrichment or pre-incubation is necessary to improve the chances of isolation. Sometimes enrichment cultures are heated prior to incubation to eliminate non-sporeforming anaerobes. However, depending on the heating regime used, 80 °C for 10 min is commonly cited, this may also eliminate the less heat resistant strains of *C. botulinum* and is therefore often omitted.

After enrichment in a medium such as cooked meat broth at 30 °C for 7 days, the culture is streaked on to fresh horse-blood or egg yolk agar and incubated anaerobically for 3 days. Characteristic smooth colonies, 2–3 mm in diameter with an irregular edge and showing lipolytic activity on egg-yolk agar (type G excepted) are transferred into a broth medium to check for toxin production.

A technique has been described that simplifies this procedure by incorporating antitoxin into the agar medium so that toxin-producing colonies are surrounded by a zone of toxin–antitoxin precipitate.

Despite the development of a range of *in vitro* immunoassay procedures for toxin, the mouse neutralization test (Figure 7.4), remains the most sensitive (a typical lethal dose of toxin for a mouse is a few picograms). However, the distressing nature of the test guarantees its eventual replacement as soon as immunoassay amplification systems have been sufficiently improved.

A suspect toxin extract is divided into three portions: one, to serve as control, is heated at 100 °C for 10 min to destroy any toxin present; a second is treated with trypsin to activate any protoxin that may be there; and a third is untreated. Each of the portions is injected intraperitoneally into 2 mice and the mice observed over 4 days for the development of typical symptoms of laboured breathing and the characteristic 'wasp-waist' appearance. The presence of toxin is confirmed by protection of mice with polyvalent antitoxin and the toxin type can be identified using monovalent antisera.

7.5.5 Association with Foods

Four common features are discernible in outbreaks of botulism.

(1) The food has been contaminated at source or during processing with spores or vegetative cells of *C. botulinum*.

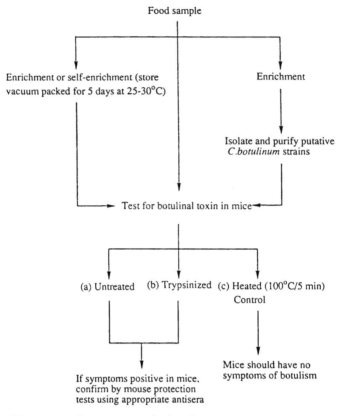

Figure 7.4 *Mouse neutralization assay for botulinum toxin*

(2) The food receives some treatment that restricts the competitive microflora and, in normal circumstances, should also control *C. botulinum*.

(3) Conditions in the food (temperature, pH, E_h, a_w) are suitable for the growth of *C. botulinum*.

(4) The food is consumed cold or after a mild heat treatment insufficient to inactivate toxin.

Since low-acid canned foods can fulfil all the above criteria, it has been necessary for the canning industry to introduce stringent process control measures to ensure safety (see Chapter 4). When canned foods are produced as a small-scale, domestic activity however, greater variability and less rigorous control are clearly potential sources of problems. In the United States, where home-canning is more widely practised than elsewhere, inadequately processed products, particularly vegetables, are the most common cause of botulism. Between 1899 and 1981 there were 522 outbreaks associated with home-canned products, including 432 involving vegetables. This compares with 55 outbreaks over the same period caused by commercially canned products; the majority occurring before 1925.

A variety of foods have been associated with botulism in the UK and they are frequently home-produced rather than commercial products [Table 7.5(a)]. Fortunately the incidence is extremely low, though slightly higher rates have been reported in some other European countries [Table 7.5(b)].

Table 7.5a *Foodborne botulism in the United Kingdom*

Year	Number of deaths/cases	Food vehicle	(Home produced)	C. botulinum toxin type
1922	8/8	Duck paté	(No)	A
1932	1/2	Rabbit and pigeon broth	(Yes)	?
1934	0/1	Jugged hare	(Yes)	?
1935	4?/5?	Vegetarian nut brawn	(Yes)	A
1935	1/1	Minced meat pie	(Yes)	B
1949	1/5	Macaroni cheese	(Yes)	?
1955	0/2	Pickled fish	(?)	A
1978	2/4	Canned salmon	(No)	E
1987	0/1	Rice and vegetables, shelf-stable airline meal	(No)	A
1989	1/27	Hazelnut puré added to yoghurt	(No)	B
1998	1/2	Bottled mushrooms	(Yes)	B
2003	1/1	Polish sausage	(Yes)	B
2004	0/2	Travel/Hummus	(?)	—
2005	0/1	Polish preserved pork	(Yes)	—

Adapted from *Eurosurveillence*, vol.4, Jan 1999

Table 7.5b *Botulism in Europe (number of cases/year)*

	1988	1989	1990	1991	1992	1993	1994	1995	1996	1997	1998[a]
Belgium	0	2	1	0	1	1	0	0	1	3	1
Denmark	0	0	0	0	0	0	0	2	0	0	1
England and Wales	0	27	0	0	0	1	1	0	0	0	2
Finland	0	0	0	0	0	0	0	0	0	0	0
France	4	6	11	3	5	10	13	7	5	8	NA
Germany	39	15	15	23	4	17	13	11	12	9	19
Greece	0	0	0	0	0	0	0	0	0	0	0
Italy	53	54	45	12	26	39	26	41	58	32	26
Scotland	0	0	0	0	0	0	0	0	0	0	0
Spain	8	8	10	5	12	9	7	6	7	9	11
Sweden	0	0	1	2	0	0	2	1	1	0	0
The Netherlands	0	0	0	0	0	0	0	0	0	0	0

[a] January – October. NA = not available
Data for Austria, Ireland and Portugal not available
Adpated from *Eurosurveillance*, vol. 4, Jan 1999

Fish can be contaminated with *C. botulinum*, particularly type E, from the aquatic environment and uncooked fish products have been responsible for several outbreaks of type E botulism.

Smoked fish consumed without reheating has generally been hot-smoked so control of *C. botulinum* depends on microbial inactivation by heat plus the inhibitory effects of salt, smoke constituents and surface drying. With the advent of refrigeration, the severity of the salting and smoking stages has been reduced in line with the perceived consumer preference for a less strongly flavoured product. In the early 1960s, two outbreaks of type E botulism in North America associated with vacuum-packed, hot-smoked fish caused considerable alarm and led to Canada banning the importation of all types of packaged fish. A similar outbreak in Germany in 1970 was caused by smoked trout from a fish farm. At first it was feared that vacuum packing, an emerging technology at that time, was responsible by providing an anaerobic environment in which *C. botulinum* could flourish.

It transpired that the problem was compounded of several factors. The salting and smoking treatments had been insufficient to eliminate *C. botulinum* or inhibit its growth during storage. A minimum salt concentration (in the water phase) of 3% and an internal temperature not less than 63 °C during smoking are recommended. The product had also been subjected to severe temperature/time abuse allowing *C. botulinum* to grow and produce toxin. The product should have been stored at temperatures below 4 °C. Finally, vacuum packing had improved the product shelf-life by inhibiting the normal spoilage microflora of bacteria and moulds which would have indicated that the product was inedible.

Fish products that are consumed raw after a fermentation process have also caused occasional problems, for example I-sushi (see Chapter 9). In 1986 in the Canadian Northwest Territories an outbreak of type E botulism was recorded after consumption of a meal comprising raw fish, seal meat and fermented seal flipper. The latter had been prepared by packing the product in a plastic bucket, covering with seal fat and leaving it outside the house to ferment. The process differed from normal in that the product was stored for 7 days instead of the usual three and the weather had been unseasonably warm. It was claimed that the seal flipper had an unusual taste and subsequent investigation established the presence of *C. botulinum* type E in the product. In Europe, the Norwegian fermented trout rak-orret has also been responsible for outbreaks of botulism.

The long association of botulism with meat products in Europe has already been noted and inadequate curing of meats still gives rise to occasional problems in some European countries. Outbreaks of botulism in the UK are relatively infrequent. The largest outbreak this century occurred in 1989 when 27 people fell ill and one died. In this outbreak the vehicle was hazelnut yoghurt. The pH of yoghurt is too low for toxin production *in situ*, but the toxin (type B) had been produced in the hazelnut puree which was inadequately heat processed.

Soil contamination is a major source of *C. botulinum* in foods and one to which vegetables, particularly root crops, are inevitably prone. Three outbreaks of type A botulism in the United States have been attributed to potato salad where cooked or partly cooked potatoes had been stored for several days at ambient temperatures and under anaerobic conditions before further processing. In 1987 an airline passenger in Europe contracted type A botulism from a prepacked vegetable salad. Important features in these outbreaks were temperature abuse and anaerobiosis created by vacuum packing or wrapping in aluminium foil. In 2006 a number of cases were reported in the USA and Canada caused by pasteurised carrot juice, presumably as a result of inadequate refrigeration of the product.

7.6 CLOSTRIDIUM PERFRINGENS

7.6.1 Introduction

Clostridium perfringens, formerly *welchii*, has been known as a cause of the serious wound infection, gas gangrene, since 1892 when it was first described by the American bacteriologist Welch. Although accounts appeared shortly after, in 1895 and 1899, linking it with outbreaks of gastroenteritis in St. Bartholomew's Hospital, London, it was not until the mid-1940s that outbreaks associated with school meals in England (1943) and pre-cooked chicken dishes in the USA (1945) firmly established *C. perfringens* as a cause of food poisoning.

The species is classified into five types, designated A–E, based on the production of four major exotoxins, α, β, ε, and ι, and eight minor ones (Table 7.6). *C. perfringens* type A which is responsible for food poisoning and gas gangrene produces only the α major toxin which has lecithinase (phospholipase C) activity. Its ability to hydrolyse lecithin and some other phospholipids plays an important role in the pathogenesis of gas gangrene;

Table 7.6 *Classification of* Clostridium perfringens *based on major exotoxin production*

Type	Toxin			
	α	β	ε	ι
A	+	–	–	–
B	+	+	+	–
C	+	+	–	–
D	+	–	+	–
E	+	–	–	+

by attacking cell membranes it produces local tissue disruption in the wound and its absorption into the circulation causes a serious toxaemia. It does not however have any role in the food poisoning syndrome.

C. perfringens type C which produces α and β toxins causes enteritis necroticans, a more severe, but far more rare, enteric disease in which the β toxin damages the intestinal mucosa causing necrosis. Illness is preventable by active immunization against the β toxin. Outbreaks were reported in Germany in 1946 and 1949, but it is nowadays particularly associated with Papua New Guinea where it is known as pigbel. Symptoms of abdominal pain and bloody diarrhoea develop several days after a high-protein meal, often pork consumed on festive occasions. Low levels of intestinal proteases are a predisposing factor in victims. This could arise from a poor diet low in protein, as with the European outbreaks after the Second World War, and may be compounded in Papua New Guinea by protease inhibitors consumed with other foods in the diet such as sweet potatoes.

C. perfringens type A ranks below *Salmonella* as a cause of outbreaks of bacterial food poisoning in the UK. Total cases numbered less than one-tenth the number of salmonella cases between 1980 and 2005. Reported outbreaks of *C. perfringens* food poisoning have declined in recent years from an average total of 28 per year in the period 1992-1998 to an average of 9 p.a. in the period 1999-2005. The total numbers of people made ill in these outbreaks are shown in Figure 7.5. These do not include sporadic cases which the Infectious Intestinal Disease Study in England recognised as being "quite common".

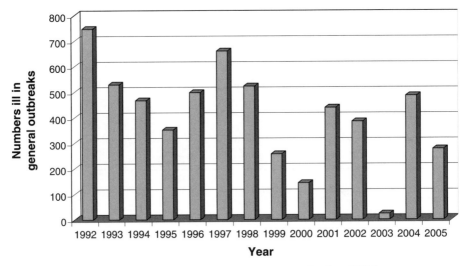

Figure 7.5   Clostridium perfringens *food poisoning in England and Wales*

7.6.2 The Organism and its Characteristics

Clostridium perfringens is a Gram-positive, rod-shaped anaerobe which forms oval subterminal spores. It differs from most other clostridia in that the relatively large rods (1×3–$9\,\mu m$) are encapsulated and non-motile. Though a catalase-negative anaerobe, *C. perfringens* will survive and occasionally grow in the presence of oxygen.

Growth occurs over the temperature range 12 to 50 °C although it is very slow below about 20 °C. At its temperature optimum, 43–47 °C, growth is extremely rapid with a generation time of only 7.1 min at 41 °C. Vegetative cells show no marked tolerance to acid (minimum pH 5, optimum 6.0–7.5), have a minimum a_w for growth of 0.95–0.97, depending on the humectant, and will not grow in the presence of 6% salt.

The heat resistance of vegetative cells is comparable to that of non-sporeforming bacteria with D values at 60 °C in beef of a few minutes D values of spores at 100 °C show a wide inter-strain variation with recorded values from 0.31 min to more than 38 min. This may, in part, be due to differences in the culture methods used since some workers included lysozyme in their media to improve the recovery of heat-damaged spores.

Distribution of type A *C. perfringens* is widespread in the environment. In soil, where it can be found at levels of 10^3–$10^4\,g^{-1}$, it persists much longer than types B,C,D, and E which are obligate animal parasites and of more limited distribution. It can be isolated from water, sediments, dust, raw and processed foods and is a common inhabitant of the human gastrointestinal tract. Spore counts of 10^3–$10^4\,g^{-1}$ are common in faeces from healthy individuals and surveys in Japan and the UK have found levels of up to 10^8–$10^9\,cfu\,g^{-1}$ in healthy elderly people in long-stay care.

7.6.3 Pathogenesis and Clinical Features

C. perfringens food poisoning is generally a self-limiting, non-febrile illness characterized by nausea, abdominal pain, diarrhoea and, less commonly vomiting. Onset is usually 8 to 24 h after consumption of food containing large numbers of the vegetative organism; the median count of *C. perfringens* in foods implicated in UK outbreaks is $7 \times 10^5\,g^{-1}$ and the required ingested dose has been variously estimated at 10^6–$10^8\,cfu$. In otherwise healthy individuals, medical treatment is not usually required and recovery is complete within 1–2 days, although occasional fatalities occur in the very old or debilitated.

Ingested vegetative cells that survive the stomach's acidity pass to the small intestine where they grow, sporulate and release an enterotoxin. The enterotoxin is synthesized by the sporulating cells, although low levels of production have been observed in vegetative cultures. The toxin is closely associated with the spore coat, but is not thought to be an important structural component, and is released into the intestinal lumen on lysis of the sporangium.

Toxin production can also occur *in vitro*. Low levels have been detected in foods, including a sample involved in an outbreak, and a few reported cases with incubation periods less than 2 h suggest that, on occasion, ingestion of pre-formed toxin may cause illness. This is generally held to be rare however.

The enterotoxin is a 35 kDa protein with an isoelectric point of 4.3. It is inactivated by heating in saline at 60 °C for 10 min and is sensitive to some proteolytic enzymes. It acts like cholera toxin by reversing the flow of Na^+,Cl^-, and water across the gut epithelium from absorption to secretion, though it does so by a different mechanism. Rather than increase the level of intracellular cyclic nucleotides, it acts at the cell membrane. It first binds to specific protein receptors on the epithelial cell and is then inserted into the cell membrane producing morphological changes within a few minutes. It changes cell permeability, inhibits synthesis of cell macromolecules, and produces pores in the membrane of the cell which eventually dies as a result of membrane damage. Illness tends to be relatively mild and short-lived since the toxin affects primarily the cells of the villus tip, which are the oldest intestinal cells and therefore replaced much sooner than younger cells.

Diagnosis of *C. perfringens* food poisoning is normally based on a number of factors:

(i) case history and symptoms;
(ii) large numbers ($> 10^6 g^{-1}$) of *C. perfringens* spores in the patient's faeces;
(iii) large numbers of vegetative cells of the same serotype in the incriminated food ($> 10^6 g^{-1}$);
(iv) presence of enterotoxin in faeces.

Faecal count data should be treated with some caution since the level of excretion can be very high in aged, healthy, institutionalized patients.

7.6.4 Isolation and Identification

In the investigation of outbreaks, enrichment culture is rarely necessary since *C. perfringens* will invariably be present in high numbers in implicated foods or clinical samples. Similarly, in routine quality assurance of foods, there is generally little value in being able to detect very low numbers in view of its ubiquity in the environment. In the examination of foods, the total count (vegetative cells plus spores) is determined but with faecal specimens a spore count, obtained after heating a suspension at 80 °C for 10 min, is also performed.

The most commonly employed selective plating media used to enumerate *C. perfringens* employ antibiotic(s) as the selective agent and sulfite reduction to produce black colonies as the differential reaction. The most popular combinations are tryptose/sulfite/cycloserine (TSC) medium and oleandomycin/polymyxin/sulfadiazine/perfringens (OPSP), incubated anaerobically for 24 h at 37 °C. A better diagnostic reaction is obtained if pour plates are used since colonies on the agar surface of spread plates can appear white.

Suspect colonies can be confirmed by the absence of motility, their ability to reduce nitrate to nitrite, lactose fermentation, and gelatin liquefaction. A traditional confirmatory test, the Nagler reaction, which looks for the production of α-toxin and its neutralization by antitoxin on lactose egg-yolk agar is less favoured now because of its lower specificity and the occurrence of lecithinase-negative strains.

Serotyping based on capsular antigens is employed for epidemiological purposes. Most isolates from outbreaks can be serotyped and this can be usefully supplemented by typing with bacteriocins, particularly when serotyping is not possible.

A number of methods are available for the detection of enterotoxin. Traditional biological tests such as the ligated ileal loop and mouse challenge have been superseded by more sensitive, rapid, convenient and humane serological techniques. A commercially available kit employs reverse passive latex agglutination. The name derives from the fact that in a standard agglutination assay, soluble antibody reacts with a particulate antigen such as bacterial cells. In a reversed passive latex agglutination assay, a soluble antigen reacts with antibody attached to latex particles. These play no part in the reaction and are therefore passive, but they do provide a visual signal when they cross-link as a result of the antigen–antibody reaction.

7.6.5 Association with Foods

For an outbreak of *C. perfringens* food poisoning, the typical scenario includes the following events:

(i) a meat dish containing spores of *C. perfringens* is cooked;
(ii) the spores survive the cooking to find themselves in a genial environment from which much of the competitive flora has been removed;
(iii) after cooking, the product is subjected to temperature/time abuse, such as slow cooling or prolonged storage at room temperature. This allows the spores to germinate and multiply rapidly to produce a large vegetative population;

(iv) the product is either served cold or reheated insufficiently to kill the vegetative cells. Some of the ingested cells survive through into the small intestine where they sporulate and produce enterotoxin.

From the above outline it is clear that *C. perfringens* food poisoning is more likely to occur where food is being prepared some time in advance of consumption and that adequate refrigeration is the key to its control.

Most cases (>70% in the United States and >87% in England and Wales) are associated with meat products such as stews, meat gravies, roast joints and pies. This is partly due to the frequent association of the organism with meats, but major contributory factors are the low redox potential, mode of preparation and consumption which can give *C. perfringens* the opportunity to multiply to dangerous levels.

Cured meats are rarely involved in *C. perfringens* food poisoning. This is a fine example of the hurdle concept in action (see Section 3.4); individual preservative factors such as salt content, nitrite level and heat processing are insufficient on their own to assure safety but effectively control growth of *C. perfringens* in combination.

Most outbreaks occur in connection with institutional catering such as schools, old people's homes and hospitals. The association between *C. perfringens* and hospital food in particular goes back a long way, to the outbreaks at St. Bartholomew's Hospital in the 1890s, but here at least there are signs of progress. In the UK, the 1991 Richmond Report on the safety of food noted that outbreaks in hospitals fell from a peak of more than 20 per year in the 1970s to about half this number in the 1980s: a decline attributed to improvements in facilities and staff training.

7.7 ENTEROBACTER SAKAZAKII

7.7.1 Introduction

Enterobacter sakazakii is an emerging opportunistic pathogen associated particularly with sporadic life threatening infections in low birth-weight infants, though it can cause disease in all age groups. The first recorded cases occurred in 1958 in Europe and the United States. Up until 2005 there had been about 75 cases of *Ent. sakazakii* infection reported worldwide. In many cases, particularly those reported up to the mid-1980s, the source of the infection was unknown but in more recent years contaminated infant milk formula has been recognised as the principal source of the organism in clinical infections.

7.7.2 The Organism and its Characteristics

Enterobacter sakazakii is a Gram-negative, motile member of the Enterobacteriaceae. It was originally classified as a strain of *Enterobacter cloacae* distinguishable from other members of the species by production of a yellow water diffusible pigment on tryptone soy agar. In 1980 it was renamed *Enterobacter sakazakii* in honour of the eminent Japanese bacterial taxonomist Riichi Sakazaki.

It is a typical mesophile and can grow between 6° and 47 °C. As with many species, its heat resistance varies between strains. Some workers have described the heat sensitivity as similar to other Enterobacteriaceae while others have reported slightly higher resistance. A typical published D value measured in reconstituted dried milk at 60 °C was 2.5 minutes with a z value of 5.8 °C. *Ent. sakazakii* appears to be relatively more resistant to low a_w stress than other Enterobacteriaceae and this may be a significant factor in its transmission.

7.7.3. Pathogenesis and Clinical Features

Ent. sakazakii is the recognised cause of severe infections in infants characterised by meningitis, cerebritis, bacteraemia and necrotising enterocolitis. Infection is associated with a high mortality rate of 50% or more and severe long term, irreversible sequelae occur in most survivors. These include quadriplegia and impaired sight or hearing. The most common predisposing factors for infection are low birth-weight or premature birth.

7.7.4. Isolation and Identification

The severity of the illness and an apparently low infectious dose have led to very stringent criteria applying to the presence/absence of *Ent. sakazakii* in powdered infant formulae. Thus cultural detection follows a similar scheme to *Salmonella* detection involving pre-enrichment, and selective enrichment stages prior to the use of selective agars. Agar media can be selective and diagnostic for Enterobacteriaceae in general followed by biochemical testing to confirm the identity of isolates, although selective media have also been developed to detect a key biochemical characteristic of the organism such as its ability to produce α-glucosidase.

A variety of molecular methods have also been developed for the identification and typing based on PCR and ribotyping, pulsed-field gel electrophoresis (PFGE) and random amplification of polymorphic DNA (RAPD)

7.7.5. Association with Foods

Ent. sakazakii appears widespread in the environment and has been isolated from water, soil and vegetation as well as the contents of household vacuum cleaners. Powdered infant formula foods have been established as a common cause of infection although occasional cases in adults and infants not exposed to powdered infant formula indicate that this is not invariably the vehicle.

Pasteurisation is an effective control measure; conventional pasteurisation conditions of 72 °C for 15 seconds would produce more than a 10 log reduction in the number of survivors, assuming a D_{60} of 2.5 minutes and $z = 5.8$ °C. It seems that the organism is most likely to enter the product as a result of post pasteurisation contamination.

Surveys of powdered infant formulae have shown contamination rates ranging between 0 and 14% but generally, when it occurs, levels of contamination are low ranging from 0.36 cfu/100g to 66 cfu/100g. It may be that the infectious dose is very low in the very vulnerable patients affected, but poor hygienic practices during reconstitution and prolonged storage of the reconstituted product allowing bacterial multiplication have been identified as significant risk factors.

7.8 *ESCHERICHIA COLI*

7.8.1 Introduction

Since 1885, when it was first isolated from childrens' faeces and described by the German bacteriologist Theodor Escherich, scientific attention has been lavished on *Escherichia coli* to such an extent that it is today probably the best understood free-living organism.

E. coli is an almost universal inhabitant of the gut of humans and other warm-blooded animals where it is the predominant facultative anaerobe though only a minor component of the total microflora. Generally a harmless commensal, it can be an opportunistic pathogen causing a number of infections such as Gram-negative sepsis, urinary tract infections, pneumonia in immunosuppressed patients, and meningitis in neonates. Its common occurrence in faeces, ready culturability, generally non-pathogenic character, and survival characteristics in water led to the adoption of *E. coli* as an indicator of faecal contamination and the possible presence of enteric pathogens such as *S.* Typhi in water. This usage has been transferred to foods where greater circumspection is required in interpreting the significance of positive results.

Strains of *E. coli* were first recognized as a cause of gastroenteritis by workers in England investigating summer diarrhoea in infants in the early 1940s. Until 1982, strains producing diarrhoea were classified into three types based on their virulence properties: enteropathogenic *E. coli* (EPEC), enteroinvasive *E. coli* (EIEC), and enterotoxigenic *E. coli* (ETEC). They are not very common causes of foodborne illness in developed countries, but an important cause of childhood diarrhoea in less developed countries.

ETEC is also frequently associated with so-called traveller's diarrhoea. However since 1982, enterohaemorrhagic *E. coli* (EHEC) particularly associated with serotype O157:H7 has been recognized as the cause of a number of outbreaks of haemorrhagic colitis and haemolytic uraemic syndrome, particularly in North America, where foods such as under-cooked ground meat, raw milk and fresh produce have been implicated. An exponential rise in isolations of O157:H7 was reported in Canada between 1982 and 1986 and a study in the UK between 1985 and 1988 suggested that the increased reporting of isolations there (118 in England and Wales and 86 in Scotland) represented a real increase. The number of cases in the UK continued to increase until 1997 and has fluctuated between 600 and 1000 isolations per year since then. Other European countries have also reported increased isolation rates.

Two further types of *E. coli* are recognized as causes of diarrhoea, primarily in children. Termed enteroaggregative *E. coli* (EaggEC) and diffusely adherent *E. coli* (DAEC), they have characteristic patterns of adherence to Hep-2 cells in culture. The emergence of these numerous pathotypes of *E. coli* is thought to reflect the plasticity of the organism's genome. The acquisition, loss or rearrangement of genetic elements introduces new pathogenicity and virulence characterisitics and the different pathotypes represent strains sharing common virulence determinants.

7.8.2 The Organism and its Characteristics

Escherichia is the type genus of the Enterobacteriaceae family and *E. coli* is the type species of the genus. It is a catalase-positive, oxidase-negative, fermentative, short, Gram-negative, non-sporing rod. Genetically, *E. coli* is very closely related to the genus *Shigella*, although characteristically it ferments the sugar lactose and is otherwise far more active biochemically than *Shigella* spp. Late lactose fermenting, non-motile, biochemically inert strains of *E. coli* can however be difficult to distinguish from *Shigella*.

E. coli can be differentiated from other members of the Enterobacteriaceae on the basis of a number of sugar-fermentation and other biochemical tests. Classically an important group of tests used for this purpose are known by the acronym IMViC. These tested for the ability to produce:

(i) indole from tryptophan (I);
(ii) sufficient acid to reduce the medium pH below 4.4, the break point of the indicator methyl red (M);
(iii) acetoin (acetylmethyl carbinol) (V); and
(iv) the ability to utilise citrate (C).

Table 7.7   *The IMViC tests*

	Indole	*Methyl Red*	*Voges Proskauer*	*Citrate*
Escherichia coli	+	+	−	−
Shigella	V	+	−	−
Salmonella Typhimurium	−	+	−	+
Citrobacter freundii	−	+	−	+
Klebsiella pneumoniae	−	−	+	+
Enterobacter aerogenes	−	−	+	+

Since production of mixed acids and acetoin are alternative pathways for the metabolism of pyruvate, most species of the Enterobacteriaceae are either VP positive or methyl red positive. In the IMViC tests, most strains of *E. coli* are indole and methyl red positive and VP and citrate negative (Table 7.7). The tests are still used for identification purposes but nowadays usually as part of the larger range of tests available in modern miniaturized test systems.

$\quad$*E. coli* is a typical mesophile growing from 7–10 °C up to 50 °C with an optimum around 37 °C, although there have been reports of some ETEC strains growing at temperatures as low as 4 °C. It shows no marked heat resistance, with a D value at 60 °C of the order of 0.1 min, and can survive refrigerated or frozen storage for extended periods. A near-neutral pH is optimal for growth but growth is possible down to pH 4.4 under otherwise optimal conditions. The minimum a_w for growth is 0.95.

$\quad$A serotyping scheme for *E. coli* based on lipopolysaccharide somatic O, flagellar H, and polysaccharide, capsular K antigens was proposed by Kauffman in the 1940s. As currently applied in the O:H system, principal serogroups are defined by O antigens and then subdivided into serotype on the basis of H antigens. Strains of each category of pathogenic *E. coli* tend to fall within certain O:H serotypes, so the scheme plays an important role in detecting pathogens as well as in epidemiological investigations.

7.8.3   Pathogenesis and Clinical Features

There are four major categories of diarrhoeagenic *E. coli* based on distinct, virulence properties.

7.8.3.1   Enterotoxigenic E. coli *(ETEC).*   Illness caused by ETEC usually occurs between 12 and 36 h after ingestion of the organism. Symptoms can range from a mild afebrile diarrhoea to a severe cholera-like syndrome of watery stools without blood or mucus, stomach pains and vomiting. The illness is usually self-limiting, persisting for 2–3 days, although in developing countries it is a common cause of infantile diarrhoea where it can cause serious dehydration.

The ingested organism resists expulsion from the small intestine with the rapidly flowing chyme by adhering to the epithelium through attachment or colonization factors in the form of fimbriae on the bacterial cell surface. These can have different morphology and be either rigid (6–7 nm diameter) or flexible (2–3 nm diameter) structures composed of 14–22 kDa protein subunits. They are mannose resistant, *i.e.* they mediate haemagglutination in the presence of mannose, and particular colonization fimbriae are restricted to certain O:H serotypes. They are encoded on plasmids which frequently also encode for the diarrhoeagenic toxins.

Two toxin types are produced: the heat-stable toxins (ST), which can withstand heating at 100 °C for 15 min and are acid resistant, and the heat-labile toxins (LT) which are inactivated at 60 °C after 30 min and at low pH. LTI bears a strong similarity to cholera toxin; it consists of five B subunits (M_r 11.5 kDa) which are responsible for binding of the toxin to the epithelial cells and an A subunit (M_r 25 kDa) which is translocated into the epithelial cell where it activates adenylate cyclase. The subsequent increase in cAMP levels then inhibits Na^+, Cl^- and water absorption by the villus cells and stimulates their loss from intestinal crypt cells thus leading to profuse watery diarrhoea. LTII toxin produced by certain ETEC strains has similar biological activity to LTI but does not cross react with antiserum to LTI or cholera toxin.

Two types of ST have been recognized; the most common, ST_A, is a low molecular weight, poorly antigenic polypeptide of less than 20 amino acids produced from a 72 amino acid precursor. Its resistance to heat, low pH and proteolytic digestion probably derive from its compact three-dimensional structure which contains at least 3 disulfide linkages. It acts by stimulating the production of cGMP by guanylate cyclase in epithelial cells. The mechanism of action of ST_B, which can be distinguished from ST_A by its inability to produce fluid secretion in the intestines of suckling mice, is not known but does not appear to operate through the stimulation of cyclic nucleotide production.

7.8.3.2 Enteroinvasive E. coli *(EIEC).* Infection by EIEC results in the classical symptoms of an invasive bacillary dysentery normally associated with *Shigella*. Like *Shigella*, EIEC invades and multiplies within the epithelial cells of the colon causing ulceration and inflammation, though EIEC strains do not produce Shiga toxin. Clinical features are fever, severe abdominal pains, malaise and often a watery diarrhoea which precedes the passage of stools containing blood, mucus, and faecal leukocytes. Invasiveness is determined by a number of outer membrane proteins which are encoded for on a large plasmid ($\approx$ 140 MDa). The infective dose of EIEC appears to be substantially higher than for *Shigella* and this is thought to be a reflection of the organism's greater sensitivity to gastric acidity.

7.8.3.3   Enteropathogenic E. coli *(EPEC)*.   When the properties of ETEC and EIEC were established it was noted that these strains were rarely of the same serotypes first associated with *E. coli* diarrhoea in the 1950s. Subsequent investigation of some of these earlier strains in most cases failed to demonstrate the property of enteroinvasiveness or the ability to produce ST or LT and yet they retained the ability to cause diarrhoea in volunteers.

Symptoms of EPEC infection, malaise, vomiting and diarrhoea with stools containing mucus but rarely blood, appear 12–36 h after ingestion of the organism. In infants, the illness is more severe than many other diarrhoeal infections and can persist for longer than two weeks in some cases. Pathogenesis is related to the ability of EPEC strains to adhere closely to the enterocyte membrane and produce the so-called attaching and effacing lesions. This is a complex and fascinating process mediated by the genes encoded on a 35 kb pathogenicity island called the locus of enterocyte effacement (LEA). Binding to the enterocytes occurs in three stages: non-intimate association mediated by pili, attachment or signal transduction, and then intimate contact. During this process the bacteria facilitate their own binding by producing a series of changes in the underlying enterocytes. A bacterial type III secretion system translocates another LEA encoded protein, Tir, into the enterocyte where it is incorporated into the cell's membrane. There it acts as a receptor for an outer membrane bacterial protein, intimin, which mediates close contact. The attachment stage is accompanied by increased levels of intracellular Ca^{2+}, release of inositol phosphates and activation of tyrosine kinase, an enzyme which phosphorylates tyrosine residues on intracelluar proteins. Following this the enterocytes accumulate filamentous actin as they form pedestal-like surface structures on which the bacteria rest. This results in deformation and loss of some microvilli; events which are thought to cause diarrhoea by disrupting the balance between absorption and secretion in the small intestine.

7.8.3.4   Enterohaemorrhagic E. coli *(EHEC)*.   EHEC, sometimes also known as Verotoxin-producing *E. coli* (VTEC), was first described in Canada where in some areas it rivals *Campylobacter* and *Salmonella* as the most frequent cause of diarrhoea. *E. coli* O157:H7 is the most common EHEC serotype reported, although others do occur. Non-motile (H negative) O111 and O157 are more common in Australia for example. EHEC has attracted attention not only because foodborne transmission is more common than with other diarrhoeagenic *E. coli*, but because the illness it causes can range from a non-bloody diarrhoea, through haemorrhagic colitis, to the life threatening conditions haemolytic uraemic syndrome (HUS) and thrombotic thrombocytopaenic purpura (TTP).

Haemorrhagic colitis is typically a self-limiting, acute, bloody diarrhoea lasting 4–10 days. Symptoms start with stomach cramps and watery diarrhoea 1–2 (sometimes 3–8) days after eating the contaminated food and, in most cases, progress over the next 1–2 days to a bloody diarrhoea with severe abdominal pain. It can be distinguished from inflammatory colitis by the usual lack of fever and absence of leukocytes in the stools. It affects mainly adults, with a peak incidence in the summer months, and can be life-threatening in the elderly.

Haemolytic uraemic syndrome is characterized by three features, acute renal failure, haemolytic anaemia (reduction in the number of red blood cells) and thrombocytopaenia (a drop in the number of blood platelets), sometimes preceded by a bloody diarrhoea. It is most common in children among whom it is the leading cause of acute renal failure in western Europe and North America. Approximately 10% of children under 10 with symptomatic *E. coli* O157 infection go on to develop HUS; half will require kidney dialysis and the mortality rate is generally 3–5%. In 70 cases seen in London between 1980 and 1986 the fatality rate was 6%, with 13% of cases showing some long-term kidney-damage. In one outbreak in a North American nursing home, the fatality rate among the 55 affected residents was 31%.

Thrombotic thrombocytopaenic purpura is a less common complication which is largely confined to adults. It is related to HUS but causes less kidney damage and includes fever and neurological symptoms resulting from blood clots in the brain.

Attachment is an important factor in virulence and O157:H7 strains possess the LEA pathogenicity island and adhere by a mechanism similar to EPEC, characterised by intimate attachment of the bacteria to the epithelial cells and effacement of the underlying microvilli.

EHEC strains produce the cytotoxin Verotoxin (so-called because of its ability to kill Vero (African Green Monkey Kidney) cells). Studies have revealed the presence of at least two toxins VTI and VTII which because of their similarity to Shiga toxin (see Section 6.6) have also been called Shiga-like toxins, SLTI and SLTII. It has been proposed that the nomenclature for these toxins be rationalised as Shiga family toxins so that the prototype toxin Shiga toxin is designated Stx, and SLTI and II become Stx 1 and Stx 2 respectively. Stx 1 bears the closest resemblance to Shiga toxin; it cross reacts with antisera to Shiga toxin, is also composed of A (M_r 32 kDa) and B (M_r 7.7 kDa) subunits and the B units are structurally identical. The B units bind specifically to the glycolipid receptor, globotriaosylceramide (Gb_3), on the eukaryotic cell surface and the susceptibility of the kidney in O157 infections may be due to higher levels of these receptors in kidney glomeruli. Following binding, the toxin is internalized by endocytosis and the A subunit activated. It then hydrolyses the *N*-glycosidic bond of a specific adenosine residue in 28S rRNA thus stopping protein synthesis in the cell. Stx 2 also comprises an A and B subunit but these are larger than in Stx 1 (M_r 35 kDa and 10.7 kDa respectively) and do not cross-react immunologically, though they do share a 60% amino acid sequence homology, with Shiga toxin. Both toxins have been shown to be phage encoded in a number of strains.

7.8.4 Isolation and Identification

Selective techniques for *E. coli* mostly exploit the organism's tolerance of bile and other surfactive compounds, a consequence of its natural habitat, the gut. Aniline dyes and the ability of many strains to grow at temperatures around 44 °C are also used as selective agents.

The first selective and differential medium was that originally devised by MacConkey in 1905. It has been variously modified since but its essential characterics have remained unchanged. Bile salts (and sometimes the aniline dye, crystal violet) act as inhibitors of Gram-positive and some fastidious Gram-negative bacteria. Lactose is included as a fermentable carbohydrate with a pH indicator, usually neutral red. Strong acid producers like *Escherichia, Klebsiella*, and *Enterobacter* produce red colonies, non-lactose fermenters such as *Salmonella, Proteus*, and *Edwardsiella*, with rare exceptions produce colourless colonies. MacConkey agar is not however strongly selective and will support the growth of a number of non-Enterobacteriaceae including Gram-positives such as enterococci and staphylococci.

Eosin/methylene blue agar is a popular selective and differential medium in North America. The aniline dyes eosin and methylene blue are the selective agents but also serve as an indicator for lactose fermentation by forming a precipitate at low pH. Strong lactose fermenters produce green-black colonies with a metallic sheen.

A biochemical feature of *E. coli* increasingly being used in diagnostic media is β-glucuronidase activity, which is possessed by around 95% of *E. coli* strains but by only a limited number of other bacteria. A fluorogenic or chromogenic glucuronide is incorporated into a conventional medium and enzyme activity detected by the production of colour or fluorescence. Most widely used is the fluorogen 4-methylumbelliferyl-β-D-glucuronide (MUG) which is hydrolysed to produce fluorescent 4-methylumbelliferone.

Suspect colonies from selective and differential media can be confirmed by further biochemical testing.

Detection of *E. coli* O157:H7 is based on phenotypic differences from most other serotypes: its inability to ferment sorbitol on MacConkey sorbitol agar and absence of β-glucuronidase activity in most strains. Presumptive *E. coli* O157:H7 from these tests must then be confirmed serologically for which a latex agglutination kit is commercially available.

Identification of diarrhoeagenic *E. coli* can be based on detection of their associated virulence factors. For example, procedures are available to detect the ST and LT of ETEC serologically, and the *LTI* and *Stx* genes in ETEC and EHEC using gene probes and the polymerase chain reaction (PCR).

7.8.5 Association with Foods

Faecal contamination of water supplies and contaminated food handlers have been most frequently implicated in outbreaks caused by EPEC, EIEC and ETEC. A number of foods have been involved, including a coffee substitute in Romania in 1961, vegetables, potato salad, and sushi. In the United States, mould-ripened soft cheeses have been responsible for outbreaks in 1971, associated with EIEC in which more than 387 people were affected, and in 1983, caused by ETEC (ST). *E. coli* would not be expected to survive well in a fermented dairy product with a pH below 5 but, where contamination is associated with mould-ripening, the local increase in pH as a result of lactate utilization and amine production by the mould would allow the organism to grow.

Outbreaks caused by EHEC serotype O157:H7 have mostly involved undercooked ground meat products and occasionally raw milk. Cattle seem to be an important reservoir of infection and O157:H7 has been isolated from 0.9–8.2% of healthy cattle in the UK. Other surveys have isolated the serotype from 3.7% (6/164) samples of retailed fresh beef and a significant percentage (1–2%) of other fresh meat products such as pork, poultry and lamb.

There have been a number of very large outbreaks around the world and their public impact has often been dramatic. Six hundred people became ill and four children died in a major US outbreak in 1993 caused by undercooked beef hamburgers. This caused a major public outcry over meat hygiene and resulted in, amongst other things, the introduction of new meat-labelling regulations.

In August 1997, a cluster of cases in Colorado prompted the largest food recall in US history when more than 12 000 tons of ground beef were recalled.

A large outbreak in Scotland in 1996 had a similar impact in the UK. Nearly 500 were affected and 20 elderly patients died. The cause was thought to be cross-contamination of cooked meats from raw meat in a butcher's shop and the resultant enquiry produced a tightening of regulation.

The failures that led to these outbreaks were generally simple breakdowns of basic food hygiene. With both raw milk and ground beef products, the primary cause has been a failure to heat process/cook the products adequately. While it is true that intact cuts of meat such as steaks can often be consumed safely when the interior is undercooked, this is because microbial contamination is usually a surface phenomenon. Comminution of the meat, however, will mix surface contaminants into the middle of products and they will therefore need cooking throughout to ensure microbial safety. The USDA has produced regulations specifying that the centre of beef hamburgers should reach on cooking: 71.1 °C (160 °F) instantaneously for consumers and 68.3 °C (155 °F) for 16 seconds in food service operations.

Outbreaks of EHEC have been reported with other foods. Lettuce has been associated on several occasions and unpasteurised apple juice was the vehicle in a large outbreak in the US. In the summer of 1996, an epidemic in Japan involved over 9000 cases and 12 deaths in children. The largest outbreak during the epidemic, in Sakai City, involved 5700 people and was associated with contaminated radish sprouts, and the same vehicle was implicated in a further outbreak the following year. Alfalfa sprouts were also implicated in an outbreak in the US.

Outbreaks caused by acidic foods such as apple juice and fermented sausages, and laboratory studies with mayonnaise, remind us of the potential for bacteria to survive for prolonged periods at pH values that do not permit growth, particularly when the product is refrigerated. EHEC does appear to have a more marked ability to survive at low pH values than some other bacteria and this may also account for the relatively low infectious dose, 2–2000 cells, recorded in outbreaks.

7.9 *LISTERIA MONOCYTOGENES*

7.9.1 Introduction

L. monocytogenes is the only important human pathogen among the six species currently recognized within the genus *Listeria*, although *L. seeligeri*, *L. welshimeri*, and *L. ivanovii* have occasionally been associated with human illness. It was first described by Murray in 1926 as *Bacterium monocytogenes*, the cause of an infection of laboratory rabbits where it was associated with peripheral blood monocytes as an intracellular pathogen, and it has since been established as both an animal and human pathogen. As an important veterinary problem, it causes two main forms of disease: a meningoencephalitis most common in adult ruminants such as sheep and cattle, and a visceral form more common in monogastrics and young ruminants which attacks organs other than the brain causing stillbirth, abortion and septicaemia. Listeriosis in sheep increased in Britain from 86 recorded incidents in 1979 to 423 in 1988. This was partly due to the increased size of the national flock over that period but has also been attributed to changes in silage-making techniques towards greater use of big-bale ensilage. In big-bale ensilage, the silage is made in large plastic bags rather than in a single large clamp. Rupture of the bag or inadequate sealing at its neck can allow mould growth to occur on the lactic acid present increasing the silage pH to a value at which *L. monocytogenes* can flourish.

Human listeriosis is described in more detail in Section 7.9.3 below. The recent widespread concern it has caused is largely attributable to the realization that food is a major source of the infection (a possibility first suggested as long ago as 1927), the psychrotrophic character of the organism, and the high mortality rate of the illness. Reported incidence of human listeriosis increased in several countries during the 1980s, but remains generally low when compared to other foodborne infections such as salmonellosis. For example, in England and Wales reported cases of listeriosis peaked in the late 1980s at around 300 per year while reports of *Salmonella* and *Campylobacter* infections numbered nearly 27 500 and 29 000 respectively. Reported cases of listeriosis dropped in 1990 and 1991 to 118 and 131 respectively; a decrease attributed to the effect of Department of Health advice to the immuno compromised and pregnant to avoid soft cheeses and to reheat certain chilled foods adequately and to withdrawal of contaminated paté from a single manufacturer (Figure 7.6). Reported incidence of human listeriois in England and Wales remained at around 100 cases p.a. until 2001 when numbers increased to more than 200 cases p.a. in 2003 and 2004. Unlike the earlier peak in 1988, this was marked by an increase in non-pregnancy associated cases. In the United States, the Center for Disease Control (CDC) has estimated an annual incidence of around 1700 cases resulting in 450 adult deaths and 100 foetal and postnatal deaths.

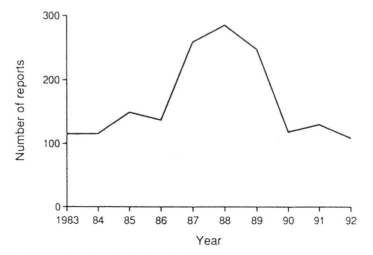

Figure 7.6 *Human listeriosis in England and Wales*

7.9.2　The Organism and its Characteristics

L. monocytogenes is a Gram-positive, facultatively anaerobic, catalase-positive, oxidase-negative, non-sporeformer. The coccoid to rod shaped cells (0.4–$0.5\,\mu m \times 0.5$–$2.0\,\mu m$) cultured at 20–$25\,°C$ possess peritrichous flagella and exhibit a characterisitic tumbling motility. Colonies on tryptose agar viewed under oblique illumination have a characteristic blue–green sheen.

　　L. monocytogenes elaborates a $58\,kDa$ β-haemolysin, listerolysin O, which acts synergistically with the haemolysin produced by *Staphylococcus aureus* to give enhanced haemolysis on blood agar. This reaction forms the basis of a useful diagnostic test to distinguish *L. monocytogenes* from *L. innocua*, and is known as the CAMP test after Christie, Atkins, and Munch-Peterson who first described the phenomenon with group B streptococci.

　　L. monocytogenes will grow over a wide range of temperature from 0–$42\,°C$ with an optimum between 30 and $35\,°C$. Below about $5\,°C$ growth is extremely slow with lag times of 1 to 33 days and generation times from 13 to more than $130\,h$ being recorded. The thermal survival characteristics of *L. monocytogenes* have received considerable attention following an outbreak in the United States associated with pasteurized milk and the suggestion that the organism could survive commercial pasteurization conditions. Despite some conflicting data in the literature, it appears that the heat resistance of *L. monocytogenes* is similar to that of other non-sporeforming Gram-positives with a typical D_{60} of a few minutes and a D_{70} of a few seconds. It was proposed that *L. monocytogenes* cells in contaminated milk are protected from heat by their intracellular location within milk leucocytes but subsequent studies have failed to demonstrate any significant effect.

　　Models of the thermal inactivation of *L. monocytogenes* in milk have indicated that conventional HTST pasteurization achieves a reduction of 5.2 log cycles in the number of survivors; an acceptable safety margin assuming low numbers of the organism present on the incoming milk.

　　Growth of all strains is inhibited at pH values below 5.5 but the minimum growth pH is dependent on both strain and acidulant and has been variously reported as between 5.6 and 4.4. *L. monocytogenes* is also quite salt tolerant being able to grow in 10% sodium chloride and survive for a year in 16% NaCl at pH 6.0.

　　The organism is ubiquitous in the environment. It has been isolated from fresh and salt water, soil, sewage sludge, decaying vegetation, and silage. Its prolonged survival in the environment has been demonstrated in one study where the level of *L. monocytogenes* in sewage sludge sprayed on to agricultural land remained unchanged for more than 8 weeks. Asymptomatic human and animal carriage is also common with reports of isolation of the organism from the faeces of, among others, cattle, pigs, sheep, chickens, turkeys, ducks, crustaceans, flies and ticks. In a study of faecal carriage in human population groups, it was isolated from 4.8% of healthy slaughterhouse workers, 1.2% of hospitalized adults, 1% of patients with diarrhoea, and 26% of household contacts of listeriosis patients.

7.9.3 Pathogenesis and Clinical Features

Its ubiquity in the environment suggests that human exposure to *L. monocytogenes* must be frequent. Incidence of infection is however low since invasive infection will result only if a susceptible individual is exposed to a sufficiently high dose of a virulent strain.

Estimates of the minimum infective dose are always fraught with difficulty and this is certainly the case with *L. monocytogenes*. It is thought to be relatively high since foods implicated in outbreaks have been found to contain numbers in excess of 10^3 cfu g^{-1}.

Incubation periods for the disease have varied from 1 day to as long as 90 days with a typical incubation period of a few weeks; a situation which makes the identification of food vehicles difficult if not often impossible.

Symptoms of the disease, which is most likely to develop in pregnant women, the very young or elderly and the immunocompromised, can vary from a mild, flu-like illness to meningitis and meningoencephalitis.

In pregnant women, it most commonly features as an influenza-like illness with fever, headache and occasional gastrointestinal symptoms, but there may be an associated transplacental foetal infection which can result in abortion, stillbirth, or premature labour.

Listeriosis in the newborn can be an early-onset syndrome, which occurs at birth or shortly afterwards, or a late-onset disease appearing several days to weeks after birth. Early-onset illness results from *in utero* infection, possibly through the aspiration of infected amniotic fluid, and is characterized by pneumonia, septicaemia and widely disseminated granulomas (abscesses). Meningitis is rare.

In the late-onset syndrome, meningitis is more common, 93% (39 of 42) of late-onset cases in Britain between 1967 and 1985 had evidence of infection of the central nervous system. Infection may occur from the mother during passage through the birth canal, but some may also be acquired after delivery. A study in the UK found a lower mortality rate for late-onset disease (26%) than for early-onset listeriosis (38%).

Listeriosis in non-pregnant adults is usually characterized by septicaemia, meningitis and meningoencephalitis, but can also include endocarditis. It is particularly associated with those with an underlying condition which leads to suppression of their T cell mediated immunity, so that malignancies or immunosuppression (after renal transplantation, for example) are often predisposing factors. Although not a common infection in AIDS patients, its incidence is around 300 times that in the general population. Other conditions such as alcoholism, diabetes and cirrhosis can act as predisposing factors, but illness does occur in otherwise healthy individuals who account for about 18% of adult cases in England and Wales.

Adult listeriosis has a high mortality rate, figures calculated using data for 1989 gave values around the world of between 13 and 34%. Early treatment with antibiotics, normally ampicillin, with or without an aminoglycoside, or chloramphenicol, is essential but in the most severe forms, the prognosis remains poor.

L. monocytogenes is a facultative intracellular pathogen which like *Mycobacterium, Brucella,* and others can survive and multiply in cells of the monocyte–macrophage system. The organism penetrates the gut either by crossing the Peyer's patches or by invading enterocytes. This process of endocytosis is promoted by a number of virulence factors including internalin, an 800 amino acid bacterial surface protein encoded by a chromosomal gene *inlA*, other products of the so-called *inl* family of genes, and p60, a 60 kDa extracellular protein. Internalization results in the bacteria being enclosed in a phagosome. In order to multiply intracellularly the organism must survive in the phagosome and escape rapidly before it fuses with a lysosome, a process which would kill the bacteria. Pathogenic *L. monocytogenes* produce listeriolysin O (LLO) a 58 kDa haemolysin which breaks down the lipid bilayer of the phagosomal membrane allowing the bacteria to escape from the phagosome. Most *L. monocytogenes* infecting epithelial-type cells do this and are released into the cytoplasm. Only 10% of those invading monocytes are successful, the remaining 90% being destroyed.

During intracellullar replication, actin polymerizes around the bacterial surface, a process which propels the bacterial cell around the host cell and into adjacent cells, thereby spreading the organism while avoiding the host's immune system. The bacterial cells reach the mesenteric lymph nodes and are disseminated around the body *via* the blood. The liver plays an important role in eliminating the organism and controlling the infection. Infection of hepatocytes causes an intense inflammatory reaction. Polymorphonuclear cells (neutrophils) destroy infected hepatocytes forcing them to release the bacteria which are in their turn destroyed. Where infection is not controlled in the liver, it can be further disseminated by the blood to the central nervous system or placenta causing more severe illness.

7.9.4   Isolation and Identification

Low-temperature enrichment at 4 °C is the traditional technique for isolating *L. monocytogenes* from environmental samples, but the increased interest in routine isolation of the organism from foods has led to its replacement by more rapid, selective enrichment procedures based on antibiotic cocktails as selective agents and incubation at near-optimal growth temperatures.

Selective agars have likewise relied on a combination of selective agents such as lithium chloride, phenylethanol and glycine anhydride and antibiotics. Identification of presumptive *Listeria* colonies was based on microscopic examination of plates illuminated from below at an incident angle of 45° (Henry illumination), when they appear blue–grey to blue–green. Some media avoid the use of this technique by incorporating aesculin and ferric ammonium citrate so that *Listeria* colonies appear dark brown or black as a result of their ability to hydrolyse aesculin.

Confirmation of *L. monocytogenes* requires further biochemical testing including sugar-fermentation tests to distinguish it from other *Listeria* species and, in particular, the CAMP test to differentiate *L. monocytogenes* from *L. innocua*. Specific miniaturized test kits have been produced to simplify this procedure including one which replaces the CAMP test, which is not always easy for the inexperienced to interpret, with one for acrylamidase activity (*L. monocytogenes*, negative; *L. innocua*, positive). Enzyme-linked immunosorbent assay (ELISA) and gene probe kits are also available.

L. monocytogenes may be serotyped according to a scheme based on somatic and flagellar antigens. This is of limited epidemiological value since the majority of human cases of listeriosis are caused by just three of the thirteen serotypes identified (1/2a, 1/2b, and 4b). Phage typing and molecular typing techniques can however be used to assist in epidemiological investigations.

7.9.5 Association with Foods

Its widespread distribution in the environment and its ability to grow on most non-acid foods offer *L. monocytogenes* plenty of opportunity to enter the food chain and multiply.

The transmission of listeriosis by food was first convincingly demonstrated in an outbreak that occurred in the Maritime Provinces of Canada in 1981. The outbreak involved 41 cases in all. Of the 34 perinatal cases, there were 9 stillbirths, 23 neonatal cases with a mortality rate of 27%, and 2 live births of healthy infants. The mortality rate in adult cases was 28.6%. Coleslaw was implicated as the result of a case control study and *L. monocytogenes* serotype 4b (the outbreak strain) was isolated from a sample of coleslaw in a patient's refrigerator. It was not possible to isolate the organism at the manufacturer's plant but it transpired that a farmer who supplied cabbages to the manufacturer also kept sheep, two of whom had died of listeriosis. The cabbage had been grown in fields fertilized by fresh and composted manure from the sheep and the harvested cabbages had been stored in a large shed through the winter – factors thought to account for the introduction of the organism and its multiplication to dangerous levels.

Raw vegetables, in the form of a garnish containing celery, tomatoes and lettuce, were also implicated on epidemiological grounds in an outbreak that occurred in eight Boston hospitals in 1979.

Surveys in the UK, the United States, Australia and elsewhere have reported a high frequency of isolation of *L. moncytogenes* from meats and meat products, where serotype 1 generally predominates. A number of sporadic cases of listeriosis have been associated with products such as pork sausage, turkey frankfurters, cook-chill chicken, and chicken nuggets.

L. monocytogenes is relatively resistant to curing ingredients and has been found in a range of delicatessen meats such as salami, ham, corned beef, brawn and paté. In an Australian survey 13.2% of samples were found to be positive, largely as a result of cross-contamination in the shop. In Britain in 1989/90, high levels on vacuum-packed ham and on paté, from which serotype 4b was isolated, prompted the recall of both products from the market. Pork tongues in aspic were identified as the original source of a large outbreak in France caused by serotype 4b. Between March and December 1992, 279 cases were reported with 63 deaths and 22 abortions.

Dairy products such as raw and pasteurized milk and soft cheeses have been associated with a number of major outbreaks of listeriosis. The overall incidence of *L. monocytogenes* in raw milk derived from surveys in Australasia, Europe and the United States averages at around 2.2%, although one Spanish study reported an incidence in excess of 45%. Pasteurized milk was responsible for an outbreak in Massachusetts in 1983 involving 42 adult and 7 perinatal cases with an overall mortality rate of 29%. The milk had come from farms where bovine listeriosis is known to have occurred at the time of the outbreak. It was the absence of evidence of improper pasteurization at the dairy that gave rise to the concern that *L. monocytogenes* might display marked heat resistance in some instances (see Section 7.9.2 above).

Soft cheeses are also frequently contaminated with *L. monocytogenes*. In 1985 there was an outbreak in California in which a Mexican-style soft cheese which had been contaminated with raw milk was the vehicle. One hundred and forty-two cases were recorded comprising 93 perinatal and 49 adult cases with an overall mortality rate of 34%. This outbreak served to focus attention on soft cheeses and there have since been other incidents identified in which they have been implicated, including a major outbreak covering the period 1983–87 with 122 cases and 31 deaths associated with the Swiss cheese Vacherin Mont d'Or.

This association with soft cheeses appears to be due to the cheese ripening process. *L. monocytogenes* survives poorly in unripened soft cheeses such as cottage cheese but well in products such a Camembert and Brie. During the ripening process, microbial utilization of lactate and release of amines increase the surface pH allowing *Listeria* to multiply to dangerous levels. There have also been two European outbreaks of listeriosis in 1998 and 2003 and a major product recall in the United States in 2004 associated with butter, hitherto considered a relatively low risk food.

7.10 MYCOBACTERIUM SPECIES

7.10.1 Introduction

The genus *Mycobacterium* consists largely of harmless environmental organisms but is best known as the cause of two of the most feared and ancient of human diseases, tuberculosis (TB) and leprosy. TB, described by John Bunyan as 'Captain of these men of death', can sometimes be foodborne and is therefore of more concern to us here.

There is archaeological evidence to suggest that TB was endemic in much of the world from ancient times but with the rise in urbanisation between the 18th and 20th centuries it became epidemic in many areas, killing millions. Death rates in Europe and the United States peaked in the 19th century when it has been estimated that 30% of all deaths under the age of 50 in Europe were due to TB. By the late 20th century, a combination of improved social conditions, childhood immunization, screening and effective chemotherapy had reduced the incidence of TB in the developed world to the point where public health officials talked confidently of eliminating the disease altogether. This optimism proved unfounded as we have seen increasing numbers of cases since the late 1980s in groups such as AIDS patients and the socially disadvantaged, as well as the emergence of drug resistant strains. In the world's poorer countries, tuberculosis has always remained an important cause of morbidity and mortality. In 1990 the WHO and International Union against Tuberculosis and Lung Disease estimated that one-third of the world's population was infected with the tubercle bacillus and there were 7–8 million new cases each year.

Human illness is primarily associated with *Mycobacterium tuberculosis* which is thought to account for 98% of cases of pulmonary TB and 70% of non-pulmonary forms. It is spread person to person by aerial transmission of droplets produced by an infected person coughing, sneezing or spitting. *Mycobacterium bovis* is very closely related to *Myco. tuberculosis* but causes tuberculosis in cattle and other animals as well as in humans. It too is spread by respiratory aerosols between animals, and from animals to humans, but can also be transmitted to humans by milk and, to a lesser extent, by meat from tuberculous animals.

Mycobacterium paratuberculosis causes paratuberculosis, otherwise known as Johne's disease, in cattle and it has been suggested that it may be implicated in the etiology of Crohn's disease in humans. This remains to be established, but if so, consumption of infected milk may be a possible route of transmission.

7.10.2　The Organism and its Characteristics

Mycobacterium species are generally non-fastidious, Gram-positive, non-sporeforming, pleomorphic aerobes 1–4 µm in length. *Myco. bovis* is mesophilic and is not heat-resistant, being readily killed by normal milk pasteurization conditions.

A special feature of mycobacteria is the chemical composition of their cell walls. These have a high lipid content made up of esterified mycolic acids, complex branched-chain, hydroxy lipids with the general formula $R^1CHOH.CHR^2.COOH$ (where R^1 and R^2 are very long aliphatic chains), and, as a result, the wall is very hydrophobic and waxy. This confers a number of important properties on the organisms. For example, uptake of nutrients from aqueous solution is impeded making them very slow growing so that it often takes more than a week for growth to be apparent on solid media. They are also very resistant to drying and therefore can persist and remain infectious in the environment for long periods. The cell wall is more resistant to degradation by lysosomal enzymes in phagocytes enabling the pathogenic mycobacteria to survive and grow in macrophages. Its hydrophobic nature also makes the cells rather difficult to stain. However, once stained they are very resistant to decolourization and have the characteristic diagnostic property of 'acid fastness'. This was first noted by Ehrlich in 1882 and is detected using the Ziehl–Neelsen staining procedure in which cells are stained with hot carbol fuchsin, and mycobacteria, if present, will resist subsequent decolourization with acid alcohol.

7.10.3　Pathogenesis and Clinical Features

Most forms of tuberculosis are chronic taking months or even years before recovery or death. The commonest clinical signs include fever, chills and weight loss, but other symptoms present depending on the organs involved. The tissue damage produced is not a direct result of microbial activity as the infecting organism does not produce toxins, but is a consequence of the body's immune response to the organism.

In foodborne tuberculosis, *M. bovis* enters the body through the intestinal tract and the primary infection usually occurs at the mesenteric lymph nodes. The bacteria are engulfed by macrophages and are then isolated in nodules called tubercles or granulomas which are mainly composed of a dense accumulation of activated macrophages and lymphocytes. For many people this is as far as the infection proceeds, the development of the tubercle is checked by surrounding it with a fibrous wall and it then calcifies to a yellow gritty mass. In others, however, illness ensues when the tubercle liquefies causing local tissue necrosis and releasing the bacteria to spread infection around the body.

7.10.4 Isolation and Identification

Tuberculous lesions can be identified in meat animals by post-mortem inspection of carcasses but disease can also be identified in the live animal (and humans) using the tuberculin test. In this, the animal exhibits delayed hypersensitivity to injection of tuberculin, a protein preparation from *Myco. bovis*. In clinical specimens mycobacteria can be identified directly on the basis of their acid-fast reaction in the Ziehl–Neelson stain when the organisms appear red and the surrounding tissue blue. The organisms can be cultured on simple media but are very slow growing.

7.10.5 Association with Foods

In 1900 at the London Congress on Tuberculosis, Robert Koch caused consternation when he concluded that the risk of transmission of bovine tuberculosis to humans was so slight that he did not deem it advisable to take any measures against it. The impact of this from the world's leading bacteriologist and discoverer of the tubercle bacillus can only be imagined – especially on John McFadyean who was due to talk on the same subject two days later! Nonetheless, when his turn came he felt compelled to 'offer some criticism on the pronouncement of one, the latchet of whose shoes I am not worthy to unloose'. He pointed out that in post-mortem examinations of hundreds of children in London and Edinburgh, primary infection appeared to have occurred through the intestines in approximately 28% of cases, and that 2% of all cows in Britain had tuberculosis of the udder and were excreting the bacillus in their milk.

Pooling of milk increased the incidence of *Myco. bovis* so that in the 1920s and 1930s the organism could be isolated from 5–12% of milk samples and the high rates of TB in children due to *Myco. bovis* were attributed to the consumption of unpasteurized milk. In North America, compulsory pasteurization regulations were introduced in a number of the large cities from about 1910 and had a marked effect reducing the incidence of bovine TB in children. In the UK there was considerable resistance to the introduction of milk pasteurization, but the available evidence suggests that its later and more gradual introduction had a similar effect. By 1944, all London's milk was pasteurized and the death rate from abdominal tuberculosis in children was 4% of what it had been in 1921 when there was no pasteurization. In contrast, the death rate in 1944 in rural areas, where pasteurization was less extensively practised, was 10 times the London rate.

Though there has been considerable recent concern about the world-wide resurgence of TB, the contribution of foodborne transmission is probably insignificant in the developed world where it is effectively controlled by the testing and elimination of infected cattle, rigorous meat inspection and milk pasteurization. This may not be true in many developing countries such as those of Africa where the extent of human tuberculosis caused by *Myc. bovis* is not known and there is wide-spread consumption of unpasteurized milk and, in some areas, raw meat products.

7.11 *PLESIOMONAS SHIGELLOIDES*

7.11.1 Introduction

Plesiomonas shigelloides is the only species of the genus whose name is derived from the Greek word for neighbour; an allusion to its similarity to *Aeromonas*. Its position as a causative agent of foodborne illness also bears some similarity to *Aeromonas*. It is not normally recovered from human faeces, except in Thailand where a carriage rate of 5.5% has been reported. The association with diarrhoea is largely based on its isolation from patients suffering from diarrhoea in the absence of any other known pathogens and the strongest of this evidence has come with isolation from several patients in the same outbreak. However volunteer feeding trials have failed to demonstrate a causal link.

7.11.2 The Organism and its Characteristics

A member of the family Enterobacteriaceae (previously classified in the Vibrionceae), *P. shigelloides* is a short, catalase-positive, oxidase-positive, Gram-negative rod. It is motile by polar, generally lophotrichous flagella in contrast to *Aeromonas* and *Vibrio* which are monotrichous. It grows over a temperature range from 8–10 °C to 40–45 °C with an optimum at around 37 °C. It is not markedly heat resistant and is readily eliminated by pasteurization treatments. Growth is possible down to pH 4.5 and the maximum salt concentration it will tolerate is between 3 and 5% depending on other conditions.

The organism is ubiquitous in surface waters and soil, more commonly in samples from warmer climates. Carriage in cold-blooded animals such as frogs, snakes, turtles, and fish is common and it has been isolated from cattle, sheep, pigs, poultry, cats and dogs. It is not normally part of the human gut flora.

7.11.3   Pathogenesis and Clinical Features

Cases of *P. shigelloides* infection are more common in warmer climates and in travellers returning from warmer climates. The usual symptoms are a mild watery diarrhoea free from blood or mucus. Symptoms appear within 48 h and persist for several days. More severe colitis or a cholera-like syndrome have been noted with individuals who are immunosuppressed or have gastrointestinal tumours.

Little is known of the pathogenesis of *P. shigelloides* infections. Motility appears to be an important factor and evidence has been presented for an enterotoxin causing fluid secretion in rabbits' ligated ileal loops.

7.11.4   Isolation and Identification

The relatively recent growth of interest in *P. shigelloides* is reflected in the use of 'second-hand' media in its isolation. Alkaline peptone water and tetrathionate broth have both been used for enrichment culture of *P. shigelloides* at 35–40 °C and salmonella–shigella and MacConkey agars have been used as selective plating media. Selective plating media have been developed such as inositol/brilliant green/bile salts, *Plesiomonas* agar. Isolates can be readily confirmed on the basis of biochemical tests.

7.11.5   Association with Foods

Fish and shellfish are a natural reservoir of the organism and, with the exception of one incident where chicken was implicated, they are the foods invariably associated with *Plesiomonas* infections. Examples have included crab, shrimp, cuttle fish and oysters.

7.12   *SALMONELLA*

7.12.1   Introduction

Most salmonellas are regarded as human pathogens, though they differ in the characteristics and the severity of the illness they cause. Typhoid fever is the most severe and consequently was the earliest salmonella infection to be reliably described. This is credited to Bretonneau, the French physician who is also regarded as the founder of the doctrine of the aetiological specificity of disease. During his life, he published only one paper on typhoid, or 'dothinenterie' as he called it, in 1829, and his treatise on the subject was only published in 1922 by one of his descendants.

In 1856, the English physician William Budd concluded that each case of typhoid is epidemiologically linked to an earlier case and that a specific toxin is disseminated with the patients faeces. To support his proposition he demonstrated that treating the excreta of victims with chlorinated lime (bleaching powder) reduced the incidence of typhoid. The typhoid bacillus was first observed by the German bacteriologists Eberth and Koch in 1880 and four years later Gaffky succeeded in its cultivation. The paratyphoid bacilli, responsible for the clinically similar condition, paratyphoid fever, were first isolated by Achard and Bensaud (1896) and by Gwyn (1898), and confirmed as culturally and serologically distinct from the typhoid bacillus by Schottmüller in 1901. Other salmonellas were isolated during the same period; Salmon and Smith (1885) isolated *Bacillus cholerae-suis* from pigs with hog cholera, a disease now known to be viral in origin, and similar bacteria were isolated from cases of foodborne infection and animal disease. The genus *Salmonella* was finally created in 1900 by Lignières and named in honour of D.E. Salmon, the American veterinary pathologist who first described *Salmonella cholerae-suis*.

Salmonellas are now established as one of the most important causes of foodborne illness worldwide. In Europe in 1989 the annual incidence of salmonellosis was around 50 per 100 000 inhabitants in most countries, though actual figures varied from below 10 in the case of Luxembourg to more than 120 in Hungary and Finland. Data collected by the European surveillance system Enter-net reported an increase in the annual total for salmonellosis in 12 European countries from 41 870 in 1995 to 55 278 in 1997. Since the contribution of England and Wales to these figures was 29 314 in 1995 and 32 596 in 1997 it suggests that the effectiveness of data collection and/or food hygiene are far from uniform across these countries. In the United States in 1998–2001 the incidence was 15.1 per 100 000.

On the basis of DNA/DNA hybridization, the genus *Salmonella* was recognized to contain a single species, *S. enterica* (formerly known as *S. cholerae-suis*), which comprises seven subspecies. One of these subspecies, which is relatively unimportant as a cause of human infection and accounts for less than 1% of *Salmonella* serovars, has been proposed for elevation to species status as *S. bongori*.

The Kauffman–White serotyping scheme has proved the most useful technique for differentiating within the genus. This describes organisms on the basis of their somatic (O) and flagellar (H) antigens, and by capsullar (Vi) antigens (possessed by *S. typhi*, *S. dublin* and occasional strains of *S. paratyphi* C). In 1941 the scheme contained 100 serovars and the number has since risen to the current level of more than 2400.

The taxonomic nomenclature of the genus is rather different from that of other genera. Many of the different serovars were named as if they were distinct species. The earliest to be described were given species epithets derived from the disease they caused, either in humans (*S. typhi, S. paratyphi* A and B), or in animals (*S. typhimurium, S. cholerae-suis*, or *S. abortusovis*). Limitations in this approach led to the use of serovar names based on the geographical location of the first isolation, for example *S. dublin, S. montevideo, S. minneapolis*, and even *S. guildford*. This has some advantage over the use of long serological formulae but since 1966 has only been applied to serovars of subspecies I (*S. enterica* subsp. *enterica*) which accounts for more than 59% of the 2400 serovars known and the vast majority (>99%) of human isolates.

To introduce some taxonomic rectitude the non-italicized serovar name is used after the species name so that *S. typhimurium* becomes *S. enterica* subsp. *enterica* ser. Typhimurium or, more concisely, *Salmonella* Typhimurium. By retaining the old serovar name much of the potential for confusion inherent in other schemes is reduced. In the case of other subspecies which comprise mainly isolates from the environment and cold-blooded animals, the serovar formula is used after the name of the subspecies, e.g. *Salmonella fremantle* would be *S. enterica* subsp. *salamae* ser.42;g,t:-.

7.12.2 The Organism and its Characteristics

Salmonellas are members of the Enterobacteriaceae. They are Gram-negative, non-sporeforming rods (typically $0.5\,\mu m$ by $1-3\,\mu m$) which are facultatively anaerobic, catalase-positive, oxidase-negative, and are generally motile with peritrichous flagella.

Growth has been recorded from temperatures just above 5 °C up to 47 °C with an optimum at 37 °C. Salmonellas are heat sensitive and are readily destroyed by pasteurization temperatures. *S.* Senftenberg 775 W is the most heat resistant serotype at high a_w and has a D_{72} in milk of 0.09 min (*S.* Typhimurium $D_{72} = 0.003$ min). Heat resistance has been shown to be enhanced by sub-lethal heat shocking at 48 °C for 30 min and can also be markedly increased in low a_w media, for example *S.* Typhimurium has a D_{70} of 11.3–17.5 h in chocolate sauce. In frozen foods, numbers of viable salmonella decline slowly, the rate decreasing as the storage temperature decreases.

The minimum a_w for growth is around 0.93 but cells survive well in dried foods, the survival rate increasing as the a_w is reduced. The minimum pH for growth varies with the acidulant from 5.4 with acetic acid to 4.05 with hydrochloric and citric acids. Optimal growth occurs around pH 7.

It was noted in Section 7.12.1 above that the most important technique for sub-dividing the genus is the serotyping scheme of Kauffman and White. This does not provide a complete account of the antigenic structure of each salmonella, but does provide a workable scheme using antigens of diagnostic value. In the case of the more common serotypes such as *S.* Typhimurium and *S.* Enteritidis a more discriminating scheme of classification is required for epidemiological purposes and this is provided by phage typing.

This was first applied to *S.* Typhi where most strains could be classified into one of 11 phage types using a set of phages that acted only on bacteria possessing the Vi antigen. A high degree of correlation has been observed between phage type and epidemic source. Similar successful phage typing-schemes have been developed for, among others, *S.* Typhimurium, which employs 36 phages to distinguish at least 232 definitive types currently recognized, *S.* Enteritidis and *S.* Virchow.

Biotyping according to biochemical characteristics has sometimes proved useful in epidemiological investigations where it can supplement phage typing or subdivide a large group of otherwise untypable strains. This has proved most useful for *S.* Typhimurium where Duguid's scheme based on 15 biochemical tests has identified 184 full biotypes.

Plasmid profiling based on the isolation and separation of plasmids by electrophoresis on agarose gels has also met with some success as an epidemiological tool. One notable example of its use was in the early 1980s when it was used to identify a strain of *S.* Muenchen responsible for an outbreak in the United States where the food vehicle was marijuana. The plasmid profile was sufficiently distinctive and stable to allow the outbreak strain to be distinguished from strains of other serotypes and non-outbreak strains of *S.* Muenchen. A number of other molecular typing techniques described in Chapter 10 have been used with *Salmonella* including pulsed field gel electrophoresis (PFGE).

Salmonellas are primarily inhabitants of the gastrointestinal tract. They are carried by a wide range of food animals, wild animals, rodents, pets, birds, reptiles, and insects, usually without the display of any apparent illness. They can be disseminated *via* faeces to soil, water, foods and feeds and thence to other animals (including humans).

Most salmonellas infect a range of animal species but some serotypes are host adapted such as *S.* Enteritidis PT4, *S.* Pullorum and *S.* Gallinarum in poultry and *S.* Cholerae-suis in pigs. In these cases direct animal-to-animal transmission can be more important and vertical transmission may occur – parents infecting offspring. For example, *S.* Enteritidis PT4 can pass from breeding flocks to newly hatched broiler and egg-laying chicks *via* transovarian infection of the egg or its shell.

7.12.3 Pathogenesis and Clinical Features

Salmonellas are responsible for a number of different clinical syndromes grouped here as enteritis and systemic disease.

7.12.3.1 Enteritis. Gastrointestinal infections are predominantly associated with those serotypes which occur widely in animals and humans. They can range in severity from asymptomatic carriage to severe diarrhoea and are the most common type of salmonellosis.

At any one time human illness is usually associated with a limited number of serotypes; in the UK only about 200 serotypes may be reported in any one year. Currently *S*. Enteritidis, and *S*. Typhimurium, are the most common, accounting for about three-quarters of laboratory reports. Other relatively more common serotypes are *S*. Virchow, *S*. Infantis and *S*. Newport.

The incubation period for salmonella enteritis is typically between 6 and 48 h. The principal symptoms of mild fever, nausea and vomiting, abdominal pain and diarrhoea last for a few days but, in some cases, can persist for a week or more. The illness is usually self-limiting but can be more severe in particularly susceptible groups such as the very young, the very old and those already ill. One example of this is the outbreak which occurred in the Stanley Royd Hospital in the UK in 1984 where about 350 patients and 50 staff were affected and 19 of the patients died.

Ingested organisms, which survive passage through the stomach acid, adhere to the epithelial cells of the ileum *via* mannose-resistant fimbriae. They are then engulfed by the cells in a process known as receptor mediated endocytosis. The ability of salmonellas to enter non-phago-cytic cells is a property essential to their pathogenicity. Our understanding of the molecular basis of this process has increased considerably with the discovery that it is largely encoded on a 35–40 kb region of the chromosome, described as a pathogenicity island. This region of the DNA encodes a complex secretion system for the proteins required in the signalling events which subvert the host cell and ultimately lead to bacterial uptake. Known as a type III or, in some cases, a contact dependent secretion system, such systems are also present in a number of other enteropathogens such as *Shigella*, *Yersinia*, enteropathogenic and enterohaemorraghic *E. coli*. Phylogenetic analysis and their base composition suggest that these regions of DNA may have been acquired from another micro-organism as a block; an event which clearly marks an important evolutionary step towards pathogenicity. Endocytosed salmonellas pass through the epithelial cells within a membrane-bound vacuole, where they multiply and are then released into the lamina propria *via* the basal cell membrane. This prompts an influx of inflammatory cells leading to the release of prostaglandins which activate adenylate cyclase producing fluid secretion into the intestinal lumen. The picture is a little more complex than this since there are at least four other pathogenicity islands also contributing to the overall pathogenicity of the organism.

As a general rule, the infectious dose of salmonella is high, of the order of 10^6 cells, but this will vary with a number of factors such as the virulence of the serotype, the susceptibility of the individual and the food vehicle involved. A number of outbreaks have occurred where epidemiological evidence points to an infective dose as low as 10–100 cells. This appears to be particularly associated with more susceptible individuals such as children and the elderly, and with fatty foods such as cheese, salami and chocolate. In an outbreak in Canada where the vehicle was cheddar cheese it was found to contain 1.5–9.1 cells per 100 g. It seems likely that the high fat content in some foods affords the bacteria some protection from stomach acidity. A low infective dose (<200) was also indicated in a waterborne outbreak in the early 1970s. In this case fat was clearly not a factor, but the more rapid transit of water through the stomach may have served a similar purpose.

After symptoms have subsided, carriage of the organism and its passage in high numbers in the stools may occur for a few weeks, or occasionally months.

7.12.3.2 Systemic Disease.   Host-adapted serotypes are more invasive and tend to cause systemic disease in their hosts; a feature which is linked to their resistance to phagocytic killing. In humans, this applies to the typhoid and paratyphoid bacilli, *S.* Typhi, and *S.* Paratyphi A, B, and C, which cause the septicaemic diseases, enteric fever.

Typhoid fever has an incubation period of anything from 3 to 56 days, though it is usually between 10 and 20 days. Invasive salmonellas penetrate the intestinal epithelium and are then carried by the lymphatics to the mesenteric lymph nodes. After multiplication in the macrophages, they are released to drain into the blood stream and are then disseminated around the body. They are removed from the blood by macrophages but continue to multiply within them. This eventually kills the macrophages which then release large numbers of bacteria into the blood stream causing a septicaemia. In this, the first phase of the illness, the organism may be cultured from the blood. There is a slow onset of symptoms including fever, headache, abdominal tenderness and constipation and the appearance on the body of rose red spots which fade on pressure.

During the second stage of the illness, the organism reaches the gall bladder where it multiplies in the bile. The flow of infected bile reinfects the small intestine causing inflammation and ulceration. The fever persists but with the onset of a diarrhoea in which large numbers of the bacteria are excreted with the characteristic 'pea soup' stools and, to a lesser extent, with the urine. In more serious cases, haemorrhage of the ulcers may occur and perforation of the intestine leading to peritonitis. In milder cases, the ulcers heal and fever falls with recovery after 4–5 weeks.

Unlike the more localized enteric infections, typhoid is usefully treated with antibiotics such as chloramphenicol, ampicillin and amoxycillin.

After remission of symptoms, a carrier state can persist for several months and occasionally years as parts of the gall bladder are colonized and bacteria are discharged intermittently with the bile into faeces. This occurs more commonly in women and the elderly and there have been a number of typhoid carriers who have achieved some notoriety as a result of their condition and its consequences. These include the 'Strasbourg Master Baker's Wife', the 'Folkestone Milker', and, probably best known of all, 'Typhoid Mary'. Mary Mallon worked as a cook in a number of households and institutions in the New York area at the beginning of the 20th century. She first attracted the attention of the authorities when she disappeared after an outbreak of typhoid fever in a family for whom she had been working. When she was eventually tracked down by following a trail of outbreaks in places she worked, she was forcibly detained by the New York City Health Department for three years. Despite an undertaking not to work as a cook or handle food on her release, she disappeared again, assumed a false name, and started work as a cook. In 1915 she was working at a New York hospital when a typhoid outbreak occurred in which 25 people were affected and two died. She failed to return from leave, but was later found and held at a hospital on North Brother Island until her death, from a stroke, in 1938, aged 70.

Nowadays chronic carriers can be treated with antibiotics, but in particularly recalcitrant cases cholecystectomy (surgical removal of the gall bladder) is necessary.

A number of non-human adapted serotypes such as *S*. Blegdam, *S*. Bredeny, *S*. Cholerae-suis, *S*. Dublin, *S*. Enteritidis, *S*. Panama, *S*. Typhimurium, and *S*. Virchow can also be invasive in susceptible individuals. They can cause less severe forms of enteric fever and septicaemia, and focal infections at a wide variety of sites around the body such as the heart, appendix, gall bladder, peritoneum, lungs, urinary tract, brain, meninges and spleen. Localization is more likely to occur at sites where there is pre-existing disease or damage and some sites of infection are associated with particular population groups such as meningitis in infants, pneumonia in the elderly, and osteomyelitis in patients with sickle-cell anaemia.

7.12.4 Isolation and Identification

Methods for the isolation and identification of salmonellas in foods have arguably received more attention than those for any other foodborne pathogen. Using traditional cultural techniques, a five-stage procedure has emerged as the widely accepted norm. This is outlined in Figure 7.7.

Pre-enrichment in a non-selective medium increases the recovery rate of salmonellas by allowing the repair of cells which have been sublethally

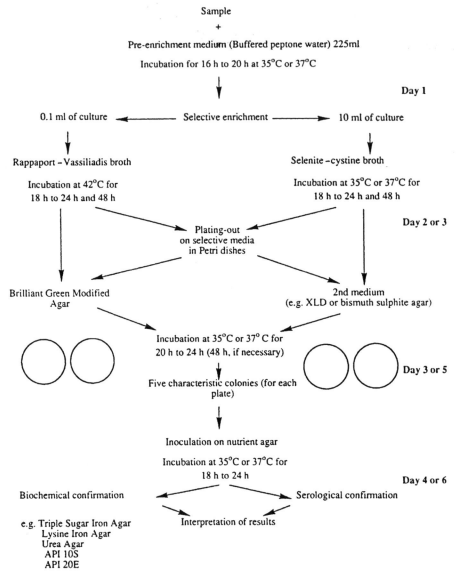

Figure 7.7   *Traditional cultural protocol for isolation of* Salmonella *from food*

damaged. Such damage can result from any exposure to adverse conditions that might occur during food processing, such as chilling, freezing, or drying, and increases the cell's sensitivity to selective agents used in media in subsequent stages of the isolation procedure. Failure to include a resuscitation step could therefore result in the non-detection of cells that might recover and cause infection if the food is mishandled.

The selective enrichment stage is intended to increase the proportion of salmonella cells in the total microflora by allowing them to proliferate while restricting growth of other micro-organisms present. To this end a number of different media have been proposed employing selective agents such as bile, brilliant green, malachite green, tetrathionate and selenite. The most widely used are selenite–cystine broth, which contains cystine to stimulate growth of salmonellas; Muller–Kauffman tetrathionate broth, containing tetrathionate, brilliant green, and bile; and Rappaport–Vassiliadis (RV) broth, which contains malachite green, magnesium chloride and a slightly reduced pH as selective factors. Since they differ in their selectivity, two broths are usually used in parallel; commonly a combination of the less selective selenite–cystine broth and one of the others.

From the selective enrichment broths, cultures are streaked on to selective and differential solid media. Once again it is usual to use two different media in parallel. The selective agents used are bile salts or deoxycholate and/or brilliant green and the diagnostic reaction is usually provided by the inability of most salmonellas to ferment lactose and/or the production of hydrogen sulfide. In choosing the media to use, it is advisable to select two based on different diagnostic reactions to ensure that atypical strains, for instance lactose-positive ones, will not be missed.

Presumptive salmonellas from selective plating media must be confirmed by biochemical testing and serologically by agglutination with polyvalent O antisera.

The whole protocol is rather complex and lengthy, requiring at least four days for a negative result. In view of this, a number of procedures have been described which attempt to simplify the procedure and reduce the elapsed time involved. Two of these employ the motility of salmonellas which means that they would fail to detect non-motile salmonellas (incidence $<0.1\%$).

In one, a conventional pre-enrichment culture is inoculated into an elective medium, salmonellas swim into a compartment containing a selective medium and from there into one containing a diagnostic medium. A diagnostic medium giving the appropriate colour change is then tested for ability to agglutinate latex particles coated with salmonella antibodies. A positive result indicates a presumptive salmonella, which must then be confirmed by conventional serological and biochemical testing using a sub-culture from the diagnostic medium. With this technique, presumptive identification of a salmonella is obtained within 42 h compared with 3–4 days by the traditional cultural method.

In another system, salmonella detection is by formation of an immunoprecipitate as *Salmonella* antibodies diffusing down through a medium meet salmonellas swimming up from a chamber containing a selective medium.

Impedance–conductance techniques (see Chapter 10) have been successfully applied to the detection of salmonellas. The original medium of

Easter and Gibson comprises a modified selenite–cystine broth containing dulcitol and trimethylamine oxide (TMAO). Salmonellas are able to ferment dulcitol and reduce TMAO to the base trimethylamine. This increases the conductivity of the medium and provides the basis for detection. The detection time is reduced if the samples are pre-enriched in a medium containing dulcitol and TMAO to induce the relevant enzymes. In a comparison using 2586 samples of milk powder, this method was found to be as effective as a traditional cultural method but with considerable savings of time and labour. With a 24 h pre-enrichment step, *Salmonella*-negative samples can be detected within 48 h.

A number of modifications to the original medium and protocol have been described. These include the incorporation of a *Salmonella*-specific bacteriophage in a parallel sample to demonstrate that observed changes in electrical properties are in response to salmonella; the replacement of dulcitol with mannitol or deoxyribose in order to detect dulcitol-negative salmonellas; and the use of detection media based on lysine decarboxylase activity.

ELISA and gene probe kits for the detection of salmonellas are also available, but like all the techniques described, they require a certain threshold concentration of salmonellas. One approach to avoid or curtail the enrichment steps that this usually entails is immunomagnetic separation. *Salmonella* antibodies are attached to magnetic particles which are added to a liquid culture containing salmonellas which are then captured by the antibodies. The beads with adhering *Salmonella* cells can then be readily separated from the culture with a magnet, achieving a substantial enrichment in minutes. Their presence can then be confirmed using conventional media or one of the more rapid techniques.

7.12.5 Association with Foods

Salmonellosis is described as a zoonotic infection since the major source of human illness is infected animals. Transmission is by the faecal–oral route whereby intestinal contents from an infected animal are ingested with food or water. A period of temperature abuse which allows the salmonellae to grow in the food and an inadequate or absent final heat treatment are common factors contributing to outbreaks.

Meat, milk, poultry, and eggs are primary vehicles; they may be undercooked, allowing the salmonellas to survive, or they may cross-contaminate other foods that are consumed without further cooking. Cross-contamination can occur through direct contact or indirectly *via* contaminated kitchen equipment and utensils.

Human carriers are generally less important than animals in the transmission of salmonellosis. Human transmission can occur if the faecally contaminated hands of an infected food handler touch a food which is then consumed without adequate cooking, often after an intervening period in which microbial growth occurs. This was the cause of a major outbreak affecting an international airline in 1984. The outbreak involved 631 passengers and 135 crew and was due to contamination of an aspic glaze by a member of catering staff who returned to work after illness but was still excreting *Salmonella* Enteritidis PT4.

Direct person-to-person spread by the faecal–oral route is also possible but is usually restricted to institutional outbreaks such as occur in hospitals, old people's homes, and nurseries.

Food animals may acquire salmonella infection on the farm from wild birds and rodents, but the principal sources are other animals, which may be symptomless excreters, and contaminated feeding stuffs (Figure 7.8). Measures that can be taken to minimize transmission between animals on the farm include good animal husbandry, protection of feeds and water from contamination, hygienic disposal of wastes, and maintenance of a generally clean environment. Transfer of *Salmonella* between animals is particularly associated with situations where animals may be stressed and crowded such as during transport, at markets, and when in lairage at the slaughterhouse. It is best minimized by avoidance of overcrowded conditions, ensuring a clean environment, and otherwise limiting the stress to animals on such occasions.

An important factor in maintaining the cycle of *Salmonella* infection in food animals has been the practice of using animal by-products as animal feeds such as meat and bone meal. The heat process that these materials undergo in their conversion to feeds should destroy any salmonellas present. Nevertheless they are subject to post-process contamination either in the plant or on the farm by contact with unprocessed material or with bird and rodent faeces. The importance of animal feeds

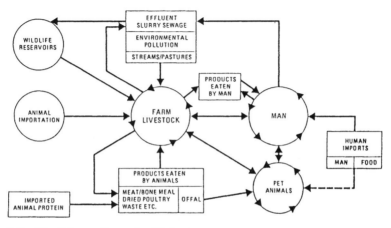

Figure 7.8 *The* Salmonella *cycle of infection*
(Reproduced with permission from WHO, 1983)

as a source of *Salmonella* should decline following the 1996 ban on feeding mammalian-derived protein to all farm animals.

In the UK, the major source of *Salmonella* infection is poultry and poultry products. Here the problem is not confined to horizontal transfer of the organism between animals but also includes vertical transmission of host-adapted serotypes from the breeding flocks to their progeny. Particularly noteworthy in this respect is *S.* Enteritidis PT4 which has been responsible for the rise in salmonellosis since 1985 (Figure 7.9). Isolations of *S.* Enteritidis increased 14-fold between 1981 and 1988, while those of *S.* Typhimurium less than doubled, and *S.* Enteritidis is now the commonest serotype recorded. Poultry was the food most commonly implicated in outbreaks of salmonellosis in 1986 and 1987 but in 1988 and 1989, eggs were the most frequent vehicle. This remained the case in 1995 and 1996. Most outbreaks were associated with raw eggs in products such as home-made mayonnaise and ice cream or, in one instance, a 'body-building' drink.

Contamination of eggs with salmonellas is a long-recognized problem but in most cases this was due to contamination of the eggshell exterior with faecal material in the hen's cloaca or after laying in the nest or battery. The shell could then contaminate the contents when the egg was broken. This is a particular problem when breaking large quantities of

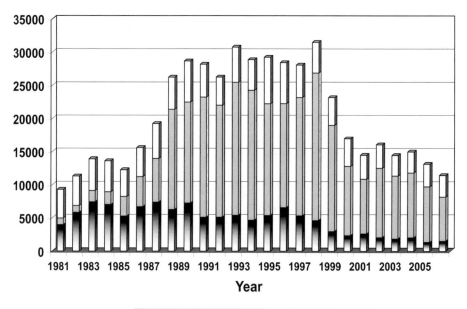

Figure 7.9   Salmonella *in humans. England and Wales 1981–2005.*

eggs where it is difficult to avoid some contamination with shell fragments. During the Second World War, there were a number of outbreaks attributed to dried whole egg powder and a survey conducted in 1961/2 found 16% of frozen whole egg samples to contain salmonellas. This led to the introduction in 1963 of regulations requiring that liquid whole egg be pasteurized at 64.4 °C for 2.5 min. The prescribed heat treatment also inactivates the yolk enzyme α-amylase and provides the basis of a simple test to ensure that the regulations have been complied with.

In the more recent cases however, contamination of the yolk of intact hen's eggs has also been indicated. In the UK and in other European countries, particularly Spain, this problem has been associated with *S.* Enteritidis PT4 but other phage types (PT8 and PT13a) have been reported to cause similar problems in the United States. It is thought that these organisms infect the bird's ovaries and oviduct and thereby contaminate the egg contents. The temperatures reached in the yolk during mild cooking procedures such as 'soft boiling' or light frying are probably insufficient to kill the organism and the fat content of the yolk may protect the organism from gastric acidity.

The precise extent of this problem is difficult to determine, but one survey found *Salmonella* in the contents of one in a thousand eggs from flocks associated with human illness. When one considers that 30 million eggs are eaten daily in the UK, then eggs are clearly an important source of human infection.

The massive increase in salmonella infections which started in 1985 prompted a number of new biosecurity measures to exclude salmonella and reduce the number of infections on egg and poultry farms. Compulsory bacteriological monitoring of all commercial egg-laying and breeding flocks was introduced and if *S.* Enteritidis was isolated the birds were required to be slaughtered. This practice has now been stopped although breeder flocks infected with either *S.* Enteritidis or *S.* Typhimurium are still subject to compulsory slaughter. General hygiene practices on farm were introduced or improved but probably the single most significant intervention was the introduction of vaccination of broiler breeder flocks (1994) and commercial laying flocks (1996) against *S.* Enteritidis. As a result of these measures, surveillance studies have shown the contamination rates in retail frozen poultry in the UK declined from 79% in 1984 to 41% in 1994 to 11% in 2001. The equivalent figures for chilled poultry were 54%, 33% and 4% respectively. The prevalence of *S.* Enteritidis and/or *S.* Typhimurium in laying hen flocks in the UK in 2004-2005 was found to be 8% compared to an EU average of 20.3%. Figures for individual member states ranged from zero in Norway, Sweden and Luxembourg to more than 50% in Spain, Poland and the Czech Republic.

The positive effect of these interventions can be seen in the substantial decline in human *Salmonella* infections since 1997. In particular infections caused by *S.* Enteritidis PT4 decreased from 10,056 in 1998 to 2693 in 2003. Hidden by the overall figures however there was an almost doubling of infections caused by *S.* Enteritidis non PT4 from 3548 in 2000 to 7065 in 2003. This was attributed to imported raw shell eggs used largely in catering.

Salmonella Typhimurium definitive phage type (DT) 104 emerged initially in cattle in the UK but has since been reported in poultry, sheep, pigs and horses. It is now the second most prevalent salmonella in humans in England and Wales with reported isolations increasing more than 16 fold to 4006 in 1996. Ninety six percent of isolates in 1996 were multiresistant, displaying resistance to four or more antimicrobials. Twenty one percent were **R**-type ACSSuSpT which have chromosomal resistance genes to Ampicillin, Chloramphenicol, Streptomycin, **S**ulphonamide, **Sp**ectinomycin and **T**etracyclines. Since 1994 there has been an increasing number of isolates with additional resistances to trimethoprim, nalidixic acid and to quinolone antibiotics such as ciprofloxacin. Quinolones are used in the treatment of salmonellosis and typhoid fever in humans and the emergence of resistance may be linked to their veterinary use since 1993 to combat salmonellosis in cattle, pigs and poultry. Increased isolation rates of multiresistant DT104 have also been reported from other European countries, such as Germany where DT104 accounted for more than 10% of 10 000 human isolates, and the United States. The organism also appears to be unusually virulent since hospitalization and fatality rates respectively twice and ten times those of other foodborne salmonella infections have been reported.

Though the primary source appears to be foods of animal origin such as poultry and unpasteurized milk, cross-contamination has led to a bewildering variety of foods being implicated in DT104 outbreaks. These include chicken drumsticks, tuna and salmon sandwiches, ham, cod roe, scotch eggs, unpasteurized milk, roast beef, apple crumble and coleslaw.

Raw milk will inevitably contain *Salmonella* and any slight nutritional advantage it may have over pasteurized milk is far outweighed by the very real risk of salmonellosis (and campylobacteriosis). Outbreaks in a number of countries have been associated with pasteurized milk that has been inadequately processed or subject to post-process contamination. *Salmonella* is unable to grow in dried milk but is able to survive and resume growth when the milk is reconstituted. *S.* Ealing was responsible for an outbreak in the UK in 1985 where the vehicle was a dried baby-milk. The organism had contaminated the insulation surrounding a spray drier and penetrated the drying chamber itself through a small defect in the chamber wall. Rigorous cleaning and disinfection were unable to eliminate the contamination from the spray drier which was eventually decommissioned.

Fish and fish products are only occasionally associated with salmonellosis, although fish meal for animal feed often contains *Salmonella* as a result of contamination from rodents and birds. Filter-feeding shellfish harvested from polluted waters and frozen precooked prawns have been identified as higher risk products.

Since birds, rodents, insects, infected food handlers or infected foods can all contaminate foods directly or indirectly, potential food vehicles for salmonella are numerous. Contaminated cocoa beans which had been processed into chocolate were responsible for outbreaks of *S.* Eastbourne in the United States and Canada, of *S.* Napoli and *S.* Montevideo in England, and *S.* Typhimurium in Scandinavia. Although the production of chocolate involves a heating stage, this was insufficient to kill all the salmonellas present, possibly as a result of a protective effect from the cocoa butter.

Desiccated coconut is used in a range of confectionary products and was identified as a hazard following cases of typhoid and salmonellosis in Australia. In 1959/60, a survey of desiccated coconut imports into the UK from Sri Lanka revealed that 9% of samples contained *Salmonella*. In response, the introduction and enforcement of regulations in Sri Lanka to improve production hygiene have now reduced the contamination rate dramatically.

Other plant products such as salad vegetables have been associated with occasional outbreaks of typhoid and salmonellosis. Use of polluted irrigation water or human and animal manure as fertilizer can be important contributory factors in such cases.

7.13 *SHIGELLA*

7.13.1 Introduction

The genus *Shigella* was discovered as the cause of bacillary dysentery by the Japanese microbiologist Kiyoshi Shiga in 1898. It consists of four species *Sh. dysenteriae*, *Sh. flexneri*, *Sh. boydii* and *Sh. sonnei*, all of which are regarded as human pathogens though they differ in the severity of the illness they cause. *Sh. dysenteriae* has been responsible for epidemics of severe bacillary dysentery in tropical countries but is now rarely encountered in Europe and North America where *Sh. sonnei* is more common. *Sh. sonnei* causes the mildest illness, while that caused by *Sh. boydii* and *Sh. flexneri* is of intermediate severity.

Although *Shigella* is relatively inactive biochemically when compared with *Escherichia* species, studies of DNA relatedness have demonstrated that they do in fact belong to the same genus. The separate genera are retained however, because, unlike *Escherichia*, most strains of *Shigella* are pathogenic and a redesignation might cause confusion with potentially serious consequences.

Laboratory reports of *Sh. sonnei* infections in England and Wales rose to 9830 in 1991 compared to 2319 and 2228 in the previous two years. In the United States annual reports over recent years have ranged between 300 000 and 450 000. Shigellas are spread primarily person-to-person by the faecal–oral route although foodborne outbreaks have been recorded. Some experts consider that the problem of foodborne shigellosis is greatly underestimated.

7.13.2   The Organism and its Characteristics

Shigellas are members of the family Enterobacteriaceae. They are non-motile, non-sporeforming, Gram-negative rods which are catalase-positive (with the exception of Shiga's bacillus, *S. dysenteriae* serotype 1), oxidase-negative, and facultative anaerobes. They produce acid but usually no gas from glucose and, with the exception of some strains of *S. sonnei*, are unable to ferment lactose; a feature they share with most salmonellas.

Shigellas are generally regarded as rather fragile organisms which do not survive well outside their natural habitat which is the gut of humans and other primates. They have not attracted the attention that other foodborne enteric pathogens have, but such evidence as is available suggests that their survival characteristics are in fact similar to other members of the Enterobacteriaceae. They are typical mesophiles with a growth temperature range between 10–45 °C and a heat sensitivity comparable to other members of the family. They grow best in the pH range 6–8 and do not survive well below pH 4.5. A number of studies have reported extended survival times in foods such as flour, pasteurized milk, eggs and shellfish.

The species are distinguished on the basis of biochemical tests and both serotyping and phage typing schemes are available for further subdivision of species.

7.13.3   Pathogenesis and Clinical Features

Shigellas cause bacillary dysentery in humans and other higher primates. Studies with human volunteers have indicated that the infectious dose is low; of the order of 10–100 organisms. The incubation period can vary between 7 h and 7 days although foodborne outbreaks are commonly characterized by shorter incubation periods of up to 36 h. Symptoms are of abdominal pain, vomiting and fever accompanying a diarrhoea which can range from a classic dysenteric syndrome of bloody stools containing mucus and pus, in the cases of *Sh. dysenteriae*, *Sh. flexneri* and *Sh. boydii*, to a watery diarrhoea with *Sh. sonnei*. Illness lasts from 3 days up to 14 days in some cases and a carrier state may develop which can persist for several months. Milder forms of the illness are self-limiting and require no treatment but *Sh. dysenteriae* infections often require fluid and electrolyte replacement and antibiotic therapy.

Shigellosis is an invasive infection where the organism's invasive property is encoded on a large plasmid. Other details of the pathogenesis of the infection are described in Chapter 6 (Section 6.7).

7.13.4   Isolation and Identification

Lack of interest in *Shigella* as a foodborne pathogen has meant that laboratory protocols for its isolation and identification from foods are relatively underdeveloped. A pre-enrichment procedure has been described based on resuscitation on a non-selective agar before overlaying with selective media. Selective enrichment in both Gram-negative broth and selenite broth has been recommended. Selective plating media used are generally those employed for enumerating the Enterobacteriaceae or *Salmonella* although neither are entirely satisfactory.

Rapid techniques for identification based on immunoassays which detect the virulence marker antigen, and on the polymerase chain reaction to detect the virulence plasmid by DNA/DNA hybridization have also been applied.

7.13.5   Association with Foods

Foodborne cases of shigellosis are regarded as uncommon though some consider the problem to be greatly underestimated. The limited range of hosts for the organism certainly suggests that it is relatively insignificant as a foodborne problem when compared with say *Salmonella*.

In foodborne cases, the source of the organism is normally a human carrier involved in preparation of the food. In areas where sewage disposal is inadequate the organism could be transferred from human faeces by flies. Contamination during primary production of a crop was responsible for an extensive outbreak of *Sh. sonnei* which affected several European countries in 1994 and was associated with imported iceberg lettuce.

Uncooked foods which may have received extensive handling such as prawn cocktail or tuna salad have been implicated in a number of outbreaks. In one, which occurred in Cambridgeshire, England, in 1992, 107 out of 200 guests at a buffet meal developed diarrhoea and *Sh. sonnei* was isolated from 81 of 93 faecal samples taken. The organism was also isolated from two of the catering staff. Investigation revealed a strong association between illness and consumption of two prawn dishes for which both infected caterers had been involved in the preparation.

7.14   *STAPHYLOCOCCUS AUREUS*

7.14.1   Introduction

The staphylococci were first described by the Scottish surgeon, Sir Alexander Ogston as the cause of a number of pyogenic (pus forming) infections in humans. In 1882, he gave them the name staphylococcus (Greek: *staphyle*, bunch of grapes; *coccus*, a grain or berry), after their appearance under the microscope.

The first description of food poisoning caused by staphylococci is thought to be that of Vaughan and Sternberg who investigated a large outbreak of illness in Michigan believed to have been caused by cheese contaminated with staphylococci. Clear association of the organisms with foodborne illness had to wait until Barber (1914) demonstrated that staphylococci were able to cause poisoning by consuming milk from a cow with staphylococcal mastitis. In 1930, Dack showed that staphylococcal food poisoning was caused by a filterable enterotoxin.

There are currently 27 species and 7 subspecies of the genus *Staphylococcus*; enterotoxin production is principally asssociated with the species *Staph. aureus*, although it has also been reported in others including *Staph. intermedius* and *Staph. hyicus*.

As a relatively mild, short-lived type of illness, staphylococcal food poisoning is perhaps more likely to be under-reported than others. Most reported cases are associated with outbreaks and only a few sporadic cases are detected. In the United States between 1983 and 1987, staphylococci accounted for 7.8% (47) of the 600 bacterial food poisoning outbreaks that were recorded. Equivalent figures for England and Wales over the same period were 1.9% (54) out of a total of 2815 outbreaks. Outbreaks of staphylococcal food poisoning in the UK peaked during the 1950s at 150 outbreaks per year but have since declined to an annual level of 5–10 outbreaks in the period 1990 to 1996 and an average of one per year in the period 2000 to 2005.

7.14.2   The Organism and its Characteristics

Staphylococcus aureus is a Gram-positive coccus forming spherical to ovoid cells about 1 μm in diameter. Cell division occurs in more than one plane so that cells form irregular clumps resembling bunches of grapes (Figure 7.10).

Staphylococci are catalase-positive, oxidase-negative, facultative anaerobes. Their ability to ferment glucose can be used to distinguish them from the strictly respiratory genus *Micrococcus*, although there are species in both genera where this distinction is not clear cut due to low acid production by some staphylococci and production of small amounts

Figure 7.10   Staphylococcus aureus *attached to stainless steel.*
(Photo M.Lo)

Table 7.8   *Factors permitting growth and enterotoxin production by* Staphylococcus aureus

| Factor | Growth | | Enterotoxin Production | |
	Optimum	*Range*	*Optimum*	*Range*
Temperature,°C	35–37	7–48	35–40	10–45
pH	6.0–7.0	4.0–9.8	Ent. A. 5.3–6.8 others 6–7	4.8–9.0
NaCl	0.5–4.0%	0–20%	0.5%	0–20%
Water activity	0.98–>0.99	0.83–>0.99	>0.99	0.86–>0.99
Atmosphere	Aerobic	Aerobic-Anaerobic	5–20% DO_2	Aerobic-Anaerobic
E_h	>+200mV	<−200 to >+200mV	>+200mV	?

of acid under anaerobic conditions by some micrococci. Enterotoxin production is adversely affected by anaerobic conditions far more than growth.

   Staphylococcus aureus is a typical mesophile with a growth temperature range between 7 and 48 °C and an optimum at 37 °C under otherwise optimal conditions. The range of temperature over which enterotoxin is produced is narrower by a few degrees and has an optimum at 35–40 °C (Table 7.8). The organism has unexceptional heat resistance with a D_{62} of 20–65 s and a D_{72} of 4.1 s when measured in milk using log-phase cultures. Heat resistance has been shown to vary considerably though, and D values were found to increase three-fold when stationary-phase cultures were tested.

Growth occurs optimally at pH values of 6–7, with minimum and maximum limits of 4.0 and 9.8–10.0 respectively. The pH range over which enterotoxin production occurs is narrower with little toxin production below pH 6.0 but, as with growth, precise values will vary with the exact nature of the medium.

A characteristic of *Staph. aureus* which is a particularly important consideration in some foods is its tolerance of salt and reduced a_w. It grows readily in media containing 5–7% NaCl and some strains are capable of growth in up to 20% NaCl. It will grow down to an a_w of 0.83 where it has a generation time of 300 min. Once again the range over which enterotoxin production occurs is more limited with a minimum a_w recorded of 0.86.

The principal habitat of the staphylococci is the skin, skin glands and the mucous membranes of warm blooded animals. Several species are associated with particular hosts, for example *Staph. hyicus* with pigs, and *Staph. gallinarum* with chickens. *Staph. aureus* is more widespread but occurs most frequently on the skin of higher primates. In humans, it is particularly associated with the nasal tract where it is found in 20–50% of healthy individuals. It can be isolated from faeces and sporadically from a wide range of other environmental sites such as soil, marine and fresh water, plant surfaces, dust and air.

Though normally a harmless parasite of human body surfaces where it plays a useful role metabolizing skin products and possibly preventing skin colonization by pathogens, *Staph. aureus* can cause minor skin abscesses such as boils and, more seriously, as an opportunistic pathogen when the skin barrier is breached or host resistance is low.

7.14.3   Pathogenesis and Clinical Features

Food poisoning by *Staph. aureus* is characterized by a short incubation period, typically 2–4 h. Nausea, vomiting, stomach cramps, retching and prostration are the predominant symptoms, although diarrhoea is also often reported, and recovery is normally complete within 1–2 days. In severe cases dehydration, marked pallor and collapse may require treatment by intravenous infusion.

The short incubation period is characteristic of an intoxication where illness is the result of ingestion of a pre-formed toxin in the food. *Staph. aureus* produces at least 11 enterotoxins designated SEA to SEJ. To add a touch of Byzantine complexity and confuse the unwary there is no SEF and there are three variants of SEC. Toxin types A and D, either singly or in combination, are most frequently implicated in outbreaks of food poisoning. In the UK, type A is responsible for 52% of outbreaks, type D for 6%, types A and D combined for 19%, and types C and D combined for 9%. Susceptibility varies between individuals but it has been estimated that in outbreaks less than 1 μg of pure toxin has been required to elicit symptoms. The toxins are small (M_r 26–30 kDa) single-chain polypeptides which share considerable amino acid homology. With the exception of SEI each contains a single disulfide loop near the molecule's centre. As a result of their compact structure they are resistant to gut proteases and heat stable, being inactivated only by prolonged boiling. Such procedures would of course eliminate viable *Staph. aureus* from a food so it is possible for someone to become ill from eating a food which contains no viable *Staph. aureus*.

Though frequently described as enterotoxins the *Staph. aureus* toxins are strictly neurotoxins. They elicit the emetic response by acting on receptors in the gut, which stimulate the vomiting centre in the brain *via* the vagus and sympathetic nerves. If these nerves are severed then vomiting does not occur. It is not known how the toxin induces diarrhoea but it has been shown not to stimulate adenylate cyclase activity.

The *Staph. aureus* enterotoxins are now also known to be superantigens, molecules that are able to stimulate a much higher percentage of T cells than conventional antigens. What role this may play in gastrointestinal illness, if any, is not known.

7.14.4 Isolation and Identification

The most successful and widely used selective plating medium for *Staph. aureus* is the one devised by Baird-Parker in the early 1960s. It combines the virtues of a high degree of selectivity, a characteristic diagnostic reaction, and the ability to recover stressed cells. Lithium chloride and tellurite act as selective agents while egg yolk and pyruvate assist in the recovery of damaged cells. Reduction of the tellurite by *Staph. aureus* gives characteristic shiny, jet-black colonies which are surrounded by a zone of clearing, resulting from hydrolysis of the egg-yolk protein lipovitellenin. Colonies also often have an inner white margin caused by precipitation of fatty acid.

Colonial appearance on Baird-Parker (B-P) agar gives presumptive identification of *Staph. aureus* which is often confirmed by tests for the production of coagulase and thermostable nuclease.

Coagulase is an extracellular substance which coagulates human or animal blood plasma in the absence of calcium. It is not specific to *Staph. aureus* but is also produced by *Staph. intermedius* and *Staph. hyicus*. *Staph. intermedius* is unable to reduce tellurite and therefore produces white colonies on B-P agar, but *Staph. hyicus*, which is found on the skin of pigs and poultry, requires a series of further biochemical tests to distinguish it reliably from *Staph. aureus.*

The presence of coagulase can be demonstrated using EDTA-treated rabbit plasma in the tube coagulase test. More rapid test kits are available, based on the detection of bound coagulase (also known as clumping factor) and/or protein A, which reacts with the Fc part of IgG molecules. Detection is by agglutination of erythrocytes or latex particles coated with fibrinogen or plasma and colonies from selective media can be tested directly, without any intermediate sub-culturing. Coagulase production can also be detected directly in an egg yolk-free modification of B-P agar containing pig or rabbit plasma.

Detection of thermostable nuclease uses toluidine blue/DNA agar either with a boiled culture supernatant or as an overlay on heat-treated colonies on B-P agar.

Four biotypes of *Staph. aureus* are recognized but the use of biotyping is limited since nearly all of the strains isolated from human sources belong to biotype A. Phage typing schemes are used with *Staph. aureus*; most food poisoning strains belonging to serogroup III.

Since the enterotoxins will survive heat processes that eliminate the producing organism, toxin detection in a food is a more reliable indication of hazard than viable counting procedures. A number of immunoassay techniques for staphylococcal enterotoxins are available. Early immunoprecipitation techniques such as the microslide gel diffusion test are less sensitive and require lengthy extraction and concentration procedures to isolate sufficient enterotoxin for detection. ELISA techniques which will detect $0.1–1.0\,\mathrm{ng}$ toxin g^{-1} food and reverse passive latex agglutination tests with a sensitivity of $0.5\,\mathrm{ng}\,\mathrm{ml}^{-1}$ are now available and more widely used.

7.14.5   Association with Foods

The presence of small numbers of *Staph. aureus* on foods is not uncommon. It will occur naturally in poultry and other raw meats as a frequent component of the skin microflora. Similarly, it can be isolated from raw milk where levels may sometimes be elevated as a result of *Staph. aureus* mastitis in the producing herd. As a poor competitor, it normally poses no problem in these situations since it does not grow and is eliminated by cooking or pasteurization. There have however been outbreaks caused by milk products such as dried milk and chocolate milk where growth and enterotoxin production occurred in the raw milk and the enterotoxin, but not the organism, survived subsequent pasteurization. A good example of this is a large outbreak that occurred in Japan in 2000, affecting more than 13,000 people. A power cut during production of dried skimmed milk led to delays in processing that allowed *Staph.*

aureus to multiply and produce enterotoxin. The contaminated powder was then used in a number of dairy products. Though not in itself a health threat, the presence of *Staph. aureus* on raw meats does pose the risk of cross-contamination of processed food.

Contamination by food handlers is also probably a frequent occurrence in view of the high rate of human carriage. Colonization of the nose and throat with the organism will automatically imply its presence on the skin and food may also be contaminated from infected skin lesions or by coughing and sneezing. Since large numbers, typically $> 10^6 g^{-1}$, are required for the production of enough toxin to cause illness, contamination is necessary but is not alone sufficient for an outbreak to occur. In particular, temperature and time conditions must also be provided that allow the organism to grow.

Studies in the United States and the UK have found that poultry products and cold, cooked meats are the most common vehicles. Salted meats such as ham and corned beef are particularly vulnerable since the *Staph. aureus* is unaffected by levels of salt that will inhibit a large proportion of the competitive flora. Buffet meals where such meats are served are a common scenario for outbreaks as the food is necessarily prepared some time in advance and too often stored at ambient temperature or inadequately chilled.

Canned foods also offer *Staph. aureus* a congenial, competitor-free environment and post-process leakage contamination of cans has been an occasional cause of outbreaks.

Other outbreaks have been caused by hard cheeses, cold sweets, custards and cream-filled bakery products. In Japan, rice balls that are moulded by hand are the commonest vehicle while in Hungary, it is ice cream.

7.15 *VIBRIO*

7.15.1 Introduction

Historically, cholera has been one of the diseases most feared by mankind. It is endemic to the Indian subcontinent where it is estimated to have killed more than 20 million people this century. During the 19th century there were a number of pandemics of 'Asiatic cholera' spreading from the Indian subcontinent throughout Europe and the Americas. It spread inexorably across Europe at a rate of about eight kilometres a day reaching England in 1831, where it thrived in the appalling overcrowded, insanitary conditions of the burgeoning towns and cities. The approach of a second outbreak in 1848 prompted Parliament to establish the Central Board of Health which began the long task of improving sewerage and water supply systems. Similar apprehension of an approaching cholera outbreak in 1866 inspired the foundation of a similar Board in New York in the United States.

Pacini (1854) is credited with the first description of the etiological agent of cholera when he observed large numbers of curved bacilli in clinical specimens from cholera patients in Florence. His findings were not however generally accepted because of the widespread occurrence of similar but harmless vibrios in the environment. It was Robert Koch who firmly established the causal link between *Vibrio cholerae* and cholera when working in Egypt in 1886.

Koch isolated what is now known as the classical *V. cholerae* biotype which was responsible for most outbreaks of cholera until 1961. The *El Tor* biotype, first isolated in 1906 by Gotschlich from pilgrims bound for Mecca at the El Tor quarantine station in Sinai, Egypt, is responsible for the current (7th) pandemic. This started in Celebes (Sulawesi) in Indonesia in 1961, reached Africa in 1970 and the Americas in 1991. Of the 594 694 cases reported to the WHO in 1991, 391 220 were in South and Central America.

It was recognized in the 1930s that both biotypes are agglutinated by a single antiserum designated O1. Other strains of *V. cholerae* do not react with this antiserum and are termed non-agglutinable, or more correctly non-O1 strains, though some do produce cholera toxin. In 1992 a new serotype, O139, was associated with epidemic cholera in India and Bangladesh and has also been isolated from cholera patients in Thailand.

A number of other species of *Vibrio* have been recognized as pathogens causing wound and ear infections, septicaemia as well as gastrointestinal upsets (Table 7.9). In particular, *V. parahaemolyticus*, which was first shown to be an enteropathogen in 1951, is responsible for 50–70% of outbreaks of foodborne gastroenteritis in Japan. *V. fluvialis* has been isolated from sporadic cases of diarrhoea in some countries, particularly those with warm climates, although its exact role is uncertain since other enteropathogens were often present in the stool samples. *V. mimicus*,

Table 7.9 Vibrio *species associated with human diseases*

Species	Disease
V. cholerae , O1	Cholera, wound infection
V. cholerae, non-O1	Diarrhoea, gastroenteritis, wound infection, secondary septicaemia
V. mimicus	Diarrhoea, gastroenteritis, wound infection
V. parahaemolyticus	Gastroenteritis, wound infection, otitis media
V. fluvialis	Diarrhoea
V. furnissii	Diarrhoea
V. hollisae	Diarrhoea
V. vulnificus	Wound infection, primary septicaemia, secondary septicaemia
V. alginolyticus	Wound infection, otitis media
V. damsela	Wound infection

which produces diarrhoea, is distinguishable from *V. cholerae* only by its ability to produce acid from sucrose and acetoin from glucose. *V. vulnificus* does not usually cause diarrhoea but severe extra-intestinal infections such as a life-threatening septicaemia. Patients normally have some underlying disease and have eaten seafood, particularly oysters about a week before the onset of illness.

7.15.2 The Organisms and their Characteristics

Vibrios are Gram-negative pleomorphic (curved or straight), short rods which are motile with (normally) sheathed, polar flagella. Catalase and oxidase-positive cells are facultatively anaerobic and capable of both fermentative and respiratory metabolism. Sodium chloride stimulates the growth of all species and is an obligate requirement for some. The optimum level for the growth of clinically important species is 1–3%. *V. parahaemolyticus* grows optimally at 3% NaCl but will grow at levels between 0.5% and 8%. The minimum a_w for growth of *V. parahaemolyticus* varies between 0.937 and 0.986 depending on the solute used.

Growth of enteropathogenic vibrios occurs optimally at around 37 °C and has been demonstrated over the range 5–43 °C, although ≈ 10 °C is regarded as a more usual minimum in natural environments. When conditions are favourable, vibrios can grow extremely rapidly; generation times of as little as 11 min and 9 min have been recorded for *V. parahaemolyticus* and the non-pathogenic marine vibrio, *V. natrigens* respectively.

V. parahaemolyticus is generally less robust at extremes of temperatures than *V. cholerae*. Numbers decline slowly at chill temperatures below its growth minimum and under frozen conditions a 2-log reduction has been observed after 8 days at −18 °C. The D_{49} for *V. parahaemolyticus* in clam slurry is 0.7 min compared with a D_{49} for *V. cholerae* of 8.15 min measured in crab slurry. Other studies have recorded higher D values for *V. parahaemolyticus*, for instance 5 min at 60 °C produced only 4–5 log reductions in peptone/3% NaCl. Pregrowth of the organism in the presence of salt is known to increase heat resistance.

V. parahaemolyticus and other vibrios will grow best at pH values slightly above neutrality (7.5–8.5) and this ability of vibrios to grow in alkaline conditions up to a pH of 11.0 is exploited in procedures for their isolation. Vibrios are generally viewed as acid sensitive although growth of *V. parahaemolyticus* has been demonstrated down to pH 4.5–5.0.

The natural habitat of vibrios is the marine and estuarine environment. *V. cholerae* can be isolated from temperate, sub-tropical, and tropical waters throughout the world, but seem to disappear from temperate waters during the colder months. Long-term survival may be enhanced by attachment to the surfaces of plants and marine animals and a viable but non-culturable form has also been described where the organism cannot be isolated from the environment using cultural techniques even though it is still present in an infective form.

V. parahaemolyticus is primarily associated with coastal inshore waters rather than the open sea. It cannot be isolated when the water temperature is below 15 °C and cannot survive pressures encountered in deeper waters. The survival of the organism through winter months when water temperatures drop below 15 °C has been attributed to its persistence in sediments from where it may be recovered even when water temperatures are below 10 °C.

Most environmental isolates of both *V. cholerae* and *V. parahaemolyticus* are non-pathogenic. The majority of the *V. cholerae* are non-O1 serotypes and even those that are O1 tend to be non-toxigenic. Similarly 99% of environmental strains of *V. parahaemolyticus* are non-pathogenic.

7.15.3 Pathogenesis and Clinical Features

Cholera usually has an incubation period of between one and three days and can vary from mild, self-limiting diarrhoea to a severe, life-threatening disorder. The infectious dose in normal healthy individuals is large when the organism is ingested without food or buffer, of the order of 10^{10} cells, but is considerably reduced if consumed with food which protects the bacteria from stomach acidity. Studies conducted in Bangladesh indicate that 10^3–10^4 cells may be a more typical infectious dose. Individuals with low stomach acidity (hypochlorohydric) are more liable to catch cholera.

Cholera is a non-invasive infection where the organism colonizes the intestinal lumen and produces a potent enterotoxin. Details of the cholera toxin and its mode of action are given in Chapter 6 (Section 6.7). In severe cases, the hypersecretion of sodium, potassium, chloride, and bicarbonate induced by the enterotoxin results in a profuse, pale, watery diarrhoea containing flakes of mucus, described as rice water stools. The diarrhoea, which can be up to $20 \ 1 \ \text{day}^{-1}$ and contains up to 10^8 vibrios ml^{-1}, is accompanied by vomiting, but without any nausea or fever.

Unless the massive losses of fluid and electrolyte are replaced, there is a fall in blood volume and pressure, an increase in blood viscosity, renal failure, and circulatory collapse. In fatal cases death occurs within a few days. In untreated outbreaks the death rate is about 30–50% but can be reduced to less than 1% with prompt treatment by intravenous or oral rehydration using an electrolyte/glucose solution.

The reported incubation period for *V. parahaemolyticus* food poisoning varies from 2 h to 4 days though it is usually 9–25 h. Illness persists for up to 8 days and is characterized by a profuse watery diarrhoea free from blood or mucus, abdominal pain, vomiting and fever. *V. parahaemolyticus* is more enteroinvasive than *V. cholerae*, and penetrates the intestinal epithelium to reach the *lamina propria*. A dysenteric syndrome has also been reported from a number of countries including Japan.

Pathogenicity of *V. parahaemolyticus* strains is strongly linked to their ability to produce a 22 kDa, thermostable, extracellular haemolysin. When tested on a medium known as Wagatsuma's agar, the haemolysin can lyse fresh human or rabbit blood cells but not those of horse blood, a phenomenon known as the Kanagawa reaction. The haemolysin has also been shown to have enterotoxic, cytotoxic and cardiotoxic activity.

Most (96.5%) strains from patients with *V. parahaemolyticus* food poisoning produce the haemolysin and are designated Kanagawa positive (Ka+) while 99% of environmental isolates are Ka–. Volunteer feeding studies have found that ingestion of 10^7–10^{10} Ka– cells has no effect whereas 10^5–10^7 Ka+ cells produce illness. A number of other virulence factors have been described but have been less intensively studied.

V. vulnificus is a highly invasive organism that causes a primary septicaemia with a high fatality rate ($\approx 50\%$). Most of the cases of foodborne transmission identified occurred in people with pre-existing liver disease, diabetes or alcoholism. Otherwise healthy individuals are rarely affected and, when they are, illness is usually confined to a gastroenteritis. In foodborne cases, the symptoms of malaise followed by fever, chills and prostration appear 16–48 h after consumption of the contaminated food, usually seafoods, particularly oysters. Unlike other vibrio infections, *V. vulnificus* infections require treatment with antibiotics such as tetracycline.

7.15.4   Isolation and Identification

The enrichment media used for vibrios exploit their greater tolerance for alkaline conditions. In alkaline peptone water (pH 8.6–9.0) the incubation period must be limited to 8 h to prevent overgrowth of the vibrios by other organisms. Tellurite/bile salt broth (pH 9.0–9.2) is a more selective enrichment medium and can be incubated overnight.

The most commonly used selective and differential agar used for vibrios is thiosulfate/citrate/bile salt/sucrose agar (TCBS). The medium was originally designed for the isolation of *V. parahaemolyticus* but other enteropathogenic vibrios grow well on it, with the exception of *V. hollisae*. *V. parahaemolyticus*, *V. mimicus*, and *V. vulnificus* can be distinguished from *V. cholerae* on TCBS by their inability to ferment sucrose which results in the production of green colonies. *V. cholerae* produces yellow colonies. Individual species can then be differentiatcd on the basis of further biochemical tests.

V. cholerae is divided into the serogroups O1 and non-O1. O1 strains can be further classified into the classical (non-haemolytic) or El Tor (haemolytic) biotypes each of which can be subdivided by serotyping into one of three groups: Ogawa, Inaba or Hikojima. These can be further subdivided by phage typing although with the advent of molecular typing techniques this is less commonly used. Clinical strains of *V. parahaemolyticus* can be serotyped for epidemiological purposes using a scheme based on 11 thermostable O antigens and 65 thermolabile K (capsular) antigens.

7.15.5 Association with Foods

Cholera is regarded primarily as a waterborne infection, though food which has been in contact with contaminated water can often serve as the vehicle. Consequently a large number of different foods have been implicated in outbreaks, particularly products such as washed fruits and vegetables which are consumed without cooking. Foods coming from a contaminated environment may also carry the organism, for example seafoods and frog's legs. In the current pandemic in South and Central America, an uncooked fish marinade, in lime or lemon juice, *ceviche* has been associated with some cases.

V. parahaemolyticus food poisoning is invariably associated with fish and shellfish. Occasional outbreaks have been reported in the United States and Europe, but in Japan it is the commonest cause of food poisoning. This has been linked with the national culinary habit of consuming raw or partially cooked fish, although illness can also result from cross-contamination of cooked products in the kitchen. Though the organism is only likely to be part of the natural flora of fish caught in coastal waters during the warmer months, it can readily spread to deep-water species through contact in the fish market and it will multiply rapidly if the product is inadequately chilled.

The risk of *V. vulnificus* infection associated with raw oysters is most effectively controlled by cooking the product before consumption. Susceptible individuals, such as those described previously, and the immuno-suppressed should avoid consumption of uncooked shellfish.

7.16 *YERSINIA ENTEROCOLITICA*

7.16.1 Introduction

Yersinia enterocolitica is one of three species of the genus *Yersinia* recognized as human pathogens; *Y. enterocolitica* causes predominantly a gastroenteritis, while Y. *pseudotuberculosis* is associated mainly with mesenteric adenitis. In terms of their social impact, both pale into insignificance when compared to *Yersinia pestis*, responsible for the bubonic plague which killed an estimated 25% of the European population in the 14th Century.

The genus *Yersinia* is named after the French bacteriologist Alexandre Yersin who, in 1894, first described the organism responsible for the bubonic plague. It was created to accommodate former members of the genus *Pasteurella* that were clearly members of the Enterobacteriaceae. In 1964, comparison of *Bacterium enterocoliticum* with a number of closely related isolates, identified by other workers as *Pasteurella* spp., led Frederiksen to propose the creation of the new species *Yersinia enterocolitica*. Further definition within the genus has occurred with the creation of seven new species from non-pathogenic strains previously described as '*Yersinia enterocolitica-like*'. Those most commonly isolated from foods, Y. *frederiksenii*, Y. *intermedia*, Y. *kristensenii*, Y. *mollaretii* and Y. *bercovierii* can be readily distinguished from Y. *enterocolitica* on the basis of a few biochemical tests.

The importance of Y. *enterocolitica* as a cause of foodborne illness varies between countries. In England and Wales, laboratory reports of Y. *enterocolitica* infections, mostly sporadic cases, increased from 45 in 1980 to 571 in 1989 when it outnumbered cases of both *Staph. aureus* and *Bacillus* food poisoning. This represented a peak in reports which then declined to an average of 25 p.a. in the period 2000 to 2005.

Yersiniosis is most common in the cooler climates of northern Europe, particularly in Belgium, and in North America where a number of large outbreaks have been reported. It also displays a different seasonal variation from most other foodborne pathogens with a peak in reported cases occurring in the autumn and winter.

7.16.2 The Organism and its Characteristics

Yersinia enterocolitica is a member of the Enterobacteriaceae; an asporogenous, short (0.5–1.0 by 1–2 µm) Gram-negative rod which is facultatively anaerobic, catalase-positive and oxidase-negative. It can grow over a wide range of temperature, from −1 °C to +40 °C, with an optimum around 29 °C and has a number of temperature-dependent phenotypic characteristics. For example, it is non-motile at 37 °C, but motile with peritrichous flagella below 30 °C. Like other psychrotrophs, though able to grow at chill temperatures, it does so slowly and at 3 °C has been found to take 4 days to increase by 2-log cycles in broth media.

It is heat sensitive but with considerable variation between strains; measured D values in whole milk at 62.8 °C have varied from 0.7–57.6 s.

Optimal growth occurs at a pH 7–8 with a minimum (in broth at 25 °C) varying between 5.1 and 4.1 depending on the acidulant used. As the temperature decreases so the minimum growth pH increases. Growth is possible in broth media containing 5% salt but not 7% salt at 3 °C or 25 °C.

Y. enterocolitica can be isolated from a range of environmental sources including soil, fresh water and the intestinal tract of many animals. Surveys have found the organism in numerous foods including milk and dairy products, meats, particularly pork, poultry, fish and shellfish, fruits and vegetables.

Most food isolates are however non-pathogenic and are known as environmental strains. The species can be subdivided by biotyping, serotyping and phage typing and pathogenicity appears to be associated only with certain types, each with a particular geographical distribution (Table 7.10). In Europe, Canada, Japan and South Africa human yersiniosis is most frequently caused by biotype 4, serotype O3 (4/O3) and, to a lesser extent in Europe and Japan, bio-serotype 2/O9. Strains of bio-serotype 4/O3 from Europe, Canada and South Africa can be distinguished by phage typing. In the United States bio-serotype 1/O8 most commonly causes human yersiniosis, although a wider range of bio-serotypes is encountered, e.g. 1/O13a; 1/O13b; 2/O5,27.

A number of techniques other than biotyping and serotyping have been described which claim to distinguish pathogenic from environmental strains of *Y. enterocolitica* relatively simply and are therefore more within the capacities of routine laboratories. These include the ability of pathogenic strains to autoagglutinate at 37 °C, their dependency on calcium for growth at 37 °C, and their uptake of Congo red dye, and they are usually associated with the presence of the 40–48 MDa virulence

Table 7.10   *Relationship between bio-sero-phage type of* Yersinia enterocolitica *host and geographical distribution*

Biotype	Serotype	Phage type	Host	Syndrome	Country
1	O8	X	Man	Gastroenteritis	USA, Canada
2	O9	X_3	Man	Gastroenteritis	Europe, Japan
			Pigs	Healthy	Europe, Japan
3	O1	II	Chinchillas	Systemic infection	Europe
4	O3	VIII	Man	Gastroenteritis	Europe, Japan
			Pigs	Healthy	Europe, Japan
		IXA	Man	Gastroenteritis	South Africa
			Pigs	Healthy	South Africa
		IXB	Man	Gastroenteritis	Canada
			Pigs	Healthy	Canada
5	O2	XI or II	Hares	Death	Europe
			Goats		

From S.J. Walker, PhD 1986

plasmid (see below). These tests are not completely reliable due to a number of problems such as the expression of plasmid-encoded phenotype in culture, occurrence of atypical strains and the possibility of plasmid loss during isolation. A test for pyrazinamidase activity which is not plasmid mediated may offer some advantages in this respect.

7.16.3 Pathogenesis and Clinical Features

Illness caused by *Y. enterocolitica* occurs most commonly in children under seven years old. It is a self-limiting enterocolitis with an incubation period of 1–11 days and lasting for between 5 and 14 days, although in some cases it may persist for considerably longer. Symptoms are predominantly abdominal pain and diarrhoea accompanied by a mild fever; vomiting is rare. Sometimes the pain resulting from acute terminal ileitis and mesenteric lymphadenitis (inflammation of the mesenteric lymph nodes) is confined to the lower right hand side of the body and prompts a mistaken diagnosis of appendicitis and subsequent surgery. A problem of post-infection complications such as arthritis and erythema nodosum (a raised, red skin lesion) can occur in adults, the latter particularly in women. This appears to be mainly associated with serotypes O3 and O9 and is therefore more common in Europe.

Ingested cells of pathogenic *Y. enterocolitica* which survive passage through the stomach acid adhere to the mucosal cells of the Peyer's patches (gut-associated lymphoid tissue). Adhesion is mediated through bacterial outer membrane proteins that are encoded for on a 40–48 MDa plasmid possessed by all pathogenic *Y. enterocolitica*. The plasmid is essential but not the sole prerequisite for virulence since cell invasion is controlled by chromosomal genes (see p.229). The adhered cell is taken up by the epithelial cell by endocytosis where it survives without significant multiplication and can exert cytotoxic activity. Released into the *lamina propria*, it invades phagocytic cells and multiplies extracellularly producing a local inflammatory response. Damage to the absorptive epithelial surface results in malabsorption and a consequent osmotic fluid loss characterized by diarrhoea. Other plasmid-encoded characteristics are thought to contribute to this process, such as the production of outer membrane proteins that confer resistance to phagocytosis, auto-agglutination at 37 °C, and resistance to serum.

A heat-stable enterotoxin (9000–9700 Da) is produced by *Y. enterocolitica* but its role in pathogenesis, if any, is unclear. It bears some similarity to *E. coli* ST immunologically and in its ability to induce fluid accumulation in ligated ileal loops and to stimulate guanylate cyclase activity. Elaboration of enterotoxin in the gut is unlikely since production usually ceases at temperatures above 30 °C. Production in inoculated foods has been shown, but the observed incubation period is inconsistent with a foodborne intoxication. Finally, the ability to produce the toxin is not confined to pathogenic *Y. enterocolitica*, but has also been demonstrated in numerous environmental strains and a number of other *Yersinia* species as well.

7.16.4   Isolation and Identification

A large number of procedures for the isolation and detection of *Y. enterocolitica* have been developed. Enrichment procedures usually exploit the psychrotrophic character of the organism by incubating at low temperature, but this has the disadvantage of being slow with the attendant possibility of overgrowth by other psychrotrophs present. Some workers have included selective agents in their enrichment media but some strains, such as serotype O8, are reported to be sensitive to selective agents. The most commonly used enrichment media are phosphate buffered saline (PBS) or tryptone soya broth (TSB) most usually incubated at 4 °C for 21 days. *Y. enterocolitica* and related species are more alkali resistant than many other bacteria so the pH of enrichment media is sometimes adjusted to 8.0–8.3 or cultures subjected to a short post-enrichment alkali treatment.

The best results for the selective isolation of *Y. enterocolitica* from foods and enrichment broths have been obtained with cefsulodin/ irgasan/novobiocin (CIN) agar. In addition to the antibiotics, the medium contains deoxycholate and crystal violet as selective agents and mannitol as a fermentable carbon source. After incubation at 28 °C for 24 h, typical colonies of *Y. enterocolitica* appear with a dark-red centre surrounded by a transparent border. Isolates can be confirmed and biotyped by biochemical tests.

In vitro tests to distinguish between environmental and pathogenic strains of *Y. enterocolitica* have been referred to above (Section 7.16.2). Techniques using gene probes to detect the virulence-associated plasmid by a colony hybridization test have also been used with some success and offer the possibility of detecting potentially pathogenic strains in foods without the need for lengthy enrichment procedures.

7.16.5   Association with Foods

Pigs are recognized as chronic carriers of those *Y. enterocolitica* serotypes most commonly involved in human infections (O3, O5, 27, O8, O9). The organism can be isolated most frequently from the tongue, tonsils and, in the gut, the caecum of otherwise apparently healthy animals. Despite this, pork has only occasionally been shown to be the vehicle for yersiniosis, although a case control study in Belgium, which has the highest incidence of yersiniosis, implicated a national prediliction for eating raw pork. In 1988/9 an outbreak of yersiniosis in Atlanta involving 15 victims (14 children) was strongly associated with the household production of pork chitterlings.

A number of outbreaks of yersiniosis have been caused by contaminated milk including the largest hitherto recorded which occurred in 1982 in Tennessee, Arkansas and Mississippi in the United States. In this instance pigs were implicated as the original source of contamination, but not demonstrated to be carriers of the same O13 serotypes causing the infection. It was presumed that the organism was transferred from pigs, via mud, onto crates used to transport waste milk from the dairy to the pig farm. The crates were returned to the dairy and inadequately washed and sanitized before being used again to transport retail milk. As a consequence the outside of packs was contaminated with *Y. enterocolitica* which was transferred to the milk on opening and pouring. It was subsequently demonstrated that the organism involved could survive for at least 21 days on the outside of milk cartons held at 4 °C.

Contaminated water used in the production of beansprouts and in the packaging of tofu (soya bean curd) was responsible for two outbreaks in the United States in 1982.

A number of approaches to the control of yersiniosis have been proposed which are generally similar to those proposed for the control of other zoonotic infections such as salmonellosis. These include pathogen-free breeding and rearing of animals, a goal which may not be achievable in practice, and hygienic transport and slaughter practices. Work in Denmark on contamination of pork products with *Y. enterocolitica* has identified evisceration and incisions made during meat inspection as critical control points and has further shown that excision of the tongue and tonsils as a separate operation significantly reduces contamination of other internal organs.

7.17 SCOMBROTOXIC FISH POISONING

Scombrotoxic fish poisoning differs from those types of foodborne illness described above in that it is thought to be an example of where bacteria act as indirect agents of food poisoning by converting food components into harmful compounds. This view has however been questioned in recent years and discussion of scombrotoxicosis may belong more correctly in Chapter 8. Fish is almost always the food vehicle, particularly the so-called scombroid fish such as tuna, bonito and mackerel, but nonscombroid fish such as sardines, pilchards and herrings have also been implicated. In some cases canned fish has been responsible indicating that the toxic factor(s) is heat stable.

It is a chemical intoxication with a characteristically short incubation period of between 10 min and 2 h. Symptoms include a sharp, peppery

Table 7.11   *Guideline histamine levels in fish*

Histamine level Status (mg%)	Status
1	Freshly caught fish
< 5	Normal and safe for consumption
5–20	Mishandled and possibly toxic
20–100	Unsatisfactory and probably toxic
> 100	Toxic and unsafe for consumption

taste in the mouth, itching, dizziness, flushing of the face and neck, often followed by a severe headache, feverishness, diarrhoea, nausea and vomiting. A rash may develop on the face and neck and cardiac palpitations may occur.

The symptoms are those of histamine toxicity and can be alleviated with antihistamines. Histamine is produced by bacterial amino acid decarboxylases acting on histidine which occurs in high concentrations in the tissues of dark-fleshed fish. The bacteria themselves increase in numbers as a result of long storage at inappropriate temperatures and freshly caught fish have not been implicated in this type of poisoning.

When making judgements on the risk of scombrotoxic fish poisoning posed by particular products, regulatory authorities usually rely on a measure of the histamine content of the fish. In the United States, the level of histamine deemed hazardous in tuna is 50 mg%. Some guideline values published by the Health Protection Agency in the UK are presented as Table 7.11.

The problem is not as clear cut as it may at first seem, however. It has, for instance, not proved possible to reproduce the symptoms in volunteers fed histamine, and cases have also been reported where the fish contained low levels of histamine. Although histamine poisoning can occur when clearance of dietary histamine from the body is slowed by monoamine oxidase inhibitors, histamine is generally metabolized efficiently in the human gut and not absorbed *per se*. Among the explanations offered are that other biologically active amines are present in the fish which potentiate the toxicity of histamine or that algal toxins may be involved causing the release of endogenous histamine in the body.

7.18   CONCLUSION

In this chapter we have surveyed the main features of the foodborne bacterial pathogens recognized as being of current or emerging importance. As should be clear, the significance of individual organisms varies from country to country reflecting differences in both diet and culinary practices. It should also be remembered that the scene is likely to change with time. The food microbiologist must be continually vigilant in anticipating the effect that changes in dietary preferences and social behaviour, and developments in crop and animal husbandry and food processing may have on bacterial hazards.

However, as noted in Chapter 6, not all the health risks posed by food are bacteriological. In the next Chapter we will consider some other microbiological hazards that are food associated.

CHAPTER 8

Non-bacterial Agents of Foodborne Illness

We have seen that foods may act as a vehicle for viable bacteria such as *Salmonella*, or for pre-formed bacterial toxins such as botulinum toxin which can cause disease or illness once introduced into the body.

Foods may also act as vehicles for other disease causing agents such as helminths, nematodes, protozoa and viruses as well as toxic metabolites of fungi and algae. Each of these is a specialist area, and cannot be dealt with in the same detail as bacteria, but a food microbiologist should be aware of the occurrence and significance of these nonbacterial agents of foodborne illness.

8.1 HELMINTHS AND NEMATODES

The flatworms and roundworms are not normally studied by microbiologists but amongst these groups there are a number of animal parasites which can be transmitted to humans *via* food and water. These complex animals do not multiply in foods and they cannot be detected and enumerated by cultural methods in the way that many bacteria can. Their presence is normally detected by direct microscopic examination often following some form of concentration and staining procedure.

8.1.1 Platyhelminths: Liver Flukes and Tapeworms

In the context of foodborne parasites the two most important classes of the Platyhelminths (flatworms) are the Trematoda, which includes the liver fluke *Fasciola hepatica*, and the Cestoda which includes tapeworms of the genus *Taenia*. These organisms have complex life cycles which may include quite unrelated hosts at different stages. Thus the mature stage of the liver fluke develops in humans, sheep or cattle which may be referred to as the definitive host (Figure 8.1). It is a leaf-like animal, growing up to two and a half centimetres in length by one centimetre in width, which

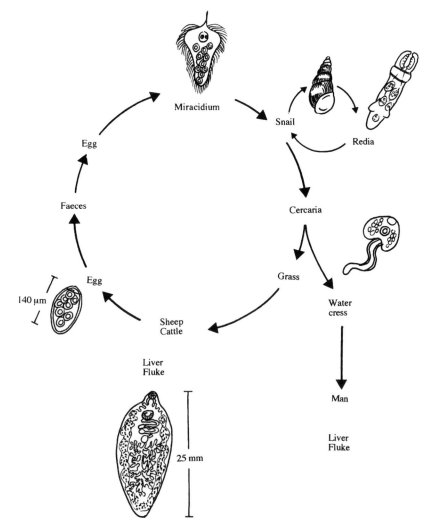

Figure 8.1 *Life cycle of the liver fluke,* Fasciola hepatica

finally establishes itself in the bile duct after entering and feeding on the liver. Having matured, it eventually produces quite large eggs ($150 \times 90\,\mu$m) which have a lid, or operculum, at one end. These are secreted in the faeces after passing from the bile duct to the alimentary tract.

The eggs hatch in water to produce a highly ciliate and motile embryo (miracidium) which cannot infect the definitive host and has to infect a species of water snail such as *Limnaea truncatula*. There are several stages in this secondary host, some leading to multiplication of the parasite, but eventually the organism is released as the final larval stage (cercaria) which encyst and may survive for up to a year depending on their environment. Cysts will only develop further if they are swallowed by an appropriate definitive host, usually cattle or sheep in which infection can cause serious economic loss, or more rarely in humans after eating raw or undercooked watercress on which the cysts have become attached.

In people the symptoms are fever, tiredness and loss of appetite with pain and discomfort in the liver region of the abdomen. The disease is known as fascioliasis and can be diagnosed by finding the eggs in the faeces or body fluids such as biliary or duodenal fluids.

Of the Cestoda, the tapeworms *Taenia solium* associated with pork and *T. saginata* associated with beef, are best known. These long, ribbon-like flatworms have humans as their definitive host but they differ in their secondary host. The larval stages of the beef tapeworm have to develop in cattle and finally infect humans through the consumption of under-cooked beef. However, the larval stages of *Taenia solium* can develop in humans or pigs. These larval stages, known as cysticerci, develop in a wide range of tissues including muscle tissue where they can cause a spotted appearance. The mature tapeworm of these species can only develop in the human intestine where it causes more severe symptoms in the young and those weakened by other diseases, than in healthy adults. The effects are fairly general and may include nausea, abdominal pain, anaemia and a nervous disorder resembling epilepsy, as well as mechanical irritation of the gut. If the latter is so severe as to cause a reversed peristalsis, so that mature segments of the tapeworm (proglottids) enter the stomach and release eggs (onchospheres), there may be an invasion of the body tissues (cysticercosis). The resulting bladder worms, or cysticerci, can especially invade the central nervous system, a situation which is often fatal. There are some species of tapeworm, such as *Diphyllobothrium latum*, which have complex life cycles involving crustacea and fish. Humans may become infected through the consumption of raw or under cooked fish.

8.1.2 Roundworms

Perhaps the most notorious of the nematodes in the context of foodborne illness, and the only one which will be dealt with here, is *Trichinella spiralis*, the agent of trichinellosis which was first recognised as a cause of illness in 1860.

This parasite has no free-living stage but is passed from host to host which can include quite a wide range of mammals including humans and pigs. Thus trichinellosis in the human population is usually acquired from the consumption of infected raw or poorly cooked pork products.

Trichinella has an intriguing life history for it is the active larval stages which cause discomfort, fever and even death. Infection starts by the

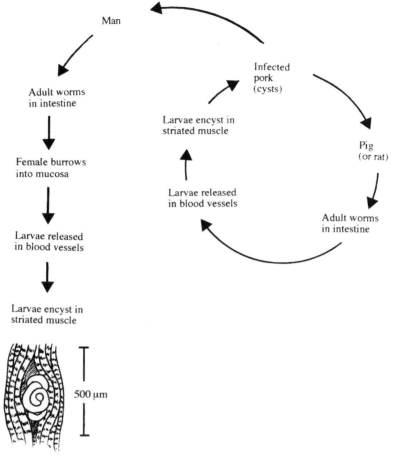

Figure 8.2   Trichinella spiralis

consumption of muscle tissue containing encysted larvae which have curled up in a characteristic manner in a cyst with a calcified wall (Figure 8.2). In this state they can survive for many years in a living host but, once eaten by a second host, the larvae are released by the digestive juices of the stomach and they grow and mature in the lumen of the intestines where they may reach 3–4 mm in length. On the assumption that uncooked human flesh is not consumed, the human host represents a dead end for nematodes such as *Trichinella*. This is unlike the situation in the Cestodes, such as *Taenia*, where passage of proglottids in human faeces can complete the cycle back to domestic animals.

The adult worms do not cause any apparent symptoms but a single female can produce more than a thousand larvae, each of which is about 100 μm long, and it is these larvae which burrow through the gut wall and eventually reach a number of specific muscle tissues in which they grow up to about 1 mm before curling up and encysting. Such cysts were first shown to contain these tiny worm larvae by a first year medical student studying at St. Bartholomew's Hospital, London in 1835. The student was James Paget who was renowned as a Norfolk naturalist and became an eminent surgeon. He had seen, and was puzzled by, some small hard white specks in the flesh of a cadaver used in a routine post-mortem dissection.

The symptoms caused by *Trichinella spiralis* occur in two phases. The period during which the larvae are invading the intestinal mucosa is associated with abdominal pain, nausea and diarrhoea. This may occur within a few days after eating heavily infested meat or after as long as a month if only a few larval cysts are ingested. The second phase of symptoms, which include muscle pain and fever, occurs as the larvae invade and finally encyst in muscle tissue.

Prevention has to be by breaking the cycle within the pig population and by adequate cooking of pork products. The United States Department of Agriculture recommends that all parts of cooked pork products should reach at least 76.7 °C. Freezing will destroy encysted larvae but in deep tissue it may take as long as 30 days at −15 °C. Curing, smoking and the fermentations leading to such products as salami do all eventually lead to the death of encysted *Trichinella* larvae.

The control of these parasites in the human food chain is effected by careful meat inspection and the role of the professional Meat Inspector, supported by legislation is very important. Badly infected animals may be recognized by ante-mortem inspection and removed at that stage. The presence of these parasites in animals usually gives rise to macroscopic changes in tissues and organs which can be recognized by meat inspection after slaughter.

Although *T. spiralis* is the most important species of *Trichinella* several others are recognized. *T. nativa* occurs in the meat of arctic carnivores, such as polar bears and walrus, and consumption of infected meat may be responsible for trichinellosis among the Inuit people. This species is particularly resistant to low temperatures and Alaskan bear meat has been shown to be infective after 35 days at −15 °C.

8.2 PROTOZOA

Amongst the protozoa only a few genera are of special concern to the food microbiologist; the flagellate *Giardia*, the amoeboid *Entamoeba* and three sporozoid (members of the phylum Apicomplexa which contains parasitic protozoa propagated by spores) genera *Toxoplasma*, *Sarcocystis* and *Cryptosporidium*. Examples are known of both enteric and systemic infections.

8.2.1 *Giardia lamblia*

Although usually associated with water, or transmission from person to person by poor hygiene, a number of outbreaks of the diarrhoeal disease caused by *Giardia lamblia*, which may also be known as *G. intestinalis* or *Lamblia intestinalis*, have been confirmed as foodborne outbreaks. The organism survives in food and water as cysts but, although it can be cultured in the laboratory, it does not normally grow outside its host. The infective dose may be very low and once ingested the gastric juices aid the release of the active flagellate protozoa, known as trophozoites, which are characterized by the possession of eight flagella and two nuclei (Figure 8.3). The organism is not particularly invasive and it is not clear how the symptoms of diarrhoea, abdominal cramps and nausea are caused but it is possible that a protein toxin is involved.

Giardia cysts have been found on salad vegetables such as lettuce and fruits such as strawberries and could occur on any foods which are washed with contaminated water or handled by infected persons not observing good hygienic practice. Confirmed foodborne outbreaks have implicated home-canned salmon and noodle salad but the difficulty of demonstrating low numbers of cysts in foods may be one reason why foods are rarely directly implicated.

Although the cysts are resistant to chlorination processes used in most water treatment systems, they are killed by the normal cooking procedures used in food preparation.

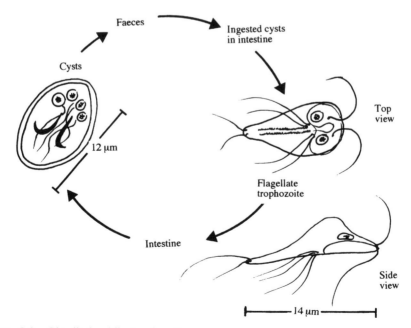

Figure 8.3 Giardia lamblia *trophozoite*

8.2.2   *Entamoeba histolytica*

Amoebic dysentery can be very widespread wherever there is poor hygiene, for it is usually transmitted by the faecal–oral route. Although outbreaks from food are also documented they are surprisingly rare. The organism is an aerotolerant anaerobe which survives in the environment in an encysted form. Indeed, a person with amoebic dysentery may pass up to fifty million cysts per day. Although most infections remain symptomless, illness may start with the passing of mucous and bloody stools, due to ulceration of the colon, a few weeks after infection and progress to severe diarrhoea, abdominal pain, fever and vomiting. *Entamoeba histolytica* infection is endemic in many poor communities in all parts of the world but there has been a steady decline in reported incidence in the United Kingdom over the past twenty years.

8.2.3   **Sporozoid Protozoa**

Cryptosporidiosis appears to be an increasing cause of diarrhoea and, although the disease is normally self-limiting, it can become a serious infection in the immunocompromised such as AIDS patients. It is very uncommon for food to be directly implicated in cryptosporidiosis but this may reflect the difficulties in detecting small numbers in foods and the low infective dose required to cause disease. Indirect evidence from epidemiological studies does suggest that certain foods such as raw sausages are a risk factor. However, as a waterborne threat *Cryptosporidium* can be very serious and caused what might have been the largest documented outbreak of gastrointestinal disease in America when an estimated 403,000 cases were reported in Milwaukee, Wisconsin in 1993.

Although it is complex, the whole life cycle of *Cryptosporidium parvum* can take place in a single host which may be human or a species of farm animal such as cattle or sheep. *Cryptosporidium meleagridis*, previously associated with birds, is also known to infect humans.

Cyclospora, closely related to *Cryptosporidium* in the phylum of protists known as the Alveolata, has been recognised since the early 1990s as a causative agent of a few gastrointestinal outbreaks associated with unprocessed fresh food products such as soft fruits and vegetables. Although there are several species associated with different animals, only *Cyclospora cayetanensis*, first recognised in 1977, is found in humans and seems to be restricted to the human host. Symptoms include non-bloody diarrhoea, loss of appetite, weight loss, stomach cramps, nausea, vomiting, fatigue and fever. Outbreaks occurring in North America during the late 1990s were often associated with the consumption of fresh raspberries imported from Guatemala. By contrast, species of *Sarcocystis* are obligately two-host parasites, the definitive host in which sexual reproduction of the parasite takes place being a carnivore such as cats, dogs or humans, and an intermediate host such as cattle, sheep or pigs in the tissues of which the asexual cysts are formed.

Two species can infect humans: *S. hominis*, which infects cattle, and *S. suihominis* from pigs. Although symptoms are usually mild, they can include nausea and diarrhoea. Beef and pork which have been adequately cooked lose their infectivity.

In the case of *Toxoplasma gondii*, the definitive host is the domestic or wild cat but many vertebrate animals including humans are susceptible to infection by the oocysts shed in their faeces. Thus herbivores can become infected by eating grass and other feedstuffs contaminated by cat faeces and, once infected, their tissues may remain infectious for life. Although foodborne infection in humans may be rare, it could occur through consumption of raw or undercooked meat, especially pork or mutton.

Toxoplasmosis is usually symptomless or associated with a mild influenza-like illness in healthy humans, but infection can be serious in immunocompromised people.

8.3 TOXIGENIC ALGAE

Although strictly speaking the term algae should now be used as a collective term for a number of photosynthetic eukaryotic phyla, for the purposes of this section the prokaryotic cyanobacteria, or blue-green algae, will also be included.

A number of planktonic and benthic algae can produce very toxic compounds which may be transported to filter-feeding shellfish such as mussels and clams, or small herbivorous fish which are food for larger carnivorous fish. As these toxins pass along a food chain they can be concentrated and it may be the large carnivorous fish which are caught for human consumption which are most toxic. In the case of shellfish, the toxins may accumulate without apparently harming the animal but with potent consequences for people or birds consuming them.

A number of distinct illnesses are now recognized including PSP (paralytic shellfish poisoning), NSP (neurotoxic shellfish poisoning), DSP (diarrhoeal shellfish poisoning), ASP (amnesic shellfish poisoning) as well as ciguatera fish poisoning. The toxins implicated in the various forms of shellfish poisoning are not only undetectable organoleptically but are also generally unaffected by cooking.

8.3.1 Dinoflagellate Toxins

Planktonic dinoflagellates may occasionally form blooms containing high numbers of organisms when environmental conditions such as temperature, light and nutrients are appropriate. *Gonyaulax catenella* and *G. tamarensis* (now both referred to the genus *Alexandrium*) are the best known of a number of dinoflagellates responsible for paralytic shellfish poisoning.

This can be a serious illness with a high mortality rate. The toxic metabolites of these algae, which include saxitoxin and gonyautoxin (Figure 8.4) block nerve transmission causing symptoms such as tingling and numbness of the fingertips and lips, giddiness and staggering, incoherent speech and respiratory paralysis. Saxitoxin, so named because it was isolated directly from the Alaskan butter clam, *Saxidomas giganteus*, before it was recognized that it is actually a dinoflagellate metabolite, is a very toxic compound having a lethal dose for the mouse of as little as 0.2 µg. Recent studies have even thrown doubt on whether saxitoxin is, in fact, a dinoflagellate metabolite for it has been shown to be produced by a species of bacteria of the genus *Moraxella* isolated from the dinoflagellates.

To control paralytic shellfish poisoning, the collection and sale of bivalves in areas affected by algal blooms is banned.

Neurotoxic shellfish poisoning is less severe and less common than paralytic shellfish poisoning and is associated with the brevitoxins, which are complex cyclic polyethers produced by the dinoflagellate *Ptychodiscus brevis* (=*Gymnodinium breve*) one of the species responsible for the phenomenon known as a 'red tide'.

A third quite different form of poisoning associated with eating shellfish which have accumulated dinoflagellates is diarrhoeal shellfish poisoning caused by lipophilic toxins such as dinophysistoxin produced by *Dinophysis fortii* (Figure 8.5). The major symptoms, which usually occur within an hour or two of consuming toxic shellfish, include diarrhoea, abdominal pain, nausea and vomiting and may persist for several days.

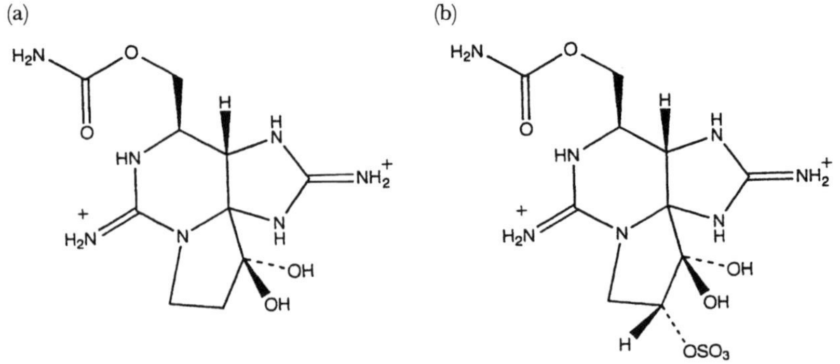

Figure 8.4   *Toxic metabolites of algae (a) saxitoxin and (b) gonyautoxin 2*

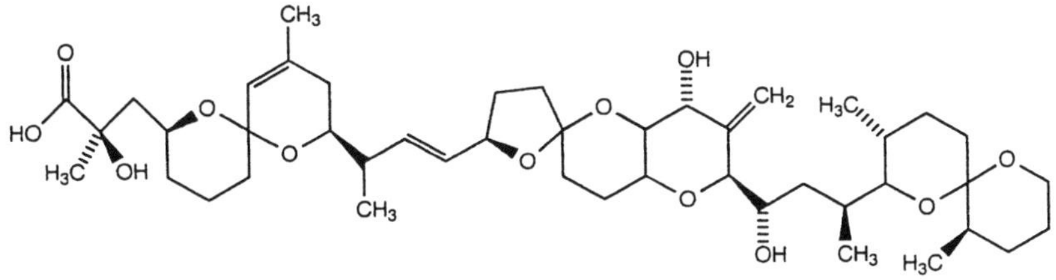

Figure 8.5 *Dinophysistoxin*

For many years a strange type of poisoning occurring after eating a number of different species of edible fish, including moray eel and barracuda, has been known as ciguatera poisoning. It can be a serious problem in tropical and subtropical parts of the world. Nausea, vomiting and diarrhoea may be accompanied by neurosensory disturbances, convulsions, muscular paralysis and even death.

After many years research it has been shown that the toxins responsible for these symptoms, which include ciguatoxin $C_{60}H_{86}O_{19}$, a polyunsaturated polycyclic ether, are produced by a benthic or epiphytic species of dinoflagellate known as *Gambierdiscus toxicus*. The toxins are concentrated along a food chain starting with the consumption of algae by herbivorous and detritus-feeding reef fish which themselves are consumed by the larger carnivorous fish caught for human consumption.

8.3.2 Cyanobacterial Toxins

Although the blue–green algae are prokaryotes, their photosynthesis, during which oxygen is liberated from water, is characteristic of that shown by the eukaryotic algae and higher plants. Several genera of freshwater cyanobacteria, especially species of *Microcystis, Anabaena* and *Aphanizomenon*, can form extensive blooms in lakes, ponds and reservoirs and may cause deaths of animals drinking the contaminated water. The presence of such cyanobacteria in public water supplies has also been implicated in outbreaks of human gastroenteritis.

The cyanoginosins, toxic metabolites of *Microcystis aeruginosa*, are cyclic polypeptides containing some very unusual amino acids (Figure 8.6) and are essentially hepatotoxins.

8.3.3 Toxic Diatoms

An outbreak of food poisoning, known as amnesic shellfish poisoning (ASP) or domoic acid poisoning, following consumption of cultivated mussels from farms in Canada, which involved more than 100 cases and three deaths, was shown to be due to a glutamate antagonist in the

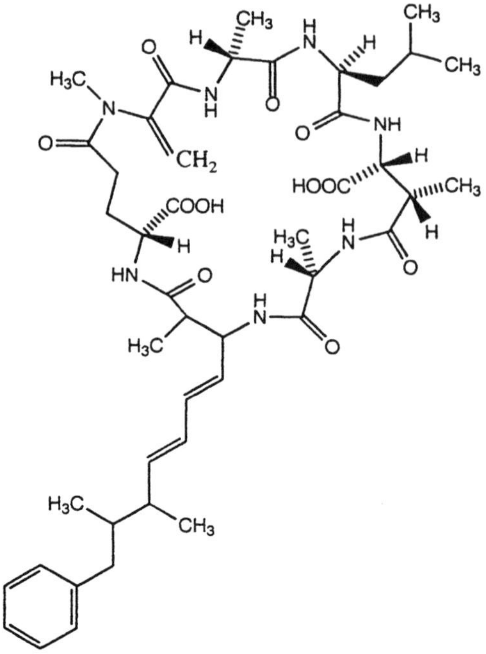

Figure 8.6 *Cyanoginosin*

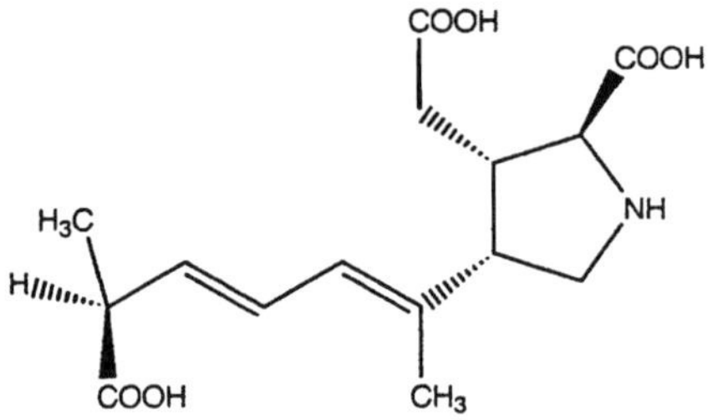

Figure 8.7 *Domoic acid*

central nervous system known as domoic acid (Figure 8.7). This compound had been produced by *Nitzschia pungens*, a chain-forming diatom of the phytoplankton which is now referred to the genus *Pseudonitzschia*.

8.4   TOXIGENIC FUNGI

The fungi are heterotrophic and feed by absorption of soluble nutrients and although many fungi can metabolize complex insoluble materials, such as lignocellulose, these materials have to be degraded by the secretion of appropriate enzymes outside the wall. A number of fungi are parasitic on both animals, plants and other fungi, and some of these parasitic associations have become very complex and even obligate. However, it is the ability of some moulds to produce toxic metabolites, known as mycotoxins, in foods and their association with a range of human diseases, from gastroenteric conditions to cancer, which concerns us here.

The filamentous fungi grow over and through their substrate by processes of hyphal tip extension, branching and anastomosis leading to the production of an extensive mycelium. Some species have been especially successful in growing at relatively low water activities which allows them to colonize commodities, such as cereals, which should otherwise be too dry for the growth of micro-organisms.

Frequently, when moulds attack foods they do not cause the kind of putrefactive breakdown associated with some bacteria and the foods may be eaten despite being mouldy and perhaps contaminated with mycotoxins. Indeed, some of the changes brought about by the growth of certain fungi on a food may be organoleptically desirable leading to the manufacture of products such as mould-ripened cheeses and mould-ripened sausages using species of *Penicillium*.

8.4.1 Mycotoxins and Mycophagy

The vegetative structures of the filamentous fungi are essentially based on the growth form of the spreading, branching, anastomosing mycelium and have a relatively limited morphological diversity. However, the structures associated with spore production and dispersal give rise to the developmental and morphological diversity of the filamentous fungi. Many are microscopic and conveniently referred to as moulds, but amongst the basidiomycetes and ascomycetes there are species producing prodigiously macroscopic fruit bodies, the mushrooms and toadstools, which have evolved as very effective structures for the production and dispersal of spores. These two aspects of fungal morphology have led to two distinct branches in the study of fungal toxins.

The mycotoxins are metabolites of moulds which may contaminate foods, animal feeds, or the raw materials for their manufacture, and that happen to be toxic to humans or their domestic animals. The study of mycotoxins, and the legislation associated with their control, are based on them being considered as adulterants of foods or animal feeds.

On the other hand, mushrooms and toadstools have provided a traditional source of food in many parts of the world for many thousands of years. Unfortunately, this group of fungi includes a number of species which produce toxic metabolites in their fruiting bodies but, because the toxins are a natural constituent of fruiting bodies deliberately ingested, usually as a result of mistaken identity, they are not considered as mycotoxins (Table 8.1). This is a somewhat arbitrary

Table 8.1 *Toxic compounds of some 'toadstools'*

Toxin	Species	Toxic effects
Coprine	*Coprinus atramentarius*	Considerable discomfort when consumed with alcohol
Illudin	*Omphalotus olearius*	Gastrointestinal irritation Vomiting
Amatoxin	*Amanita phalloides*	Liver and kidney damage, death unless treated
Orellanin	*Cortinarius orellanus*	Irreversible kidney damage, death or very slow recovery
Psilocybin	*Psilocybe cubensis*	Hallucinogenic
Muscarine	*Inocybe patouillardii*	Vomiting and diarrhoea

distinction based on human behaviour and not on the chemistry, biochemistry or toxicology of the compounds.

There are relatively few species of agarics which can be considered as deadly poisonous but they include the deathcap, *Amanita phalloides*, a quarter of a cap of which can be lethal to a healthy adult, and species of *Cortinarius* which are still foolishly mistaken for edible wild fungi (Figure 8.8). In both these cases the toxins cause irreversible damage to the liver and kidneys and death may follow several weeks after the initial consumption of the poisonous fungi.

It is worth emphasizing a major difference between the toxic metabolites of fungi and the toxins of most of the bacteria associated with food poisoning. The former are relatively low molecular weight compounds, although their chemistry may be very complex, while the latter are macromolecules such as polypeptides, proteins or lipopolysaccharides. An exception to this generalization is an unusual bacterial food poisoning associated with a traditional food produced in parts of Indonesia; a form of tempeh is made by inoculating coconut flesh with moulds such as *Rhizopus* and *Mucor*. Occasionally the process becomes contaminated with the bacterium *Burkholderia cocovenenans*, previously known as *Pseudomonas cocovenenans*, which produces at least two low molecular weight toxic metabolites, bongkrekic acid and toxoflavin (see Chapter 9).

Although there are many genera of moulds which include toxigenic species three stand out as especially important – *Aspergillus*, *Penicillium* and *Fusarium*.

8.4.2 Mycotoxins of *Aspergillus*

8.4.2.1 The Aflatoxins. In 1959 a very singular event occurred which initiated the international interest which now exists in mycotoxins. This was the deaths of several thousand turkey poults and other poultry on farms in East Anglia and, because of the implications for the turkey

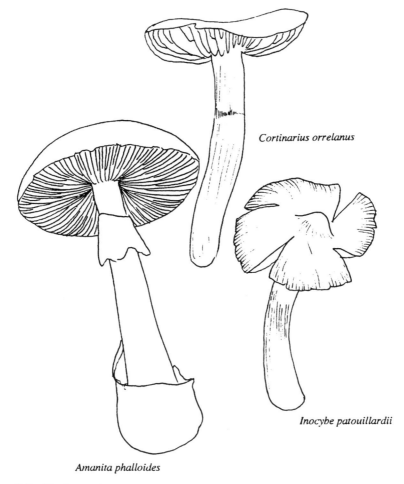

Cortinarius orrelanus

Inocybe patouillardii

Amanita phalloides

Figure 8.8 *Toxic agarics*

industry and the manufacture of pelleted feed which supported it, a considerable effort was put into understanding the etiology of this major outbreak of what was initially referred to as turkey X disease. Although the name implies a disease such as a viral infection, it was shown that the birds had been poisoned by a contaminant in the groundnut meal used as a protein supplement in the pelleted feed. The contaminant, which was called aflatoxin, fluoresces intensely under ultra-violet light and was shown to be produced by the mould *Aspergillus flavus* growing on the groundnuts.

Aflatoxin is not only acutely toxic but, for the rat, it is amongst the most carcinogenic compounds known. The demonstration of the potential carcinogenicity of aflatoxin made it possible to rationalize the etiology of diseases such as liver carcinoma in rainbow trout and

Table 8.2 *Some reports of aflatoxins during 1996 and 1997*

Commodity	Country (year) reported	Incidence (%)	Range ($\mu g \, kg^{-1}$)
Maize	Argentina (1996)	20	5–560
Peanuts	India (1996)	45	5–833
Pistachios	Netherlands (1996)	59	2–165
Wheat	Uruguay (1996)	20	2–20
Cottonseed meal	UK (1997)	71	5–25
Maize	India (1997)	45	5–666
Rice	Equador (1997)	9	6.8–40

From Pittet, *Revue Médecine Vétérinaire*, 1998, **149**, 479

hepatitis X in dogs which had been described nearly a decade earlier but had remained a mystery. Very sensitive analytical methods for aflatoxins were developed which led to the demonstration that their occurrence was widespread in many agricultural commodities, especially groundnuts and maize, much of which may be destined for human consumption.

Aflatoxins are still reported from a wide range of foods and animal feeds (Table 8.2) but, whereas the concentrations which cause acute toxic symptoms would be measured in $mg \, kg^{-1}$, today's analytical procedures make it possible for quantitative detection of $\mu g \, kg^{-1}$.

In 2005 the European Commission was still expressing anxiety about imports of pistachios, peanuts and brazil nuts from a number of producing countries because they were contaminated with unacceptable levels of aflatoxins. The aflatoxins are produced predominantly by two closely related species of mould, *Aspergillus flavus* and *A. parasiticus*, both of which are especially common in the tropics and subtropics. More recently three more species have been recognised as aflatoxigenic, *A. nomius, A. pseudotamarii* and *A. ochraceoroseus*, but the frequent reports in the early literature of the production of aflatoxins by other species, even belonging to different genera of moulds, are usually the result of artefacts or mistakes.

Initially, it was considered that aflatoxin contamination was essentially a problem of poor storage of commodities after harvest allowing the growth of storage fungi such as aspergilli and penicillia with consequent formation of mycotoxins. Indeed, conditions of high humidity and warm temperatures can give rise to the highest levels of aflatoxin in food often exceeding the upper limit initially established by the Food and Agricultural Organization (FAO) and the World Health Organization (WHO) of $30 \, \mu g \, kg^{-1}$ in foods for human consumption. It has to be recognized that these agencies faced a hard dilemma when setting these limits and this is reflected in the observation that 'clearly the group would have preferred a lower figure, but felt that the danger of malnutrition was greater than the danger that aflatoxin would produce liver cancer in man'. Meanwhile, many developed countries had set even more stringent legislative or guideline levels, some of the more recent of which are shown in Table 8.3.

It is now realized that aflatoxins are not simply a problem of poor storage, but they can be produced in the growing crop before harvest. Aflatoxigenic species of *Aspergillus* can establish an endophytic relationship with the healthy plant and produce low, but significant, amounts of aflatoxin when the plant is stressed, such as occurs during a drought.

Like many microbial secondary metabolites, the aflatoxins are a family of closely related compounds, the most toxic of which is referred to as aflatoxin B1 (Figure 8.9). The precise nature of the response to aflatoxin is dependent on species, sex and age, in general the male is more sensitive than the female. Some animals, such as the day-old duckling and the adult dog, are remarkably sensitive to the acute toxicity of aflatoxin B1 with LD_{50} values of 0.35 and $0.5\,mg\,kg^{-1}$ body weight respectively, while others, such as the adult rat and the mouse, are more resistant (LD_{50} *ca.* $9\,mg\,kg^{-1}$). Not all animals respond to the carcinogenic activity of aflatoxin but for the rat and the rainbow trout aflatoxin B1 is one of the most carcinogenic compounds known.

Table 8.3 *Some maximum tolerated levels[a] for aflatoxin in foodstuffs*

Country	Commodity	Tolerance ($\mu g\ kg^{-1}$)
Australia	Peanut products	15
Belgium	All foods	5
Canada	Nuts and nut products	15
China	Rice and other cereals	10
France	All foods	10
	Infant foods	1
India	All foods	30
United Kingdom	Nuts and nut products	2[a]
United States	All Foods	20

[a] On 16 July 1998 the European Commission set maximum tolerated levels for a number of food commodities (EC Regulation 1525/98 EC) and Member States were allowed a period of time to comply. UK Statutory Instrument, Contaminants in Food (Amendment) Regulations 1999 (SI No. 1999/1603) came into force on 30 June 1999 reflecting UK compliance with the EC regulation

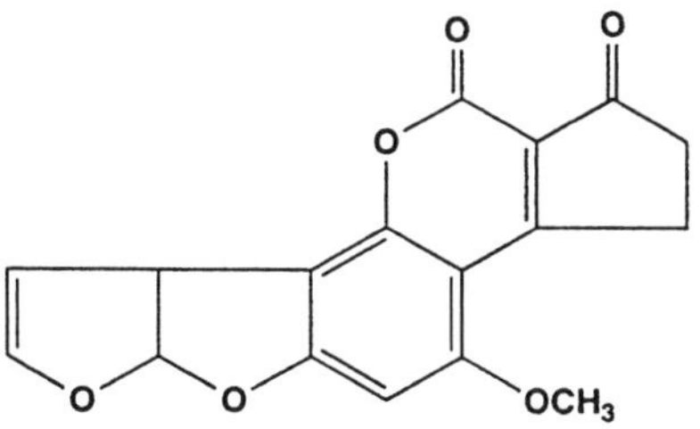

Figure 8.9 *Aflatoxin B1*

What about humans? Are they as sensitive as the dog or as resistant as the rat to the acute toxicity and does aflatoxin cause liver cancer in humans?

A particularly tragic demonstration of the acute human toxicity of aflatoxin was reported in India in 1974 when a large outbreak of poisoning occurred involving nearly 1000 people of whom nearly 100 died. From the concentrations of aflatoxins analysed in the incriminated mouldy maize it is possible to estimate that the LD_{50} of aflatoxin B1 in humans lies somewhere between that for the dog and the rat. During 2004 another large outbreak of aflatoxicosis occurred in a rural part of Kenya resulting in 317 cases and 125 deaths. Locally produced maize was shown to be the cause and a subsequent survey of 65 markets in Kenya showed that 55% of maize samples were contaminated with aflatoxin levels exceeding the Kenyan regulatory level of $20\,\mu g\,Kg^{-1}$, 35% exceeded $100\,\mu g\,Kg^{-1}$ and 7% exceeded $1,000\,\mu g\,Kg^{-1}$.

Although aflatoxin may be considered amongst the most carcinogenic of natural products for some animals, it is still not clear whether it is a human carcinogen. Liver cancer in some parts of the world, such as the African continent, is complex and the initial demonstration of a correlation between exposure to aflatoxin in the diet and the incidence of liver cancer has to be considered with caution. It is known that a strong correlation occurs between the presence of hepatitis B virus and primary liver cancer in humans and it now seems clear that these two agents act synergistically.

Although liver cancer may be attributable to exposure to aflatoxin in parts of Africa, it is necessary to ask why liver cancer is not also more prevalent in India where dietary exposure to aflatoxin also occurs. In India, cirrhosis of the liver is more common and there is still a lot to learn about the role of aflatoxin in liver cancer and liver damage in different parts of the world.

A diverse range of responses to the toxic effects of a compound may occur because the compound is metabolized in the animal body and the resulting toxicity is influenced by this metabolic activity. This is certainly the case with aflatoxin B1 from which a very wide range of metabolites are formed in the livers of different animal species (Figure 8.10). Thus the cow is able to hydroxylate the molecule and secrete the resulting aflatoxin M1 in the milk, hence affording a route for the contamination of milk and milk products in human foods even though these products have not been moulded.

The formation of an epoxide could well be the key to both acute and chronic toxicity and those animals which fail to produce it are relatively resistant to both. Those animals which produce the epoxide, but do not effectively metabolize it further, may be at the highest risk to the carcinogenic activity of aflatoxin B1 because the epoxide is known to

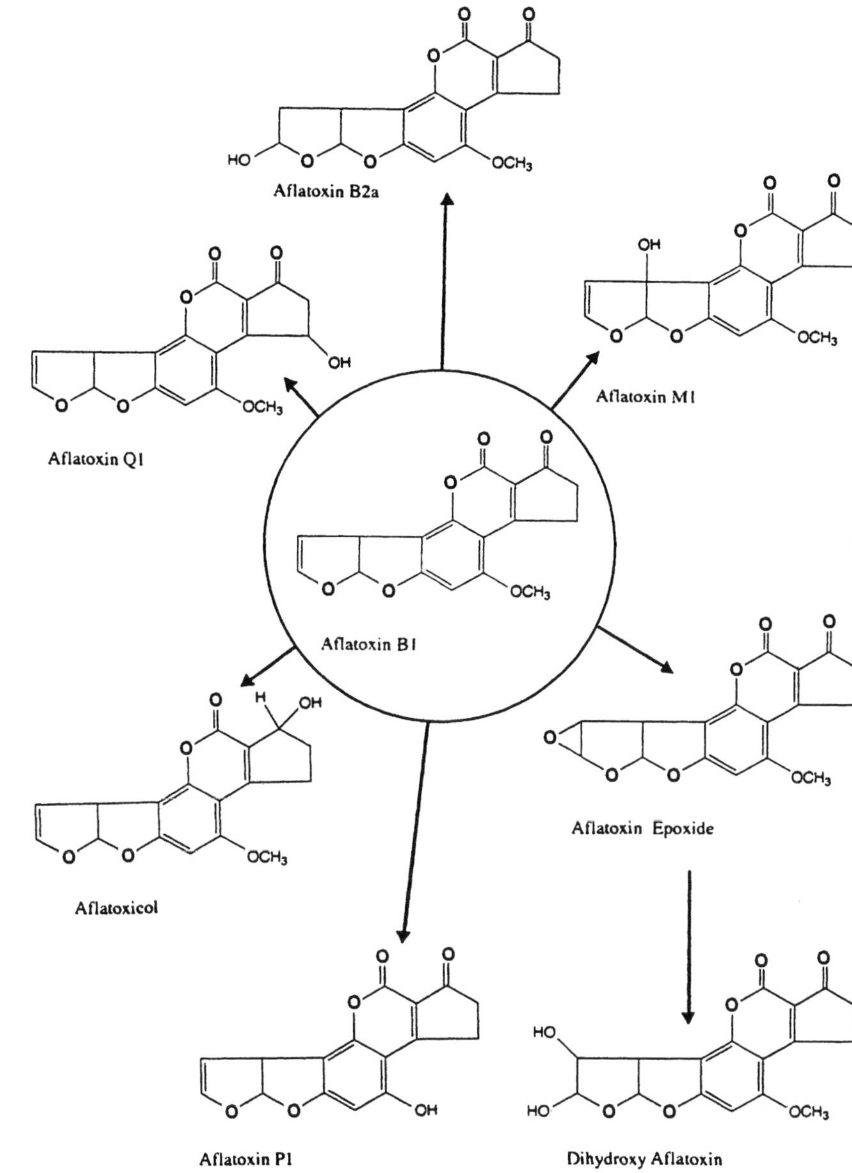

Figure 8.10 *Metabolites of aflatoxin B1*

react with DNA. Those animals which not only produce the epoxide but effectively remove it with a hydrolase enzyme, thus producing a very reactive hydroxyacetal, are most sensitive to the acute toxicity. The hydroxyacetal is known to react with the lysine residues in proteins.

It is now known that aflatoxin B1 epoxide reacts rather specifically with guanine residues of DNA at a number of hot spots, one of which is codon 249 of the *p53* gene. The product of this gene is involved in processes which normally protect against cancer and it is known that the hepatitis B virus binds to the *p53* gene product. Thus with aflatoxin B1 and hepatitis B interacting with *p53* in different ways it is easy to see that they could act synergistically.

The parent molecule may thus be seen as a very effective delivery system having the right properties for absorption from the gut and transmission to the liver and other organs of the body. It is, however, the manner in which the parent molecule is subsequently metabolized *in vivo* which determines the precise nature of an animal's response. Information available about the metabolic activity in the human liver suggests that humans are going to be intermediate in sensitivity to the acute toxicity and may show some sensitivity to the chronic toxicity of aflatoxin B1, including carcinogenicity.

Several studies have demonstrated that very young children may be exposed to aflatoxins even before they are weaned because mothers, consuming aflatoxin in their food, may secrete aflatoxin M1 in their milk. There is no doubt about the potential danger of aflatoxin in food and every effort should be made to reduce or, if possible, eliminate contamination.

8.4.2.2   The Ochratoxins.   Ochratoxin A (Figure 8.11), which is a potent nephrotoxin, was first isolated from *Aspergillus ochraceus* in South Africa, but it has been most extensively studied as a contaminant of cereals, such as barley, infected with *Penicillium verrucosum* in temperate countries such as those of northern Europe. This is because it is known to be a major aetiological agent in kidney disease in pigs and, because it is relatively stable, ochratoxin A may be passed through the food chain in meat products to humans.

A debilitating human disease known as Balkan endemic nephropathy, the epidemiology of which is still a mystery, may be associated with the presence of low levels of nephrotoxic mycotoxins, such as ochratoxin, in the diet of people who have a tradition of storing mould-ripened hams for long periods of time. The presence of ochratoxin in foods of tropical

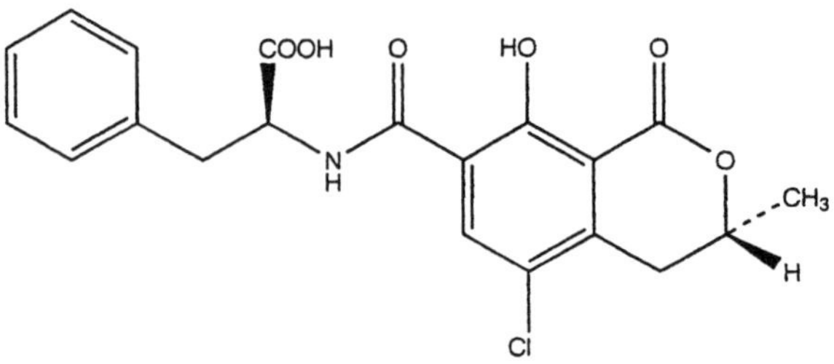

Figure 8.11   *Ochratoxin A*

and subtropical origin, such as maize, coffee beans, cocoa and soya beans is usually due to contamination by *Aspergillus* species.

It is now appreciated that ochratoxin A is quite widespread at low levels in foods and it is necessary to add wine, beer, grape juice and dried fruits (on which the black-spored *A. carbonarius* is implicated) to the list. The most recent toxicological assessment of ochratoxin indicates that it is not only an acute nephrotoxin but may also cause cancer of the kidneys and Member States of the European Union have been engaged in discussions about setting regulatory limits for this mycotoxin. Draft proposals were published by the European Commission in 1999 but it was not until 2002 that the following levels were agreed and set out in Commission Regulation (EC) No 472/2002. For raw cereal grains the maximum level is set at $5\,\mu g\,Kg^{-1}$ whereas for cereal products, and cereal grains intended for direct human consumption, a more stringent level of $3\,\mu g\,Kg^{-1}$ was set. For dried vine fruit, such as currants, raisins and sultanas, the level was set at $10\,\mu g\,Kg^{-1}$.

8.4.2.3 Other Aspergillus *Toxins.* Sterigmatocystin (Figure 8.12), a precursor in the biosynthesis of aflatoxins, is produced by a relatively large number of moulds but especially by *Aspergillus versicolor*. It is not considered to be as acutely toxic, or as carcinogenic, as aflatoxin but it is likely to be quite widespread in the environment and has been isolated from a number of human foods such as cheeses of the Edam and Gouda type which are stored in warehouses for a long period of time. In this situation the moulds grow only on the surface and sterigmatocystin docs not penetrate beyond the first few millimeters below the surface.

Cyclopiazonic acid (Figure 8.13) gets its name because it was first isolated from a mould which used to be called *Penicillium cyclopium* (now known as *P. aurantiogriseum*) but it has subsequently been isolated from *Aspergillus versicolor* and *A. flavus*. In the latter, it is formed primarily in the sclerotia and there has always been a suspicion that some of the symptoms ascribed to the ingestion of food contaminated by *A. flavus* may be due to the presence of this compound as well as to the presence of aflatoxins.

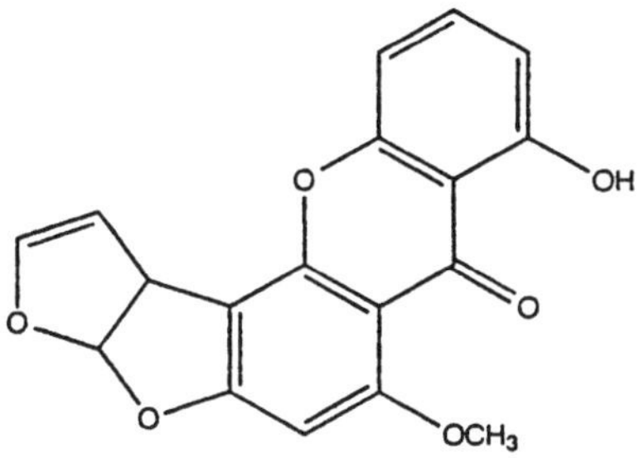

Figure 8.12 *Sterigmatocystin*

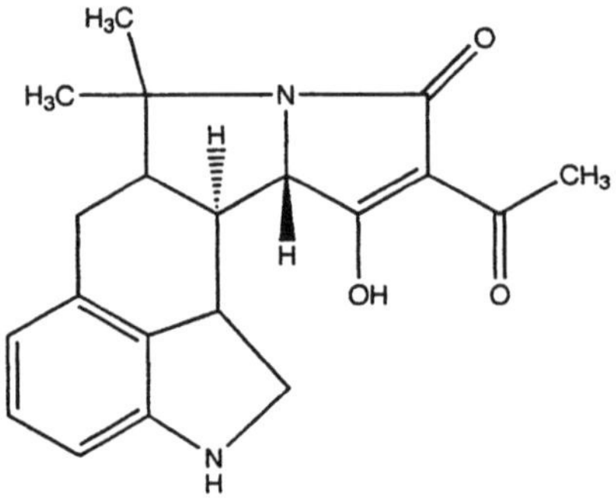

Figure 8.13 *Cyclopiazonic acid*

In parts of India a disease known as kodua poisoning occurs following the consumption of kodo millet (*Paspalum scrobiculatum*) which is both a staple food and an animal feed. *Aspergillus flavus* and *A. tamarii* have been isolated from incriminated samples of millet and both species are able to synthesize cyclopiazonic acid. Poisoning in cattle and humans is associated with symptoms of nervousness, lack of muscle co-ordination, staggering gait, depression and spasms and, in humans, sleepiness, tremors and giddiness may last for one to three days.

Some of these symptoms are reminiscent of a problem in intensively reared farm animals known as staggers in which complex indole alkaloid metabolites (tremorgens) are implicated. One of these metabolites, aflatrem, is also produced by some strains of *A. flavus*.

8.4.3 Mycotoxins of *Penicillium*

Penicillium is much more common as a spoilage mould in Europe than *Aspergillus* with species such as *P. italicum* and *P. digitatum* causing blue and green mould respectively of oranges, lemons and grapefruits, *P. expansum* causing a soft rot of apples, and several other species associated with the moulding of jams, bread and cakes. Species which have a long association with mould-ripened foods include *P. roquefortii* and *P. camembertii*, used in the mould ripened blue and soft cheeses respectively.

The mycotoxin patulin (Figure 8.14) is produced by several species of *Penicillium, Aspergillus* and *Byssochlamys* but is especially associated with *P. expansum* and was first described in 1942 as a potentially useful antibiotic with a wide spectrum of antimicrobial activity. It was discovered several times during screening programmes for novel antibiotics and this is reflected in the many names by which it is known including claviformin, clavicin, expansin, penicidin, mycoin, leucopin, tercinin and clavatin.

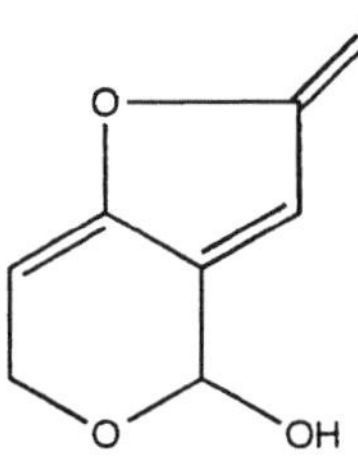

Figure 8.14 *Patulin*

It was not until 1959 that an outbreak of poisoning of cattle, being fed on an emergency ration of germinated barley malt sprouts, alerted the veterinary profession to patulin as a mycotoxin. In this instance the producing organism was *Aspergillus clavatus* but the same toxin has been implicated in several outbreaks of poisoning from such diverse materials as apples infected with *P. expansum* to badly stored silage infected with a species of *Byssochlamys*. As far as humans are concerned, it is the common association of *P. expansum* with apples, and the increasing consumption of fresh apple juice as a beverage, which has caused some concern.

Patulin is not a particularly stable metabolite. It is stable at the relatively low pH of apple juices, although it is destroyed during the fermentation of apple juice to cider. Even if there was no concern over the toxicity of patulin, the demonstration of its presence in a fruit juice is a useful indicator that very poor quality fruit has been used in its manufacture. However, at least 33 countries have now set limits on patulin in fruit and fruit juices, the most common being 50 ppb.

The nephrotoxic metabolite citrinin, produced by *P. citrinum*, was also first discovered as a potentially useful antibiotic but again rejected because of its toxicity. It is probably not as important as ochratoxin, produced by *P. verrucosum* as well as *Aspergillus ochraceus*, although it may be implicated in the complex epidemiology of 'yellow rice disease'.

8.4.3.1 Yellow Rice Disease. A complex of disorders recognized in Japan a number of times since the end of the last century has been associated with the presence of several species of penicillia and their toxic metabolites on rice. This moulded rice is usually discoloured yellow and several of the toxic metabolites implicated are themselves yellow pigments. There was an early awareness that moulds may be responsible for cardiac beriberi and in 1938 it was demonstrated that *Penicillium citreo-viride* (*P. toxicarium*) and its metabolite citreoviridin were responsible. The most toxic of the species of penicillia associated with yellow rice disease is *P. islandicum* which produces two groups of toxins, hepato-toxic chlorinated cyclopeptides such as islanditoxin, as well as the much less acutely toxic, but potentially carcinogenic, dianthraquinones such as luteoskyrin.

8.4.4 Mycotoxins of *Fusarium*

Some species of *Fusarium* cause economically devastating diseases of crop plants such as wilts, blights, root rots and cankers, and may also be involved in the post-harvest spoilage of crops in storage. The genus is also associated with the production of a large number of chemically diverse mycotoxins both in the field and in storage.

8.4.4.1 Alimentary Toxic Aleukia.   Outbreaks of this dreadful disease, which is also known as septic angina and acute myelotoxicosis, occurred during famine conditions in a large area of Russia. A particularly severe outbreak occurred during the period 1942–47 but there had been reports of the disease in Russia since the 19th century.

Studies in Russia itself demonstrated that the disease was associated with the consumption of cereals moulded by *Fusarium sporotrichioides* and *F. poae* but the nature of the toxin remained unknown. Studies of dermonecrosis in cattle in the United States showed it to be caused by a *Fusarium* metabolite called T-2 toxin (Figure 8.15) which is one of the most acutely toxic of a family of compounds called trichothecenes.

There is good evidence that T-2 toxin was a major agent in the development of alimentary toxic aleukia in humans, the first symptoms of which are associated with damage of the mucosal membranes of mouth, throat and stomach followed by inflammation of the intestinal mucosa. Bleeding, vomiting and diarrhoea, which are all associated with damage of mucosal membrane systems, were common but recovery at this stage was possible if the patient was given a healthy, uncontaminated, vitamin-rich diet. Continued exposure to the toxin, however, led to damage of the bone marrow and the haematopoietic system followed by anaemia and a decrease in erythrocyte and platelet counts. The occurrence of necrotic tissue and skin haemorrhages were further characteristics of the disease.

As well as giving rise to this sequence of acute symptoms, the trichothecenes are known to be immunosuppressive and this undoubtedly contributed to victims' sensitivity to relatively trivial infectious agents. Indeed, many people died of bacterial and viral infections before succumbing from the direct effects of the toxin itself. Unlike aflatoxin, the acute toxicity of T-2 toxin is remarkably uniform over a wide range of animal species (Table 8.4) and it is reasonable to assume that the human LD_{50} will be in the same range. Although improved harvesting and storage has eliminated alimentary toxic aleukia from Russia this

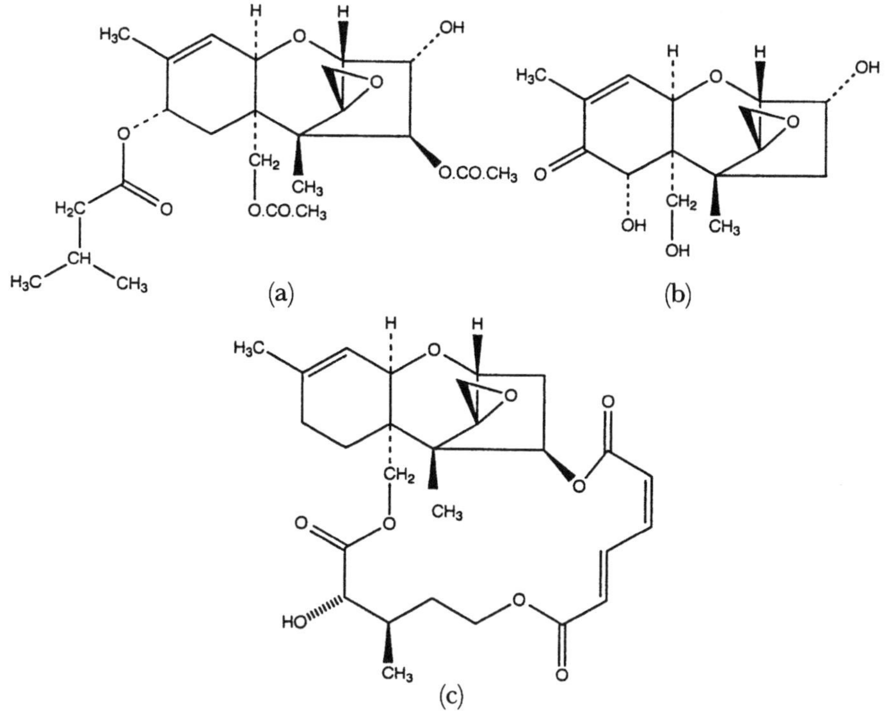

Figure 8.15 *Trichothecenes. (a) T-2 toxin, (b) deoxynivalenol, and (c) verrucarin A*

Table 8.4 *Some LD$_{50}$ values for T-2 toxin*

Species	LD$_{50}$ (mg kg^{-1})
Mouse	5.2 (intraperitoneal)
Rat	5.2 (oral)
Guinea pig	3.1 (oral)
7-Day-old chick	4.0 (oral)
Trout	6.1 (oral)

disease may still occur in any part of the world ravaged by war and famine.

Three of the most important mycotoxins, aflatoxin, ochratoxin and T-2 toxin, are immunosuppressive but react differently against the immune system. All three inhibit protein biosynthesis, aflatoxin by inhibiting transcription, ochratoxin by inhibiting phenylalanyl tRNA synthetase, and T-2 toxin by inhibiting translation through binding with a specific site on the eukaryote ribosome. One consequence of these distinct modes of activity is that mixtures of such mycotoxins are likely to be synergistic in activity and this has been shown experimentally in the case of aflatoxin and T-2 toxin. This observation is significant in the context of the probability that a food which has gone mouldy will probably be infected by several species of mould and may thus be contaminated by several different mycotoxins.

8.4.4.2 DON and Other Trichothecenes. In Japan an illness known as red-mould disease involving nausea, vomiting and diarrhoea has been associated with the consumption of wheat, barley, oats, rye and rice contaminated by species of *Fusarium*. The species most frequently incriminated was *Fusarium graminearum*, although it had been misidentified as *F. nivale*, and the trichothecene toxins isolated from them were called nivalenol and deoxynivalenol. It is now realized that *F. nivale* itself does not produce trichothecenes at all, indeed it may not even be a *Fusarium*.

Deoxynivalenol (Figure 8.15), also known as DON and vomitoxin, was also shown to be the vomiting factor and possible feed-refusal factor in an outbreak of poisoning of pigs fed on moulded cereals in the United States. Deoxynivalenol is much less acutely toxic than T-2 toxin, having an LD_{50} of $70 \, mg \, kg^{-1}$ in the mouse.

Nevertheless, it is more common than T-2 toxin especially in crops such as winter wheat and winter barley. In 1980 there was a 30–70% reduction in the yields of spring wheat harvested in the Atlantic provinces of Canada due to infections with *Fusarium graminearum* and *F. culmorum*, both of which may produce DON and zearalenone. It is not clear whether DON and other trichothecenes are as immunosuppressive as T-2 toxin but it seems prudent to reduce exposure to a minimum. Several countries have set legislative limits for DON and zearalenone in cereals and the E.C. implemented regulatory limits for the European Union in 2006. For DON they range from $200 \, \mu g \, Kg^{-1}$ in processed cereal based foods for infants and ingredients used in the manufacture of food for infants to $1750 \, \mu g \, Kg^{-1}$ in durum wheat and oats. For zearalenone they range from $20 \, \mu g \, Kg^{-1}$ in processed cereal based products for infants to $100 \, \mu g \, Kg^{-1}$ in unprocessed cereals except maize. Maximum limits for maize are likely to be implemented during 2007.

The most virulent group of trichothecenes are those with a macrocyclic structure attached to the trichothecene nucleus such as the satratoxins, verrucarins and roridins produced by *Stachybotrys atra* (Figure 8.15). This species has been implicated in a serious disease of horses, referred to as stachybotryotoxicosis, fed on mouldy hay. It seems that species of *Fusarium* do not produce such toxins.

8.4.4.3 Zearalenone. Zearalenone (Figure 8.16) is an oestrogenic mycotoxin which was first shown to cause vulvovaginitis in pigs fed on mouldy maize. Pigs are especially sensitive to this toxin and, although its acute toxicity is very low, it is common in cereals such as maize, wheat and barley being produced by *Fusarium graminearum, F. culmorum* and other species of *Fusarium*. The toxin was called zearalenone because of

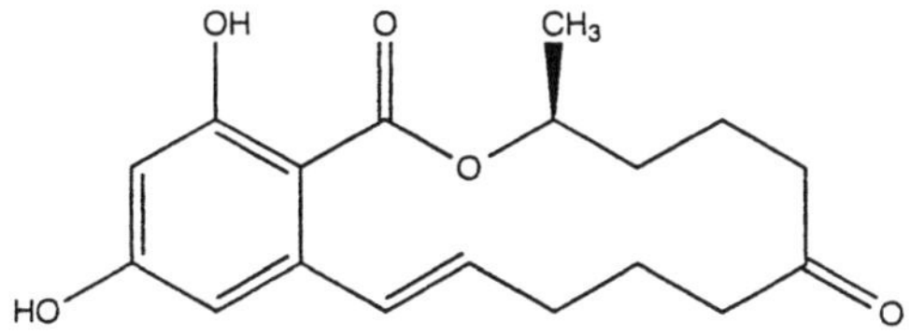

Figure 8.16 *Zearalenone*

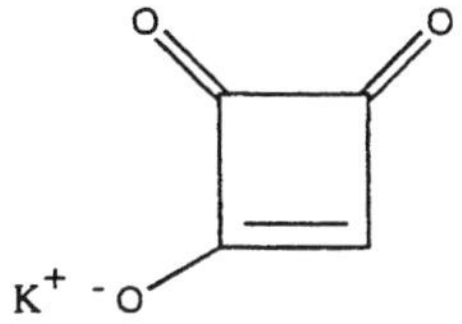

Figure 8.17 *Moniliformin*

its initial isolation from *Gibberella zeae*, the perfect stage of *F. graminea-rum*.

In gilts, the vulva and mammary glands become swollen and, in severe cases, there may be vaginal and rectal prolapse. In older animals there may be infertility, reduced litter size and piglets may be born weakened or even deformed. There is concern about the long-term exposure of the human population to such an oestrogen.

Zearalenone, and the corresponding alcohol zearalenol, are known to have anabolic, or growth promoting activity, and, although its use as a growth promoting agent is banned in some countries, it is permitted in others. This can lead to problems in international trade because zear-alenone can be detected in the meat of animals fed on diets containing it.

8.4.4.4 Oesophageal Cancer. In parts of Northern China, and the Transkei in Southern Africa, there are regions of high incidence of human oesophageal carcinoma and the epidemiology of the disease fits the hypothesis that the consumption of moulded cereals and mycotoxins are involved. *F. moniliforme* (by the strict code of biological nomenclature this should now be called *F. verticillioides*), which belongs to a distinct group of the genus which do not produce trichothecenes, seems to be the most likely fungus to be involved. Strains of this species are associated with a disease of rice which has been a particular problem in China and other, probably distinct, strains are commonly isolated from maize grown in Southern Africa and many other parts of the world. *F. moniliforme* is a very toxigenic species and its occurrence in animal feeds is associated with outbreaks of a disease known as equine le-ukoencephalomalacia in horses and liver cancer in rats.

One of the first mycotoxins to be isolated during the study of these diseases was called moniliformin (Figure 8.17) because it was presumed

to have been produced by *F. moniliforme*. It is now known that moniliformin is actually produced by strains of the related species *F. sub-
glutinans* and not *F. moniliforme*. However, it is the latter which is
especially associated with human oesophageal cancer and a number of
complex metabolites have been isolated and characterized from cultures
of this species, including fusarin C, which is mutagenic, and the fumonisins which are carcinogenic (Figure 8.18). However it would probably be
wise to be cautious about extrapolating laboratory tests demonstrating
carcinogenic activity to a human disease. Nevertheless, since the discovery of the fumonisins, reported in 1988, they have become the focus of a
considerable amount of interest. Once the analytical problems had been
overcome it was realized that they are widespread wherever maize is
grown. In a survey in the UK in 1998, 97% of the 67 samples of maize
examined were found to contain fumonisins at levels ranging from
$25\,000–27\,000\,\mu\mathrm{g\,kg}^{-1}$. Similar levels of contamination have been found

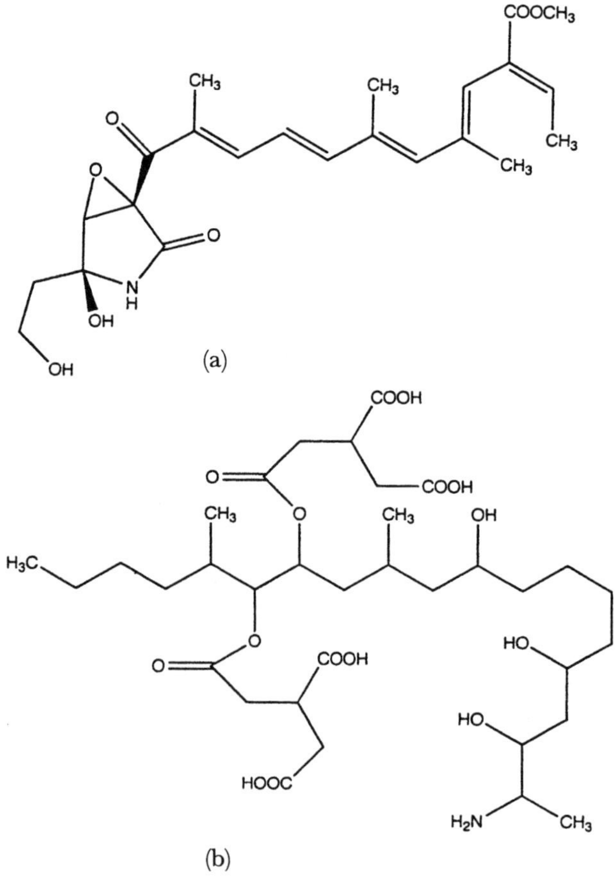

Figure 8.18   *Complex metabolites from* Fusarium moniliforme. *(a) Fusarin C and (b)
fumonisin B1*

in other parts of Europe and in South America, Africa, India and the USA.

Fumonisin B1 has been confirmed to cause equine encephalomalacia, porcine pulmonary oedema, kidney damage in rodents and hepatic cancer in rats. It is known to cause apoptosis in tissue cell cultures and has cancer promoting activity in several experimental systems. At the molecular level it is a potent competitive inhibitor of ceramide synthase, blocking the biosynthesis of complex sphingolipids and leading to the accumulation of sphinganine.

Despite all of this information it is still not clear whether the fumonisins are responsible for human oesophageal carcinoma, but clearly it is important to determine its significance to human health.

8.4.5 Mycotoxins of Other Fungi

Ergotism has been documented as a human disease since the middle ages but its aetiology remained a mystery until the mid-19th century when it was demonstrated to be caused by a fungus, *Claviceps purpurea*. This fungus is a specialized parasite of some grasses including cereals and, as part of its life cycle, the tissues of infected grains are replaced by fungal mycelium to produce a tough purple brown sclerotium which is also known as an ergot because it looks like the spur of a cockerel (Figure 8.19). The biological function of the sclerotium is to survive the adverse conditions through the winter in order to germinate in the following spring. Ergots contain alkaloid metabolites which may be incorporated into the flour, and eventually the bread, made from the harvested grain.

Ergotism, or St Anthony's fire, is infrequent in human beings. The toxicity of the ergot alkaloids is now well understood and one aspect of their activity is to cause a constriction of the peripheral blood capillaries leading, in extreme cases, to fingers and toes becoming gangrenous and necrotic. Different members of this family of mould metabolites may also have profound effects on the central nervous system stimulating smooth muscle activity.

Plant–fungal interactions can be complex and there are instances where a toxic plant metabolite is produced in response to fungal attack. Thus, when the sweet potato, *Ipomoea*, is damaged by certain plant pathogens it responds by producing the phytoalexin ipomeamarone. This antifungal agent is produced to limit fungal attack but it is also an hepatotoxin to mammals. Further complexity arises when other moulds, such as *Fusarium solani*, degrade ipomeamarone to smaller molecules such as ipomenol which can cause oedema of the lung (Figure 8.20).

A disease of sheep in New Zealand known as ryegrass staggers may cause an estimated loss of hundreds of millions of dollars in some years. It is caused by an intimate association of perennial ryegrass (*Lolium*

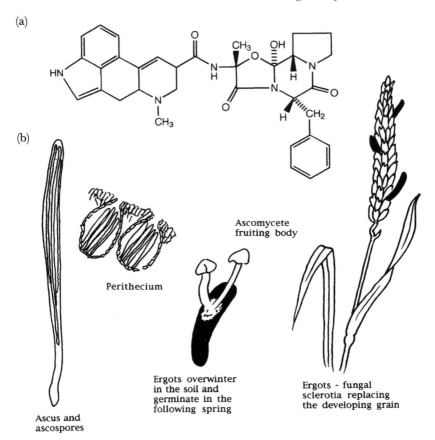

Figure 8.19 *Claviceps purpurea. (a) The alkaloid ergotamine, and (b) aspects of development*

perenne) and an endophytic fungus, *Acremonium loliae*. The endophyte–plant association results in the production of complex tremorgenic mycotoxins known as the lolitrems (Figure 8.21) which are responsible for the staggering response and possible collapse of sheep under stress. The endophyte is seedborne and completes its whole life cycle within the plant, although it can be cultured with difficulty in the laboratory.

It is possible to eliminate the endophyte by careful heat treatment of seed but, in New Zealand, the planting of endophyte-free ryegrass provides pastures which are very susceptible to insect damage such as that caused by the stem weevil *Listronotus bonariensis*. It is almost certain that the role of the endophyte in controlling insect damage is not due to the production of lolitrems so the possibility remains that a genetically engineered strain of *Acremonium loliae*, which no longer produces lolitrems, could be used to replace the wild strain in perennial ryegrass.

Although there can be no doubt about the potential for the presence of mycotoxins in food to cause illness and even death in humans, there are

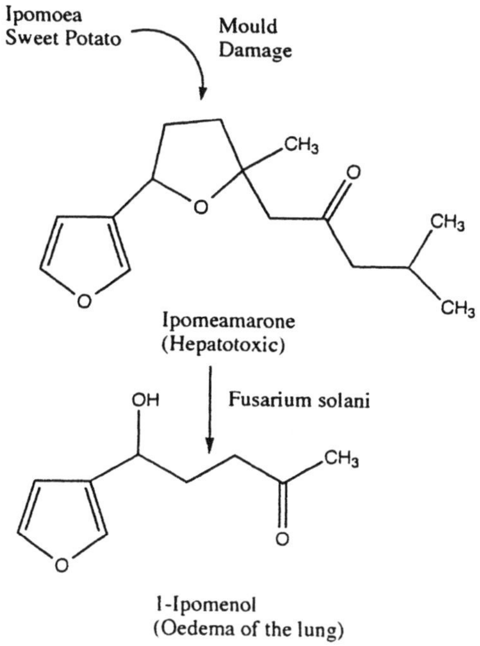

Figure 8.20 *Ipomeamarone and ipomenol*

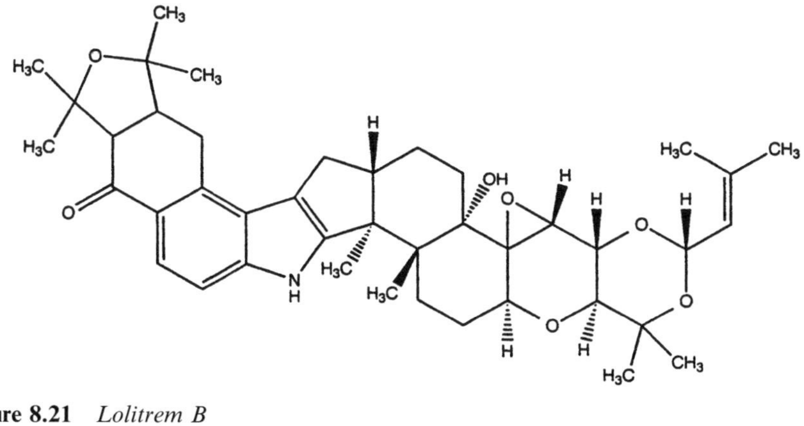

Figure 8.21 *Lolitrem B*

far more overt mycotoxicoses in farm animals throughout the world. Thus facial eczema in sheep in New Zealand, caused by the saprophyte *Pithomyces chartarum* growing on dead grass, slobbers in cattle in the United States, caused by *Rhizoctonia leguminicola* parasitic on red clover, and lupinosis of farm animals in Australia, caused by *Phomopsis leptostromiformis* growing on lupins, do not have a direct effect on humans. However, there can be no denying the impact that such outbreaks of mycotoxin poisoning have on economics through losses in productivity.

Recognition of the potential to cause harm in humans, by the imposition of maximum tolerated levels of mycotoxins such as aflatoxin, can also have a major impact on economics by rendering a commodity unacceptable in national or international trade. Thus, a major problem occurred for Turkey, the world's most important exporter of dried figs, during the Christmas of 1988.

Several European countries imposed a ban on the import and sale of dried figs following the demonstration of aflatoxin in 30% of samples of figs analysed. Fearing the loss of 50 000 jobs in the fig-drying and packing industry, Turkey was vigorous in her diplomatic efforts to have the bans lifted. This was done fairly soon after they had been imposed and an international symposium on 'Dried Figs and Aflatoxins' was held in Izmir, Turkey, in April 1989. More recently, in 1996, the first reports of aflatoxin contamination of pistachios imported into Europe caused some concern, but the producing country has taken appropriate action to reduce the level of contamination.

In 1980, nearly 66% of random samples of maize from North Carolina had concentrations of aflatoxins in excess of 20 μg kg^{-1} giving rise to an estimated loss to producers and handlers of nearly 31 million dollars. It is rare that the losses and costs arising from mycotoxin contamination can be calculated but these three isolated and very different examples indicate that on a world-wide basis they must be considerable. In at least two of these examples aflatoxin was probably formed in the commodity during growth and development in the field. Under these conditions aflatoxin formation is usually relatively low and in neither case was there any evidence of harm to human beings. However, it is when commodities are improperly stored that really high concentrations of mycotoxins may be formed and it is in these situations that human suffering can occur.

8.5   FOODBORNE VIRUSES

Viruses differ profoundly from other types of micro-organism. They have no cellular structure and possess only one type of nucleic acid (either RNA or DNA) wrapped in a protein coat or capsid. They are also extremely small, with diameters generally in the range 25–300 nm (1 nm $= 10^{-3} \mu$m), so that most are invisible using conventional light microscopy and can only be viewed with the electron microscope. Some viruses (for example HIV) are enveloped by an outer lipid membrane, but these cannot be transmitted *via* food since they are relatively fragile and are destroyed by exposure to bile and acidity in the digestive tract.

As obligate intracellular parasites, viruses cannot multiply other than in a susceptible host cell whose machinery and metabolism they hijack for the purpose of viral replication. Consequently, virus multiplication will not occur in foods which can act only as a passive vehicle in the transmission of infection.

In recent years viruses have been increasingly recognized as an important cause of foodborne illness. There are currently more than 100 human enteric viruses recognized (Table 8.5), and since these are spread by the faecal–oral route, food is one potential means of transmission. Broadly speaking there are two types of pathogenic enteric viruses which differ in their target tissues. Both enter the body *via* the gut, but gastroenteritis viruses remain confined there while others, such as the polio and hepatitis viruses, cause illness once they have migrated to other organs.

8.5.1 Polio

The genus *Enterovirus* is made up of small (28 nm), single-stranded RNA viruses, and includes poliovirus, which was at one time the only virus known to be foodborne. Polio can be a transient viraemia with an incubation period of 3–5 days and characterized by headache, fever and sore throat, but in a minority of cases it can progress to a second stage where the virus invades the meninges causing back pain and headaches. In the worst cases the virus may spread to neurons in the spinal chord causing cell destruction and various degrees of paralysis. Ascent of the infection to the brain may cause death.

Like other enterovirus infections, poliovirus is more likely to produce an asymptomatic infection in very young children. From about the turn of the century, however, improvements in hygiene and sanitation in industrialized countries meant that early infection and acquisition of immunity became less common. As a result, the disease changed from endemic to epidemic and was widely feared as it became more frequent in older children and young adults where it was likely to be much more severe.

Poliomyelitis is now virtually eradicated in developed countries due to the availability of very effective live and inactivated vaccines. At the time that mass-vaccination programmes were introduced in the 1950s, food was no longer important as a vehicle. Previously, contaminated milk had been the principal source of foodborne polio but this route of infection had been controlled by improvements in hygiene.

8.5.2 Hepatitis A and E

A similar story applies to another enterovirus, Hepatitis A, the cause of infectious hepatitis. Improvements in public hygiene and sanitation in

Table 8.5   *Human enteric viruses*

Family	Features	Viruses	Associated diseases
Adenoviridae	Icosahedral particles with fibres. 100 nm, DNA.	Group F adenovirus Serotypes 40 and 41 (AdV).	Gastroenteritis
Astroviridae	28 nm particles with surface 'star' motif. ssRNA.	Human astrovirus, 7 serotypes (HAs + V)	Mild gastroenteritis
Caliciviridae	34 nm particles with cup-shaped depressions on surface. ssRNA.	Sapovirus 5 or more serotypes	Gastroenteritis
	Less distinct surface features.	Norovirus 4–9 serotypes.	Gastroenteritis
Parvoviridae	22 nm featureless particles. ssDNA.	Parvovirus, *e.g.* Ditchling and Cockle agent.	Gastroenteritis, normally shellfish associated.
Picornaviridae	Featureless 28 nm icosahedral particles. ssRNA.	Poliovirus types 1–3.	Meningitis, paralysis fever.
		Echovirus types 1–65. Enterovirus now viruses numbered 68–71.	Meningitis, rash, diarrhoea, fever, respiratory disease.
		Coxsackie A types 1–23.	Meningitis, herpangia, fever, respiratory disease.
		Coxsackie B types 1–6.	Myocarditis, congenital heart anomalies, pleurodynia, respiratory disease, fever, rash, meningitis.
		Hepatovirus (Hepatitis A).	Infectious hepatitis.
Reoviridae	Double shelled capsids. 70–80 nm segmented as RNA.	Reovirus	No disease associations known.
	Outer shell appears as 'spokes of a wheel'. 70 nm.	Rotavirus. Mainly Group A, occasionally B and C in humans.	Gastroenteritis.
Coronaviridae[a]	Fragile, pleomorphic, enveloped particles with prominent club-shaped spikes. SsRNA.	Human enteric coronavirus (HECV).	Gastroenteritis, possibly neonatal necrotizing enterocolitis.
Unclassified		Hepatitis E virus* (Enterically transmitted, non-A, non-B hepatitis). (HEV).	Infectious hepatitis

(M. Carter)
[a] Potential agents not confirmed as human pathogens

the developed world have reduced exposure to the virus so that, when it does occur, it tends to be later in life when the illness is more severe.

The incubation period varies between two and six weeks. During this period the virus multiplies in the cells of the gut epithelium before it is carried by the blood to the liver. In the later part of the incubation period the virus is shed in the faeces. Early symptoms are anorexia, fever, malaise, nausea and vomiting, followed after a few days by symptoms of liver damage such as the passage of dark urine and jaundice.

Like other enteric viruses, hepatitis A is transmitted by the faecal–oral route. Primarily it is spread by person-to-person contact but food-and waterborne outbreaks do occur. Milk, fruits such as strawberries and raspberries, salad vegetables such as lettuce, and shellfish are common food vehicles. With the exception of those caused by shellfish, common source outbreaks are usually due to contamination by an infected food handler. The long incubation period of the illness often makes identification of the source extremely difficult. For the same reason, it is difficult to say with any accuracy what proportion of hepatitis A cases are transmitted by food, although it has been estimated that about 3% of cases in the United States are food or waterborne.

The agent of enterically transmitted non-A, non-B hepatitis has now been designated hepatitis E virus and molecular biology studies indicate it is a calici-like particle with an unusual RNA structure. It too is transmitted by the faecal–oral route and produces illness after an incubation period of 40 days.

8.5.3 Gastroenteritis Viruses

A number of different viruses have been implicated in gastroenteritis by their presence in large numbers (up to 10^8–$10^{10}\,g^{-1}$) in diarrhoeal stools. In most cases it has not proved possible to culture the virus thus preventing their full characterization. As a result, classification has been based largely on morphology and geographical origin.

Although other, better characterized, viruses such as rotavirus, calicivirus and astrovirus are also known to cause diarrhoea, it is these less well-defined agents that are responsible for most outbreaks of foodborne gastroenteritis where a virus is identified. In the United States they were originally known as Norwalk-like agents after the virus which caused an outbreak of gastroenteritis in schoolchildren in Norwalk, Ohio in 1968. In the UK, they were described as small round structured viruses (SRSVs) based on the fact that, when viewed in the electron microscope, they are particles about 25–30 nm in diameter possessing an amorphous structure lacking geometrical symmetry (Figure 8.22). They are now classified as a distinct genus, Norovirus (NoV), within the family Caliciviridae.

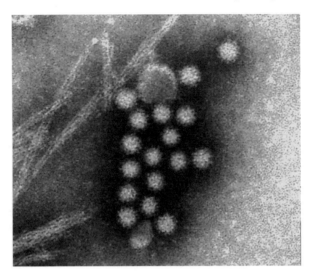

Figure 8.22   *Small round-structured viruses (SRSVs), magnification × 200 000 (Photo: H. Appleton)*

Foodborne viral gastroenteritis is characterized by an incubation period of 15–50 h followed by diarrhoea and vomiting which persists for 24–48 h. The infectious dose is not known. Studies in model systems have suggested that doses as low as one cell culture infectious unit can produce infection but in polio vaccination an oral dose of 100 000 infectious units is given to ensure a success rate of at least 90%.

The onset of symptoms such as projectile vomiting may be very sudden and unexpected and this can contribute to the further spread of illness (see below).

8.5.4   Sources of Food Contamination

The importance of viral gastroenteritis is clear from the huge under reporting revealed by the Infectious Intestinal Disease Study in England. The degree to which this is foodborne is uncertain since a considerable amount of human to human transmission must also occur. Estimates of the proportion of viral gastroenteritis which is foodborne, made in the UK, Australia and the USA, vary between 10% and 40%.

Enteric viruses may be introduced into foods either as primary con-tamination, at source where the food is produced, or as secondary contamination during handling, preparation and serving. It is possible that salad vegetables fertilized with human excrement or irrigated with sewage polluted water could be contaminated with viruses while in the field. Salads and fruits such as raspberries have been implicated in outbreaks, though in some cases this could also have been the result of secondary contamination during preparation. Evidence of unequivocal primary contamination is largely restricted to bivalve molluscan shellfish, such as clams, cockles, mussels and oysters, which have been involved in numerous outbreaks of hepatitis and gastroenteritis. In the UK between 1976 and 1987 there were several large outbreaks involving cockles from the Essex coast in which more than 2000 people were affected. Large outbreaks have also been reported from Australia and the United States, but these pale beside the outbreak of hepatitis A in Shanghai in 1988 when almost 300 000 were reported ill and contaminated clams were identified as the source of infection.

The problem arises because these shellfish are grown in shallow, inshore, coastal waters that are often contaminated with sewage. Since they feed by filtering sea water to extract suspended organic matter, they also tend to concentrate bacteria and viruses from the surrounding environment. Pumping rates can be quite substantial as an oyster will filter up to four litres of seawater per hour and concentrate microorganisms in their gut by up to a thousand-fold.

It is possible to decontaminate shellfish by relaying them in clean waters (if these can be found) or removing them to special depuration plants where they are encouraged to filter water that is recirculated and purified, usually by treatment with UV light or ozone. Depuration procedures have proved very successful for removing bacterial pathogens; coliform bacteria have been shown to be removed within 24–48 hours. However the rate at which viruses are cleansed is much slower and less predictable. This is probably due to the small size of the virion compared with the bacterial cell, the relative strength of its attachment to the gut wall and on its ability to penetrate into deeper tissues. It has been suggested that virus particles ingested by the shellfish are taken up by macrophages and transported from the gut to tissues that are remote from the depuration process, though there is little evidence for this.

The problem is compounded by the fact that some shellfish, such as oysters, are consumed without any cooking and those that are cooked receive only a mild, relatively uncontrolled heat process in order to prevent the flesh assuming the consistency of rubber. Studies on the heat inactivation of hepatitis A virus have led to the introduction of guidelines in the UK for the cooking of cockles which recommend that the internal temperature of the meat should reach 85–90 °C for 1.5 min. It is not known whether these guidelines provide an acceptable safety margin with regard to NoV since they cannot be cultured *in vitro* for their heat sensitivity to be determined.

Secondary contamination by infected food handlers is an alternative source of infection, particularly with those food items that are subject to extensive handling in their preparation and are consumed without reheating. Usually the food handler is suffering from viral gastroenteritis at the time. One outbreak in the UK provides a graphic illustration of secondary contamination and also how the sudden onset of symptoms can catch victims unaware and exacerbate the problem.

The outbreak occurred at a hotel in the UK where over 140 people were ill. One chef vomited in the changing room lavatory and then immediately returned to food preparation. Later that day he had an episode of diarrhoea. Two days later another member of the kitchen staff vomited into a bin outside the kitchen door and the following day two staff vomited in the kitchen itself. The epidemiological evidence strongly implicated the cold foods prepared by the chef who was found to be excreting the NoV 48 h after his symptoms subsided. However, it is clearly possible that several other foods may have been contaminated by droplets from the vomitus from other affected kitchen workers.

In 1982 a huge outbreak in the Twin Cities area of Minnesota was caused by a baker who was working during an episode of diarrhoeal illness. Despite claiming to have washed his hands thoroughly after a visit to the toilet, he transferred sufficient virus to the butter cream that he mixed by hand to cause an outbreak which affected at least 3000 people. Shortly after, in the same area, a second outbreak affecting 2000 people occurred in which a food handler contaminated salads during banquet catering.

There is no evidence yet of a persistent symptomless carrier state for these viruses. They are no longer apparent in patient's stools shortly after recovery but this may simply reflect the insensitivity of electron microscopy as a detection method.

8.5.5   Control

The problems of monitoring and control of foodborne viruses are very different to those posed by bacteria. Testing foods for the presence of pathogenic viruses is not possible since many cannot be cultured and the numbers present are too low to be detected by techniques such as electron microscopy. An alternative would be to use more readily cultured viruses that are shed in the faeces, such as the vaccine polio strain, as indicator organisms for the presence of pathogenic enteric viruses. However, current extraction methods are very inefficient and the culture techniques, based on observation of a cytopathic effect in cell monolayers or plaques in cell monolayers under semi-solid medium, are far more complex and expensive than bacteriological testing. Already though, techniques based on immunoassay and nucleic acid probes with the polymerase chain reaction promise to improve both the sensitivity and speed of virus detection.

An interesting approach is to use coliphage, a bacteriophage which infects the enteric bacterium *E. coli*, as a viral indicator. Coliphages do not require expensive tissue culture techniques for their enumeration since they can be detected through their ability to form plaques in a lawn culture of a suitable strain of *E. coli*. The problem of extraction of the coliphage from food remains however, and interpretation of the significance of their presence in foods is uncertain.

As with other problems of microbiological food safety, control of viral contamination is most effectively exercised at source. Primary contamination can be controlled by avoiding the fertilization of vulnerable crops with human sewage and the discharge of virus-containing effluents into shellfish-harvesting waters. Secondary contamination is even harder to detect microbiologically and can only be controlled by the strict observance of good hygienic practices in the handling and preparation of foods.

Prospects are poor for a vaccine against the gastroenteritis viruses since immunity following infection appears to be short lived. Volunteers who were made ill by ingesting a faecal extract containing the Norwalk agent became ill again a year later when given the same extract a second time. A new vaccine against hepatitis A, based on normal hepatitis A virus inactivated with formaldehyde, was licensed for use in the UK in 1992.

8.6 SPONGIFORM ENCEPHALOPATHIES

Spongiform encephalopathies (SEs) are degenerative disorders of the brain that occur in a number of species. They are recognized by the clinical appearance of the affected animal and the characteristic histological changes they produce in the brain. Microscopic examination reveals the presence of vacuoles in the neurons giving the grey matter the appearance of a section through a sponge.

Scrapie, the disease of sheep and goats, has been known since the 18th century but was first described scientifically in 1913. Its name is derived from one of the symptoms; an itching which causes the infected animal to scrape itself against objects.

The agent of scrapie and other SEs have been described as 'slow viruses' due to their long incubation periods. However, it is now thought that the infectious agent, known as a prion, is neither a bacterium nor a virus. It is invisible in the electron microscope, cannot be cultured in media or cell cultures and does not provoke the formation of specific antibodies in infected animals. It is also very resistant to heat, irradiation and chemical treatments such as formalin. In sheep, the illness is transmitted both vertically and horizontally and other animals have been infected as a result of intraperitoneal or intracerebral injection of infected tissue preparations. The evidence available suggests that these illnesses are intoxications rather than infections. The prion contains a protein PrP^{Sc} which is also a major component of the plaques formed in the brains of affected individuals. PrP^{Sc} is a modified version of the protein PrP^{C} found normally on the outer surface of neurons. Differences in its tertiary structure make it resistant to proteolytic degradation and removal when its useful life is over. It is thought that prion PrP^{Sc} finds its way to the neuron surface where its interaction with PrP^{C} leads to production of further PrP^{Sc}, a process known as 'recruitment'. This initiates a chain reaction, the accumulation of PrP^{Sc}, plaque formation, and the onset of neurological symptoms.

Four human SEs are known, Gerstmann–Sträussler–Scheinker syndrome, Fatal familial insomnia, Creutzfeldt–Jacob disease and kuru, though the last two may in fact be the same disease. Kuru, first described in 1957, is restricted to the Fore people of Papua New Guinea where it was the major cause of death among Fore women. It was shown to be transmissable to chimpanzees by intracerebral inoculation with brain extracts from dead patients and it was eventually decided that human transmission was in fact foodborne, albeit of a rather special kind. The Fore tribe have a tradition of cannibalism and it was the tribal custom for women and children to eat the brains of the dead as a mark of respect. Since this practice was suppressed in the late fifties incidence of the disease has declined.

Creutzfeldt–Jacob disease is more widespread than kuru and in the UK has an incidence of 1 per 2 000 000 inhabitants. Accidental transmission by injection of contaminated pituitary extracts, corneal grafting and implantation of contaminated electrodes has occurred but oral transmission has not been described.

Much attention has been focused on these conditions since the emergence of bovine spongiform encephalopathy (BSE), or mad cow disease, in cattle in Britain in 1985/6. The illness is thought to result from the scrapie agent crossing the species barrier and being transmitted to cattle by scrapie-infected sheep protein fed to cattle. Its emergence at this time has been linked with changes in the commercial rendering processes used in the production of animal proteins such as reduced processing temperatures and abandonment of the use of solvents to extract tallow. Control measures introduced in the UK include prohibition of the sale of bovine offal from animals more than six months old and banning the feeding of ruminant-derived protein to ruminants and the slaughter of all cattle over 30 months of age. In 1996 a ban on the use of all mammalian-derived animal protein in feed for all farm animals was introduced.

Incidence of BSE in cattle peaked in 1992 with 36,680 cases and by 2005 the annual total had declined to less than 150 cases. By and large the epidemic seems to have been confined to the UK as reported cases in other countries totalled around 600 at the end of 1997, mostly in Switzerland (256) and Ireland (224). This is a little surprising since thousands of tons of the same meat and bone meal thought to have caused BSE in the UK were also exported in the period 1985–1990, as were large numbers of British breeding cattle, nearly 58 000 to the EU alone. An alternative hypothesis is that the epidemic was caused by contaminated meat and bone meal from Africa imported into the UK.

The concern was that if the agent can cross the species barrier from sheep to cattle and be acquired through food, it could do so again–infecting humans. Initially, the concensus of expert opinion was that this was highly unlikely. There is no evidence linking human occupational exposure to potentially scrapie-infected tissues with degenerative encephalopathy, and sheep and sheep products have been eaten for longer than scrapie has been known without any evidence of it causing a similar human illness. It may be that the ease with which ingested abnormal prion protein can recruit the normal prion protein is reflected in the similarity of their amino acid compositions. If this is the case then recruitment of normal human prion by the BSE prion is likely to be relatively inefficient. The cattle and sheep prions differ at 7 positions whereas the cattle and human prions differ at more than 30 positions. The BSE agent has however been shown to cause illness in cats and some primates. Results of experiments with transgenic mice expressing PrP^C also indicate that induction of human prion produced by bovine prions is inefficient. In 1996, however, a new variant of CJD (vCJD) in humans was described which appears to be linked to BSE. It is thought that those affected acquired the agent through consuming beef products before the offal ban was imposed in 1989. The number of vCJD cases reported in the UK has increased from 1995 to a peak of 28 cases in 2000 and has since declined (Figure. 8.23).

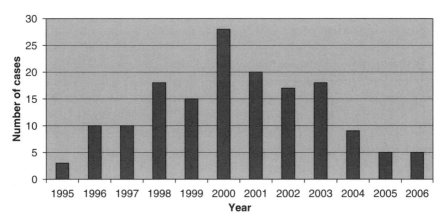

Figure 8.23 *vCJD cases in UK*
(data from Health Protection Agency)

CHAPTER 9

Fermented and Microbial Foods

9.1 INTRODUCTION

So far in this book we have been almost exclusively concerned with the negative roles that micro-organisms play in food. There is however a huge diversity of foods where microbial activity is an essential feature of their production. Some are listed in Table 9.1 and in this chapter we will describe a few of these in more detail and discuss some general features of food fermentation.

Almost without exception, fermented foods were discovered before mankind had any knowledge of micro-organisms other than as witness to the effects of their activity. It was simply an empirical observation that certain ways of storing food effected desirable changes in its character-istics (Table 9.2). Originally the most important of these changes must have been an improvement in the shelf-life and safety of a product, although these became less important in the industrialized world with the advent of alternative preservation methods such as canning, chilling and freezing. Modern technologies have in no way diminished the sensory appeal of fermented products however. This is clear from the way people rarely enthuse over grape juice or milk as some are prone to do over the vast array of wines and cheeses.

We now know that, in food fermentation, conditions of treatment and storage produce an environment in which certain types of organism can flourish and these have a benign effect on the food rather than spoiling it. The overwhelming majority of fermented foods is produced by the activity of lactic acid bacteria and fungi, principally yeasts but also, to a lesser extent, moulds. Both groups of organisms share a common ecological niche, being able to grow under conditions of low pH and reduced a_w, although only lactic acid bacteria and facultative yeasts will prosper under anaerobic conditions. As a consequence, they frequently

Table 9.1   *Some fermented foods*

Food	Ingredients	Geographical Distribution
Busa	Rice, millet, sugar	Turkey
Beer	Barley	Widespread
Cheese	Milk	Widespread
Chicha	Maize and others	S. America
Dawadawa	Locust beans	W. Africa
Gari	Cassava	Nigeria
Idli/dosa	Rice and black gram	India
Injera	Tef	Ethiopia
I-sushi	Fish	Japan
Kefir	Milk	Eastern Europe
Kenkey	Maize, sorghum	Ghana
Kimchi	Vegetables	Korea
Koko	Maize, sorghum	Ghana
Leavened bread	Wheat	Europe, N. America
Lambic beer	Barley	Belgium
Mahewu	Maize	S. Africa
Nam	Meat	Thailand
Ogi	Maize, sorghum, millet	Nigeria
Olives		Mediterranean Area
Palm wine	Palm sap	Widespread
Poi	Taro	Hawaii
Puto	Rice	Philippines
Salami	Meat	Widespread
Salt stock, cucumbers	Cucumbers	Europe, N. America
Sauerkraut	Cabbage	Europe, N. America
Sorghum beer	Sorghum	S. Africa
Sourdough bread	Wheat, rye	Europe, N. America
Soy sauce, miso	Soy beans	S.E. Asia
Tempeh	Soy beans	Indonesia
Tibi	Fruit	Mexico
Yoghurt	Milk	Widespread

Table 9.2   *Effects of food fermentation*

Raw material	Stability	Safety	Nutritive Value	Acceptability
Meats	++	+	−	(+)
Fish	++	+	−	(+)
Milks	++	+	(+)	(+)
Vegetables	+	(+)	−	(+)
Fruits	+	−	−	++
Legumes	−	(+)	(+)	+
Cereals	−	−	(+)	+

++ Definite improvement
+ Usually some improvement
(+) Some cases of improvement
− No improvement

occur together in fermented foods; in some cases members of both groups act in concert to produce a product while in others, one group plays the role of spoilage organisms. Some examples of these are presented in Table 9.3.

Table 9.3   *Yeasts and lactic acid bacteria in fermented foods*

Yeasts[1]	Lactic acid bacteria[2]	Yeasts and lactic acid bacteria
Modern European beers	Yoghurt	Sourdough bread
Bread	Sauerkraut	Kefir
Wine	Salami	Soy sauce
Cider	Cheese	African beers
		Lambic beer

[1] The presence of lactic acid bacteria in these foods is often associated with spoilage
[2] The presence of yeasts in these foods is often associated with spoilage

9.2   YEASTS

The yeasts are true fungi which have adopted an essentially single celled morphology reproducing asexually by budding or, in the case of *Schizosaccharomyces*, by fission. Although they have a simple morphology, it is probable that they are highly evolved specialists rather than primitive fungi. Their natural habitat is frequently in nutritionally rich environments such as the nectaries of plants, plant exudates, decaying fruits and the body fluids of animals. The yeasts frequently show complex nutritional requirements for vitamins and amino acids.

The yeast morphology has undoubtedly evolved several times for there are species with Ascomycete or Basidiomycete affinities and quite a number with no known sexual stage. Although a number of yeasts almost always occur as single celled organisms, quite a few can develop the filamentous structure of a typical mould. Indeed, there are a number of moulds which can take on a yeast morphology under certain conditions, usually in the presence of high nutrient, low oxygen and enhanced carbon dioxide concentrations.

A major taxonomic study of the yeasts by Kreger–van Rij (1984) describes about 500 species divided into 60 genera of which 33 are considered to be Ascomycetes, 10 Basidiomycetes and 17 Deuteromycetes. A number of yeasts, though certainly not all, are able to grow anaerobically using a fermentative metabolism to generate energy. The majority, if not all, of these fermentative yeasts grow more effectively aerobically and anaerobic growth usually imposes more fastidious nutritional requirements on them.

Although there is a large diversity of yeasts and yeast-like fungi, only a relatively small number are commonly associated with the production of fermented and microbial foods. They are all either ascomycetous yeasts or members of the imperfect genus *Candida*. *Saccharomyces cerevisiae* is the most frequently encountered yeast in fermented beverages and foods based on fruits and vegetables, an observation which is reflected in the existence of more than eighty synonyms and varieties for the species. All strains ferment glucose and many ferment other plant-associated carbohydrates such as sucrose, maltose and raffinose but none can ferment the animal sugar lactose. In the tropics *Schizosaccharomyces pombe* is frequently the dominant yeast in the production of traditional fermented beverages where a natural fermentation is allowed to occur, especially those produced from cereals such as maize and millet. *Kluyveromyces marxianus* is able to hydrolyse lactose and ferment galactose. There are a number of varieties which had previously been recognized as separate species associated with a range of different fermented milk products. *K. marxianus* var. *marxianus* (= *K. fragilis*) is the perfect state of *Candida kefir* and has been isolated from eastern European fermented milks such as koumiss and kefir. *K. marxianus* var. *bulgaricus* has been isolated from yoghurt and *K. marxianus* var. *lactis* from buttermilk, Italian cheese and fermented milks from Manchuria.

Because of its ability to grow at low water activities in the presence of high concentrations of sugar or salt, *Zygosaccharomyces rouxii* is especially associated with the fermentation of plant products in which the addition of salt is an integral part of the process. Many strains of *Hansenula anomala* and *Debaryomyces hansenii* can also grow in fairly concentrated salt solutions and the latter is frequently isolated from brined meat products and fermented sausages.

Although able to ferment carbohydrates, yeasts such as *Pichia guilliermondii* and *Saccharomycopsis fibuligera* grow best as surface pellicles and have been isolated from a number of tropical fermented products. The latter is able to break down starch and is associated with 'chalky bread'. *Geotrichum candidum* is usually considered as a filamentous mould but it has a strong affinity with the ascomycetous yeasts and is frequently isolated as part of the surface flora of fermented milk products such as cheeses. It is important to realize that, although all these species of yeasts and yeast-like fungi are thought to play a positive role in the production of a diverse range of fermented foods, they also occur as spoilage organisms in other commodities where their biochemical activities are undesirable.

One of the most important yeasts associated with spoilage is *Zygosaccharomyces bailii*. It has the ability to grow at relatively low water activities and low pH, as well as being remarkably resistant to preservatives, such as sorbic, benzoic and ethanoic acids, sulfur dioxide and ethanol, commonly used to prevent microbial spoilage of fruit juices, fruit juice concentrates, fermented beverages, pickles and sauces. *Z. bailii* is strongly fermentative and spoilage of products stored in plastic packs and glass bottles can lead to explosion of the containers. The survival of a single cell in a product containing an appropriate nitrogen source and fermentable carbohydrate can result in spoilage, so pasteurization or membrane filtration before filling, followed by stringent hygiene to prevent post-treatment contamination, are essential.

The following dichotomous key indicates how the genera discussed above differ from each other:

1. Vegetative reproduction by cross-wall formation followed by fission — *Schizosaccharomyces*[*]
1. Vegetative reproduction by budding — 2

2. Ascospores not formed — *Candida*
2. Ascospores formed — 3

3. Nitrate assimilated — *Hansenula*
3. Nitrate not assimilated — 4

4. Abundant true mycelium as well as budding — 5
4. True mycelium scarce or absent — 6

5. Asci formed exclusively on the true hyphae — *Saccharomycopsis*
5. Asci not formed exclusively on the true hyphae — *Pichia*

6. Asci dehiscent — *Kluyveromyces*
6. Asci persistent — 7

7. No conjugation preceding ascus formation — *Saccharomyces*
7. Conjugation preceding ascus formation — 8

8. Ascospores warty or with ridges — *Debaryomyces*
8. Ascospores spherical and smooth — *Zygosaccharomyces*

[*]*Schizosaccharomyces* belongs to the Archiascomycetes

9.3   LACTIC ACID BACTERIA

The term lactic acid bacteria (LAB) has no strict taxonomic significance, although the LAB have been shown by serological techniques and 16S ribosomal RNA cataloguing to be phylogenetically related. They share a number of common features: they are Gram-positive, non-sporeforming rods or cocci; most are aerotolerant anaerobes which lack cytochromes and porphyrins and are therefore catalase- and oxidase-negative. Some do take up oxygen through the mediation of flavoprotein oxidases and this is used to produce hydrogen peroxide and/or to re-oxidize NADH produced during the dehydrogenation of sugars.

Cellular energy is derived from the fermentation of carbohydrate to produce principally lactic acid. To do this, they use one of two different

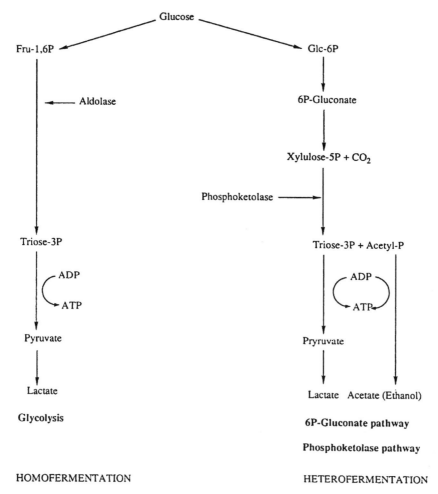

Figure 9.1   *The homo- and heterofermentation pathways*

pathways and this provides a useful diagnostic feature in their classification (Figure 9.1). Homofermenters produce lactate as virtually a single product from the fermentation of glucose. They follow the Emden–Meyerhof–Parnas (EMP) glycolytic pathway whereby the six-carbon molecule glucose is phosphorylated and isomerized before cleavage by the enzyme aldolase into glyceraldehyde-3-phosphate. This is then converted to pyruvate during which ATP is produced by substrate-level phosphorylation at two sites to give an overall yield of two molecules of ATP for every molecule of glucose fermented. In order to regenerate the NAD^+ consumed in the oxidation of glyceraldehyde-3-phosphate, pyruvate is reduced to lactate using NADH.

Heterofermenters produce roughly equimolar amounts of lactate, ethanol/acetate, and carbon dioxide from glucose. They lack aldolase and transform the hexose, glucose, into a pentose by a sequence involving oxidation and decarboxylation. The pentose is cleaved into glyceraldehyde phosphate and acetyl phosphate by the enzyme phosphoketolase. The triose phosphate is converted into lactate by the same sequence of reactions as occurs in glycolysis to give two molecules of ATP. The fate of the acetyl phosphate depends on the electron acceptors available. In the absence of alternatives, acetyl phosphate fulfils this role and is reduced to ethanol while regenerating two molecules of NAD^+ from NADH. In the presence of oxygen, NAD^+ can be regenerated by NADH oxidases and peroxidases, leaving acetyl phosphate available for conversion to acetate. This provides another site for substrate level phosphorylation and increases the overall ATP yield of heterofermentation from one to two molecules ATP per molecule of glucose dissimilated. When this is possible, the increased yield of ATP is reflected in a faster growth rate and a higher molar growth yield. The same effect can be achieved with other electron acceptors, for example fructose which is reduced to mannitol.

Heterofermenters and homofermenters can be readily distinguished in the laboratory by the ability of heterofermenters to produce carbon dioxide in glucose-containing media.

The principal genera of the lactic acid bacteria are described in Table 9.4. *Lactobacillus* is recognized as being phylogenetically very heterogeneous and this is evidenced by the broad range of %GC values exhibited within the genus. Some non-acidoduric, heterofermentative lactobacilli have been reclassified in the new genus *Carnobacterium* and there is likely to be significant further refinement of the genus in the future. Currently the lactobacilli are subdivided into three groups: obligate homofermenters, facultative heterofermenters and obligate heterofermenters. The obligate homofermenters correspond roughly to the Thermobacterium group of the Orla–Jensen classification scheme and include species such as *Lb. acidophilus*, *Lb. delbrückii* and *Lb. helveticus*. They ferment hexoses almost exclusively to lactate but are unable to ferment pentoses. The facultative heterofermenters ferment hexoses via the EMP pathway to lactate but have an inducible phosphoketolase which allows them to ferment pentoses to

Table 9.4 *Principal genera of the lactic acid bacteria*

Genus	Cell Morphology	Fermentation	Lactate isomer	DNA (mole %GC)
Lactococcus	cocci in chains	homo	L	33–37
Leuconostoc	cocci	hetero	D	38–41
Pediococcus	cocci	homo	DL	34–42
Lactobacillus	rods	homo/hetero	DL, D, L	32–53
Streptococcus	cocci in chains	homo	L	40[a]

[a] *S. thermophilus*
(Other genera that are currently included in the lactic acid bacteria, *Carnobacterium*, *Enterococcus*, *Oenococcus*, *Vagococcus*, *Aerococcus*, *Tetragenococcus*, *Alloiococcus*, *Weissella*)

lactate and acetate. They include some species important in food fermentation such as *Lb. plantarum*, *Lb. casei*, and *Lb. sake*. Obligate heterofermenters which include *Lb. brevis*, *Lb. fermentum* and *Lb. kefir* use the phosphoketolase pathway for hexose fermentation.

Leuconostoc is treated as a separate genus on morphological grounds as its members are typically irregular cocci. This is not entirely satisfactory since the vexed question, 'When does a short rod become a coccus?' often arises; for example, *Lactobacillus confusus* was originally classed as a *Leuconostoc*. It is possible to distinguish leuconostocs from most heterofermentative lactobacilli by two phenetic characteristics: their production of only D-lactate and inability to produce ammonia from arginine.

The genus *Pediococcus* also includes species of importance in food fermentations such as *P. pentosaceus* and, until fairly recently, *P. halophilus* now in a genus of its own as *Tetragenococcus halophilus*.

Nucleic acid studies of the streptococci have shown that they comprise three distinct groups worthy of genus status. The enterococci now form the genus *Enterococcus* although the faecal strains of *S. bovis* and *S. equinus* which also react with the group D antisera used in Lancefield's classical serological classification scheme are not included. What were known as Lancefield's group N streptococci, the lactic or dairy streptococci, are now members of the genus *Lactococcus* and a number of these which were considered distinct *Streptococcus* species are now classified as subspecies of *Lactococcus lactis*. The yoghurt starter *Streptococcus thermophilus* does not possess the Group N antigen and remains in the genus *Streptococcus*.

Some authors also include *Bifidobacterium* among the lactic acid bacteria although this has less justification as they are quite distinct both phylogenetically and biochemically. For example, hexose fermentation by bifidobacteria follows neither the EMP glycolytic pathway nor the phosphoketolase pathway but produces a mixture of acetic and lactic acids.

9.4 ACTIVITIES OF LACTIC ACID BACTERIA IN FOODS

9.4.1 Antimicrobial Activity of Lactic Acid Bacteria

Lactic acid bacteria are often inhibitory to other micro-organisms and this is the basis of their ability to improve the keeping quality and safety of many food products. The principal factors which contribute to this inhibition are presented in Table 9.5. By far the most important are the production of lactic and acetic acids and the consequent decrease in pH. Just how organic acids and low pH inhibit microbial growth and survival is discussed in Section 3.2.2 and will not be repeated here.

Bacteriocins are bactericidal peptides or proteins which are usually active against species closely related to the producing organism.

Table 9.5 *Factors contributing to microbial*
inhibition by lactic acid bacteria

Low pH
Organic acids
Bacteriocins
Hydrogen peroxide
Ethanol
Diacetyl
Nutrient depletion
Low redox potential

Production of bacteriocins by lactic acid bacteria has been extensively studied in recent years and a number have been described. Interest in them stems from the fact that they are produced by food-grade organisms and could therefore be regarded as 'natural' and hence more acceptable as food preservatives. A few promising candidates have been found but many others have a spectrum of activity which is too limited to be of any practical utility.

Nisin is the only bacteriocin to find practical application in the food industry to date. Produced by certain strains of *Lactococcus lactis*, it was first discovered when a nisin-producing strain caused problems in cheesemaking by inhibiting the other starter organisms present. Nisin is available commercially and has been used as a food preservative in the UK and some other countries since the early 1950s, though it was not approved for use in the United States until 1988. It differs from many other bacteriocins produced by lactic acid bacteria in having a relatively broad spectrum of activity against Gram-positive bacteria generally. In vegetative cells it acts by creating pores in the plasma membrane through which there is leakage of cytoplasmic components and a breakdown of the transmembrane potential. In Gram-negatives, the outer membrane acts as a barrier preventing nisin access to its site of action, thus making them resistant. Some Gram-negatives have been shown to become sensitive when their outer membrane has been damaged by thermal shock or by treatment with a chelating agent such as EDTA. Bacterial spores are particularly sensitive and the most important commercial applications of nisin have been to inhibit spore outgrowth in heat processed products, principally processed cheese and canned foods, but also, in some countries, in products such as clotted cream, dairy desserts, crumpets, pasteurized soups and pasteurized eggs. Nisin's value lies not just in the fact that it can improve shelf life by inhibiting spore outgrowth but its use can also permit milder heat processing regimes, allowing retention of heat sensitive properties in some products.

Nisin is an amphiphilic polypeptide containing 34 amino acids and is remarkably heat stable at acid pH. It belongs to a group of antibiotics known as lantibiotics, most of which are produced by non-lactic acid

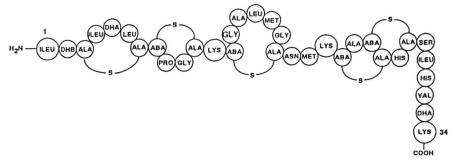

Figure 9.2 *Nisin (ABA. aminobutyric acid; DHA. dehydroalanine; DHB, dehydrobutyrine)*

bacteria and are characterized by the possession of unusual amino acids such as lanthionine (3,3′-thiodialanine) and β-methyl lanthionine (Figure 9.2). These are produced by a series of post translational modifications to a pre-propeptide which is then cleaved to remove a leader peptide.

Production of many bacteriocins appears to be a plasmid-encoded function but the gene coding for nisin has been cloned and sequenced from both chromosomal and plasmid DNA. Introduction of the ability to produce nisin into a chosen starter organism may prove useful in some fermented foods where competition from other Gram-positives needs to be controlled, although this is not desirable in cheesemaking where nisin production could inhibit the lactobacilli that contribute to cheese maturation.

Hydrogen peroxide is well known for its antimicrobial properties. Since lactic acid bacteria possess a number of flavoprotein oxidases but lack the degradative enzyme catalase, they produce hydrogen peroxide in the presence of oxygen. This will confer some competitive advantage as they have been shown to be less sensitive to its effects than some other bacteria. Accumulation of hydrogen peroxide has been demonstrated in some fermented foods but its effects are, in general, likely to be slight. Lactic acid fermentations are essentially anaerobic processes so hydrogen peroxide formation will be limited by the amount of oxygen dissolved in the substrate at the start of fermentation. It may be, however, that at this critical initial stage of a fermentation hydrogen peroxide production provides an important additional selective advantage. In milk, hydrogen peroxide is also known to potentiate the lactoperoxidase antimicrobial system (see Section 3.2.4).

Heterofermentative LAB produce ethanol, another well-established antimicrobial. It may make some contribution to the inhibition of competitors, although its concentration in lactic fermented products is generally low.

There are a number of other factors which may, like ethanol, give LAB a selective advantage in some situations. In most cases however their contribution is likely to be negligible, particularly when compared to the ability of LAB to produce lactic acid in quantities up to around 100 millimolar and a pH in the range 3.5 to 4.5.

9.4.2   Health-promoting Effects of Lactic Acid Bacteria-Probiotics

Fermented foods have long had a reputation for being positively beneficial to human health in a way that ordinary foods are not. Ilya Metchnikoff, the Russian founder of the theory of phagocytic immunity, was an early advocate of this idea based on his theories on disharmonies in nature. He held that the human colon was one such disharmony since intestinal putrefaction by colonic bacteria produced toxins which shorten life. One solution to this which he advocated in his book 'The Prolongation of Life', published in 1908, was the consumption of substantial amounts of acidic foods, particularly yoghurt. He thought that the antimicrobial activity of the lactic acid bacteria in these products would inhibit intestinal bacteria in the same way they inhibit putrefaction in foods and attributed the apparent longevity of Bulgarian peasants to their consumption of yoghurt.

Since then a number of claims have been made for lactic acid bacteria, particularly in association with fermented milks (Table 9.6). So much so, that live cultures of lactic acid bacteria (and some others such as *Bifidobacterium* spp.) consumed in foods are frequently termed 'probiotics' (Greek: for life). Much of the evidence available on these putative benefits is however inadequate or contradictory at present, and many remain rather ill defined.

Several studies have shown improved nutritional value in grains as a result of lactic fermentation, principally through increasing the content of essential amino acids. Such improvements however may be of only marginal importance to populations with a varied and well balanced diet. It has also been reported that fermentation of plant products reduces levels of antinutritional factors which they may contain such as cyanogenic glycosides and phytic acid, although this effect is often the result of other aspects of the process such as soaking or crushing rather than microbial action. Some have claimed that fermentation of milks increases the bioavailability of minerals, although this is disputed.

Table 9.6   *Beneficial effects claimed for lactic*
                acid bacteria

Nutritional improvement of foods
Inhibition of enteric pathogens
Alleviation of diarrhoea/Constipation
Hypocholesterolaemic action
Anticancer activity
Simulation of the immune system

One area where there is good evidence for a beneficial effect is in the ability of fermented milks to alleviate the condition known as lactose intolerance. All human infants possess the enzyme lactase (β-galactosi-dase) which hydrolyses the milk sugar lactose into glucose and galactose which are then absorbed in the small intestine. In the absence of this enzyme when milk is consumed, the lactose is not digested but passes to the colon where it is attacked by the large resident population of lactose-fermenting organisms producing abdominal discomfort, flatulence and diarrhoea. Only people of north European origin and some isolated African and Indian communities maintain high levels of gut β-galactosidase throughout life. In most of the world's population it is lost during childhood and this precludes the consumption of milk and its associated nutritional benefits. If however lactase-deficient individuals take milk in a fermented form such as yoghurt, these adverse effects are less severe or absent. This is not simply a result of reduced levels of lactose in the product since many yoghurts are fortified with milk solids so that they have lactose contents equivalent to fresh milk. It appears to be due to the presence of β-galactosidase in viable starter organisms, as pasteurized yoghurts show no beneficial effect. In the gut, the ingested cells become more permeable in the presence of bile and this allows them to assist the body in the hydrolysis of lactose.

The protective role of the gut's microflora has been discussed already (Section 6.5) and there is evidence that ingested lactic acid bacteria can contribute to this. Yoghurt has been shown to have a strong inhibitory effect on the growth of coliform bacteria in the stomach and duodenum of piglets and studies of human infants with diarrhoea have shown that the duration of illness was shorter in those groups given yoghurt than in control groups. However, the usual starter organisms in yoghurt, *Lactobacillus delbrueckii* subsp. *bulgaricus* and *Streptococcus thermophilus* are not bile tolerant and do not colonize the gut. They will persist in the alimentary tract and be shed in the stools only as long as they are being ingested, so that any improving effect is likely to be transient. Recently attention has focused on lactic acid bacteria such as *Lactobacillus acidophilus* and bifidobacteria such as *Bifidobacterium longum* which can colonize the gut and these organisms have been included in yoghurts and other fermented milks and some proprietory preparations. There is some evidence that such probiotic lactobacilli can shorten the duration of viral diarrhoea and may be useful in reducing antibiotic associated diarrhoea. Studies with traveller's diarrhoea however have given contradictory results.

Pathogen inhibition *in vivo* by LAB unable to colonize the gut must be by mechanisms broadly similar to those which apply *in vitro* (Section 9.4.1). With those organisms able to colonize the gut, the masking of potential attachment sites in the gut may also be involved.

Lactic acid bacteria have been reported to stimulate the immune system and various studies have described their ability to activate macrophages and lymphocytes, improve levels of immunoglobulin A (IgA) and the production of gamma interferon. These effects may contribute to a host's resistance to pathogens and to the antitumour activity noted for LAB, mainly *Lactobacillus acidophilus*, in some animal models. An additional or alternative possible mechanism proposed for the antitumour effect is the observed reduction in activity of enzymes such as β-glucuronidase, azoreductase and nitroreductase in faecal material when LAB are ingested. These enzymes, produced by components of the intestinal flora, can convert procarcinogens to carcinogens in the gut and their decreased activity is probably due to inhibition of the producing organisms by LAB.

A number of studies have indicated that probiotics may have a role in preventing and treating atopic diseases such as atopic eczema and asthma in children.

High levels of serum cholesterol are established as a predisposing factor for coronary heart disease. It has been suggested that consumption of fermented milks has a hypocholesterolaemic action and some have suggested a variety of mechanisms by which this can occur. The evidence is however weak and it has not proved possible to demonstrate this effect in a number of trials.

An alternative approach to the consumption of large numbers of probiotic bacteria is to encourage the growth of indigenous bifidobacteria and lactobacilli in the gut through consumption of prebiotics. These are defined as non-digestible food components that exert a beneficial effect on the consumer by selectively stimulating the growth and activity of certain bacteria in the colon. The most common prebiotics are polymeric forms of fructose such as inulin, a natural component of foods such as Jerusalem artichokes, leeks, onions and garlic.

9.4.3 The Malo-lactic Fermentation

LAB can decarboxylate L-malic acid to produce L-lactate in a reaction known as the malo-lactic fermentation (Figure 9.3). This process is particularly associated with wines, where malic acid can form up to half the total acid, and its effect is to reduce substantially a wine's acidity. It is particularly encouraged in wines from cool regions which tend to have a naturally high acidity and, although less desirable in wines from warmer regions, it is often promoted to provide bacteriological stability to the bottled product. It may also modify and improve the body and flavour of a wine.

A natural malo-lactic fermentation can be encouraged by refraining from sulfiting the new wine and leaving it on the yeast lees (sediment) for

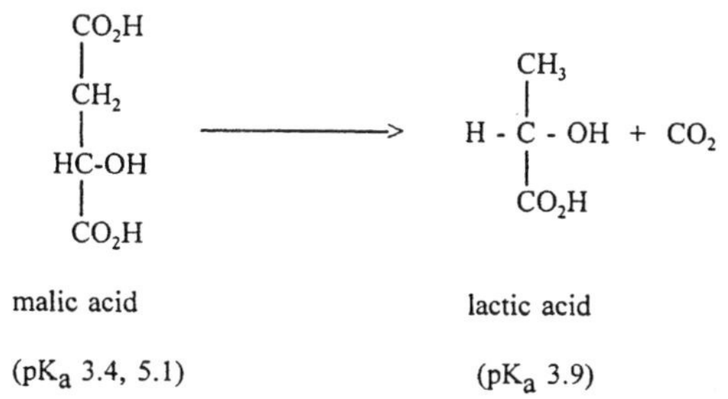

Figure 9.3   *The malo-lactic fermentation*

longer than usual. Commercial starter cultures are also available usually consisting of strains of *Oenococcus oeni*, formerly *Leuconostoc oenos*.

Until recently it was unclear how LAB derive any benefit from performing this reaction. Substrate-level phosphorylation does not occur (it is not therefore, strictly speaking, correct to call it a fermentaion) and the free energy of the decarboxylation reaction is low. It now seems that the reaction conserves energy through a proton motive force generated across the cell membrane by the transport of malate, lactate and protons.

9.5   FERMENTED MILKS

9.5.1   Yoghurt

Fermentation to extend the useful life of milk is probably as old as dairying itself. The first animals to be domesticated are thought to have been goats and sheep in the Near East in about 9000 BC. In the warm prevailing climate it is likely that their milks furnished the first fermented milks and only some time later, between 6100 and 5800 BC in Turkey or Macedonia, was the cow first domesticated.

Fermented milks which include yoghurt, buttermilk, sour cream, and kefir differ from cheese in that rennet is not used and the thickening produced is the result of acidification by lactic acid bacteria. Yoghurt whose name comes from the Turkish word 'Jugurt' is the most widely available fermented milk in the Western world today where its popularity derives more from its flavour and versatility than from its keeping properties.

It is made from milk, skimmed milk or fortified milk usually from cows but sometimes from other animals such as goats or sheep. The production process most commonly applied commercially is outlined in Figure 9.4.

The first prerequisite of any milk to be used in a fermentation process is that it should be free from antimicrobials. These could be antibiotic residues secreted in the milk as a result of mastitis chemotherapy or

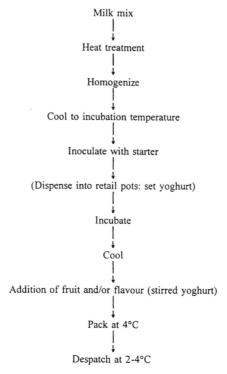

Milk mix

Heat treatment

Homogenize

Cool to incubation temperature

Inoculate with starter

(Dispense into retail pots: set yoghurt)

Incubate

Cool

Addition of fruit and/or flavour (stirred yoghurt)

Pack at 4°C

Despatch at 2-4°C

Figure 9.4   *Yoghurt production*

sanitizers carried into the milk as a result of inadequate equipment cleaning regimes at the farm or dairy. Inhibition of the starter culture would result not only in economic losses but could potentially allow pathogens to grow.

In commercial practice it is usual to supplement the solids content of the milk to enhance the final texture of the product. The SNF (solids not fat) content is increased to between 11 and 15%, compared with a level of around 8.5% in fresh milk. The simplest way of achieving this is by addition of skim- or whole-milk powder depending on whether a conventional or low-fat product is required. The properties of the product may also be improved and stabilized by the addition of small amounts of natural or modified gums which bind water and thicken the product.

If left to stand, the milk fat would separate out to form a cream layer. To prevent this, the milk is homogenized by passing it through a small orifice under pressure, typically $100–200 \, kg \, cm^{-2}$ at $50–60 \, °C$, to reduce the size of the fat globules to below $2 \, \mu m$. This improves the product's stability, increases the milk's viscosity, and also makes it appear whiter as the number of light-reflecting centres is increased.

Before addition of the starter culture, the milk is heated at $80–90 \, °C$ for about 30 min. Being well in excess of the normal pasteurization requirements for safety, this has a substantial lethal effect on the micro-flora. All but heat-resistant spores are eliminated so that the starter culture encounters little by way of competition. The heat process also improves the milk as a growth medium for the starter by inactivating immunoglobulins, expulsion of oxygen to produce a microaerophilic environment, and through the release of stimulatory levels of sulfydryl groups. Excessive heating can however lead to the production of inhib-itory levels of these compounds. Heating also promotes interactions between whey or serum proteins and casein which increase the yoghurt viscosity, stabilize the gel and limit syneresis (separation of whey).

The heat-treated milk is cooled to the fermentation temperature of 40–43 °C which is a compromise between the optima of the two starter organisms *Strep. thermophilus* (39 °C) and *Lb. delbrueckii* subsp. *bulga-ricus* (45 °C). The starter culture is added at a level of about 2% by volume to give an initial concentration of 10^6–10^7 cfu ml^{-1} composed of roughly equal numbers of the two organisms. The fermentation can be conducted in the retail pack to produce a firm, continuous coagulum, which is known as a set yoghurt, or in bulk tanks to produce a stirred yoghurt where the gel has been broken by mixing in other ingredients and by pumping into packs.

The fermentation takes about 4 h during which the starter bacteria ferment lactose to lactic acid decreasing the pH from its initial level of 6.3–6.5. The lactic acid helps solubilize calcium and phosphate ions which destabilize the complex of casein micelles and denatured whey proteins. When the pH reaches 4.6–4.7, the isoelectric point of the casein, the micelles aggregate to produce a continuous gel in which all the components are entrapped with little or no 'wheying-off'.

During fermentation growth of the streptococci is fastest in the early stages, but as the pH drops below 5.5 it slows and the lactobacilli tend to predominate. By the end of fermentation the product has a total acidity of 0.9–0.95% and the populations of the two starter organisms are roughly in balance again with levels in excess of 10^8 cfu ml^{-1}.

The relationship between the two starter organisms is one known as protoco-operation, that is to say they have a mutually favourable inter-action but are not completely interdependent. Both will grow on their own in milk but will grow and acidify the product faster when present together. Growth of the streptococcus in milk is limited by the avail-ability of peptides and free amino acids which are present in relatively low concentrations (≈ 50 mg kg^{-1}). The lactobacillus is slightly proteoly-tic and liberates small amounts of these, particularly valine, which stimulate streptococcal growth. In its turn the streptococcus produces formate, pyruvate and carbon dioxide all of which stimulate the lacto-bacillus. Formate is used in the biosynthesis of the purine base adenine, a component of RNA and DNA and *Lb. delbrueckii* subsp. *bulgaricus*

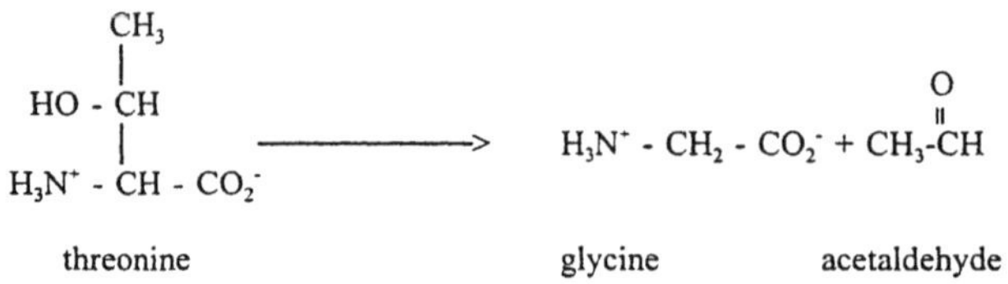

$$
\begin{array}{c}
\mathrm{CH_3} \\
| \\
\mathrm{HO - CH} \\
| \\
\mathrm{H_3N^+ - CH - CO_2^-}
\end{array}
\quad \xrightarrow{\hspace{2cm}} \quad
\mathrm{H_3N^+ - CH_2 - CO_2^-} + \mathrm{CH_3 - \overset{\overset{\textstyle O}{\|}}{C}H}
$$

threonine   glycine   acetaldehyde

Figure 9.5   *The threonine aldolase reaction*

tends to grow poorly in milk with low levels of formate, forming elongated, multinucleate cells.

Acetaldehyde (ethanal) is the most important flavour volatile of yoghurt and should be present at 23–41 mg kg^{-1} (pH 4.2–4.4) to give the correct yoghurt flavour. Its accumulation is a consequence of the fact that both starter organisms lack an alcohol dehydrogenase which would otherwise reduce the acetaldehyde to ethanol. Both will produce acetaldehyde from the glucose portion of lactose *via* pyruvate and through the action of threonine aldolase. The latter activity (Figure 9.5) is more pronounced in the lactobacillus but in the streptococcus methionine has been shown to increase levels of acetaldehyde *via* threonine. Diacetyl, an important flavour compound in many dairy products, is present at very low levels (≈ 0.5 mg kg^{-1}) but is thought to make a contribution to the typical yoghurt flavour.

When the fermentation is complete the yoghurt is cooled to 15–20 °C before the addition of fruits and flavours and packaging. It is then cooled further to below 5 °C, under which conditions it will keep for around three weeks. Yoghurt is not usually pasteurized since chill storage will arrest the growth of the starter organisms. The acidity will however continue to increase slowly during storage.

Because of its high acidity and low pH (usually 3.8–4.2), yoghurt is an inhospitable medium for pathogens which will not grow and will not survive well. It is unusual therefore for yoghurt to be involved in outbreaks of foodborne illness, although the hazelnut yoghurt botulism outbreak in the UK in 1989 (see Section 7.5.5) is a notable exception. Yoghurts are spoiled by acidoduric organisms such as yeasts and moulds. Yeasts such as the lactose-fermenting *Kluyveromyces fragilis* and, in fruit-containing yoghurts, *Saccharomyces cerevisiae* are particularly important but the yeast-like fungus *Geotrichum* and surface growth of moulds such as *Mucor*, *Rhizopus*, *Aspergillus*, *Penicillium*, and *Alternaria* can also be a problem. Advisory guidelines for microbiological quality have suggested that satisfactory yoghurts should contain more than 10^8 cfu g^{-1} of the starter organisms, <1 coliform g^{-1}, <1 mould g^{-1} and <10 yeasts g^{-1} (fruit-containing yoghurts may contain up to 100 yeasts g^{-1} and remain of satisfactory quality).

9.5.2 Other Fermented Milks

The popularity of acidophilus milk is largely due to health-promoting effects which are claimed to stem from the ability of *Lactobacillus acidophilus* to colonize the gut. It is a thermophilic homofermenter but is slow fermenting and a poor competitor and is easily outgrown. As a result, the fermentation takes longer than for yoghurt and great care must be taken to avoid contamination. In the original process whole or skimmed milk was sterilized prior to fermentation by a Tyndallization process. This involved two heating stages of 90–95 °C for up to an hour separated by a holding period of 3–4 h to allow spore germination to occur. Nowadays the same effect can be achieved more swiftly and economically by UHT processing. The milk is then homogenized, cooled to the fermentation temperature of 37–40 °C and inoculated with 2–5% of starter culture. It can take as long as 24 h to produce the required acidity of about 0.7%, after which the product is cooled to 5 °C.

In addition to the extra care required in its production, acidophilus milk suffers from a number of other drawbacks. In particular, it lacks the sensory appeal of yoghurt, being restricted to a rather sour, acidic taste. Also, the *Lb. acidophilus* cells do not survive well in the acid product, dying out after about a week's storage at 5 °C. To avoid these problems, a non-fermented sweet acidophilus milk is produced in the United States where large numbers of *Lb. acidophilus* are simply added to pasteurized milk without incubation.

In an attempt to combine the supposed virtues of acidophilus milk with those of yoghurt, a number of 'bio-yoghurts' are now produced. These contain a mixture of organisms, those able to colonize the gut such as *Lb. acidophilus* and *Bifidobacterium* spp. with *Strep. thermophilus* to provide the characteristic yoghurt flavour. However, because of their poor survival at acid pH, it is likely that the strains used are chosen for their ability to survive in the product as much as for any benefit they may have *in vivo*.

Kefir and koumiss are distinctive fermented milks produced by a mixed lactic acid bacterial fermentation and an alcoholic yeast fermentation. Kefir is further distinguished by the fact that the microflora responsible is not dispersed uniformly throughout the milk but is added as discrete kefir 'grains'. These are in fact sheets composed largely of a strong polysaccharide material, kefiran, which folds upon itself to produce globular structures resembling cauliflower florets. The outside of the sheets is smooth and is populated by lactobacilli while the inner, rougher side of the sheet carries a mixed population of yeasts and lactic acid bacteria. A large variety of different organisms have been reported as being associated with the fermentation, probably reflecting the widespread and small-scale nature of production. The morphology of the grain itself suggests that the lactic acid bacteria are responsible for its production and a capsular, homofermenter *Lactobacillus kefiranofaciens* has been shown to produce kefiran. A heterofermentative lactobacillus *Lb. kefir* is numerically very important in many grains and plays a key role in the fermentation, probably among other things contributing to the required effervescence in the product. Although less significant numerically, several yeasts have been reported including *Candida kefir*, *Saccharomyces cerevisiae* and *Sacc. exiguus*. The latter is particularly interesting because it was shown to utilize galactose preferentially in the presence of glucose and this may confer an advantage when growing in a mixed culture of organisms most of which will preferentially metabolize the glucose portion of lactose.

Kefir is produced commercially in a number of countries, most importantly in Russia and those states which comprised the old Soviet Union. In the mid-1980s production of kefir reached 12 million tonnes representing 80% of all dairy products, excluding soft cheese and sour cream. In commercial practice, milk for kefir production is homogenized and heated to 85–95 °C for between 3 and 10 min. It is cooled to 22 °C before addition of kefir grains at a level of up to 5%. The fermentation itself lasts for 8–12 h but is sometimes followed by slow cooling to around 8 °C over 10–12 h to allow for the required flavour development.

Kefir has an acidity of about 0.8% and an alcohol content which has been reported as varying between 0.01% and 1%. Ethanol levels tend to be lower in commercial products than domestically produced kefir and increase with the age of the product. In addition to the character imparted by the ethanol, lactic acid and carbon dioxide, acetaldehyde (ethanal) and diacetyl are also present as flavour components.

Koumiss is a fizzy, greyish white drink produced traditionally from mare's milk in eastern Europe and central Asia. It can have an acidity up to 1.4% and an ethanol content up to 2.5%. A mixed yeast/LAB flora is responsible for the fermentation comprising *Lb. delbrueckii* subsp. *bulgaricus* and a number of lactose fermenting yeasts. These are dispersed throughout the product and do not form discrete particles as in kefir. Cow's milk is a more convenient raw material to use nowadays and this is usually modified to resemble more closely the composition of mare's milk which has a lower fat content and higher carbohydrate levels.

Strictly speaking, buttermilk is the liquid which separates from cream during the churning of butter (see Chapter 5). However, to achieve a consistent quality product most buttermilk today is produced directly by the fermentation of skimmed or partially skimmed milk. Cultured buttermilk is an acidic refreshing drink with a distinctive buttery flavour. A mixture of starter organisms is required to produce these attributes; *Lactococcus lactis* produces most of the lactic acid, while the buttery flavour is the result of diacetyl production by so-called flavour bacteria such as strains of *Lactococcus lactis* subsp. *lactis* and *Leuconostoc mesenteroides* subsp. *cremoris*.

Most bacteria produce diacetyl and acetoin from carbohydrate *via* pyruvate. However, because of the key role pyruvate plays as an electron acceptor in LAB, it cannot usually be spared for this purpose unless an additional source other than carbohydrate or an alternative electron acceptor is available. Citrate metabolism can provide this extra pyruvate and lead to the accumulation of diacetyl as indicated in Figure 9.6. Fresh milk contains citrate but levels decline during storage so that, for the

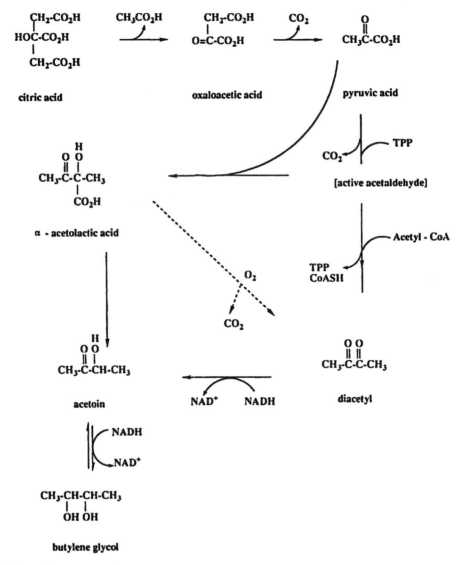

Figure 9.6 *Citrate fermentation. TPP,* thiamine pyrophosphate

production of cultured buttermilk, the milk is often supplemented with 0.1–0.2% sodium citrate to ensure good flavour development.

In the production process, pasteurized, homogenized milk is fermented at 22 °C for 12–16 h. The product contains 0.7–0.9% lactic acid and will keep for two weeks at 5 °C.

Another property of LAB valued in some fermented milks is their ability to produce a glycoprotein slime which provides a characteristic texture and viscosity to products such as Swedish *langfil* and Finnish *villi*. Like several other properties of LAB important in dairy fermentations such as the ability to ferment citrate, slime production is a plasmid-mediated characteristic and the ease with which this ability can be lost by the 'ropy' strains of *Lactococcus lactis* used in these fermentations can cause serious problems in commercial production.

9.6 CHEESE

Cheese can be defined as a consolidated curd of milk solids in which milk fat is entrapped by coagulated casein. Unlike fermented milks, the physical characteristics of cheese are far removed from those of milk. This is because protein coagulation proceeds to a greater extent as a result of the use of proteolytic enzymes and much of the water content of the milk separates and is removed in the form of whey. Typically the yield of cheese from milk is of the order of 10%.

Cheesemaking can be broken down into a number of relatively simple unit operations. Slight variations of these and the use of different milks combine to generate the huge range of cheeses available today; said to include 78 different types of blue cheese and 36 Camemberts alone. Classification of cheeses is made difficult by this diversity and the sometimes rather subtle distinctions between different types. Probably the most successful approach is one based on moisture content, with further subdivision depending on the milk type and the role of micro-organisms in cheese ripening (Table 9.7).

Cheese is a valuable means of conserving many of the nutrients in milk. In many people, it evokes a similar response to wine, playing an indispensible part in the gastronome's diet and prompting Brillat-Savarin (1755–1826) to coin the rather discomforting aphorism that 'Dessert without cheese is like a pretty woman with only one eye'. Despite this, the attraction of a well-ripened cheese eludes many people and it is sometimes hard to understand how something that can smell distinctly pedal can yield such wonderful flavours. This paradox was encapsulated by a poet, Leon-Paul Fargue, who described Camembert as 'the feet of God'.

Today cheesemaking is a major industry worldwide, producing something approaching 14 million tonnes per annum. Much is still practised

Table 9.7 *Cheese varieties and their classification*

Moisture Content	
50–80% *SOFT CHEESES*	Unripened
	Cottage, Quark, Cream, Mozzarella
	Ripened
	Camembert, Brie, Neufchatel (as made in France), Caciotta,
	Cooked
	Salt-cured or pickled
	Feta, Domiati
39–50% *SEMI SOFT*	
CHEESES	
	Ripened principally by internal mould growth
	Roquefort (milk from sheep), Stilton, Gorgonzola, Danish
	Blue
	Ripened by bacteria and surface micro-organisms
	Limburger, Brick, Trappist, Port Salut
	Ripened primarily by bacteria
	Bel Paesa, Pasta Filata, Provolone, Brick, Gouda, Edam
<39% *HARD CHEESES*	
	Without eyes, ripened by bacteria
	Cheddar, Caciocavallo
	With eyes, ripened by bacteria
	Emmental, Gruyère
<34% *VERY HARD CHEESES*	
	Asiago old, Parmesan, Romano, Grana

Based on USDA, 1978

on a relatively small scale and accounts for the rich diversity of cheeses still available. Large-scale industrialized production is increasingly important, however, and is dominated by one variety, Cheddar, which is now produced throughout the world, far removed from the small town in Somerset where it originated. Cheddar cheese is particularly valued for its smooth texture and good keeping qualities, although products sharing the name can vary dramatically in flavour. In what follows we will describe the basic steps in cheesemaking with particular reference to the manufacture of Cheddar cheese.

Cow's milk for cheese production must be free from antibiotics and sanitizing agents that might interfere with the fermentation. Although it is not compulsory, a heat treatment equivalent to pasteurization is usually applied at the start of processing. This helps to ensure a safe product and a reliable fermentation, although cheeses made from raw (unpasteurized) milk have been claimed to possess a better flavour. The milk is then cooled to the fermentation temperature which, in the case of Cheddar and other English cheeses such as Stilton, Leicester and Wensleydale, is 29–31 °C. The starter organisms used in most cheese-making are described as mesophilic starters, strains of *Lactococcus lactis* and its subspecies. Thermophilic starters such as *Lactobacillus helveticus*, *Lb. casei*, *Lb. lactis*, *Lb. delbrueckii* subsp. *bulgaricus* and *Strep. thermophilus* are used in the production of cheeses like Emmental and Parmesan where a higher incubation temperature is employed.

The role of starter organisms in cheesemaking is both crucial and complex. Their central function is the fermentation of the milk sugar lactose to lactic acid. This and the resulting decrease in pH contribute to the shelf-life and safety of the cheese and gives a sharp, fresh flavour to the curd. The stability of the colloidal suspension of casein is also weakened and calcium is released from the casein micelles improving the action of chymosin. After the protein has been coagulated, the acid aids in moisture expulsion and curd shrinkage, processes which govern the final cheese texture.

There are two different systems for uptake and metabolism of lactose in LAB. In most lactobacilli and *Strep. thermophilus*, lactose is taken up by a specific permease and is then hydrolysed intracellularly by β-galactosidase. The glucose produced is fermented by the EMP pathway which the galactose also enters after conversion to glucose-6-phosphate by the Leloir pathway (Figure 9.7). Most lactococci and some lactobacilli such as *Lb. casei* take up lactose by a phosphoenolpyruvate (PEP)-dependent phosphotransferase system (PTS) which phosphorylates lactose as it is transported into the cell. The lactose phosphate is then hydrolysed by phospho-β-galactosidase to glucose, which enters the EMP pathway, and galactose-6-phosphate which is eventually converted to pyruvate via the tagatose-6-phosphate pathway. These pathways are of practical import in cheesemaking; in the lactococci, lactose utilization is an unstable, plasmid encoded characteristic and loss of these genes can

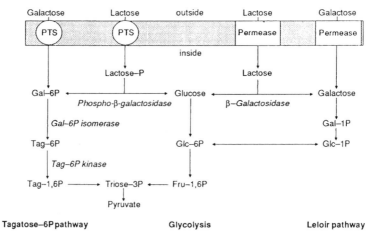

Figure 9.7 *Lactose uptake systems*

clearly have serious consequences for milk fermentation. Using transduction techniques, molecular biologists have produced strains of *Lactococcus lactis* in which this property has been stabilized by integration of the lactose utilization genes in the chromosome.

The thermophilic lactobacilli, which employ a lactose permease and β-galactosidase, metabolize the glucose produced preferentially, turning to galactose only when lactose becomes limiting. This can be a problem in some products. The accumulation of galactose can give rise to a brown discolouration during the heat processing of Mozzarella cheese. In Swiss cheeses such as Emmental, residual galactose can affect product flavour since propionic acid bacteria ferment it in preference to lactate. In doing so they produce a preponderance of acetic (ethanoic) acid which does not confer the usual nutty flavour associated with the equimolar concentrations of acetate and propionate produced by the *Propionibacterium* from lactate.

Lactic acid bacteria are nutritionally fastidious and require preformed nucleotides, vitamins, amino acids and peptides to support their growth. To grow to high cell densities and produce acid rapidly in milk, dairy starters must have proteolytic activity to overcome the limitation imposed by the low non-protein nitrogen pool in native milk. These systems are comprised of proteinases, associated with the surface of the bacterial cell wall, which can hydrolyse casein proteins. Peptidases in the cell wall degrade the oligopeptides produced down to a size that can be transported into the cell (4–5 amino acid residues) where they are further degraded and utilized. While this ability is essential to starter function, it also plays an important role in the development of cheese flavour during ripening or maturation (see below).

Citrate fermentation to diacetyl is required in some cheese varieties and starter cultures for these include species such as *Lactococcus lactis* subsp. *lactis* or *Leuconostoc cremoris*. Carbon dioxide is another product of this pathway and is important in producing the small eyes in Dutch cheese like Gouda or giving an open texture that will facilitate mould growth in blue-veined cheeses. In other cheese, such as Cheddar, this would be regarded as a textural defect.

To produce Cheddar cheese, starter culture is added at a level to give 10^6–10^7 cfu ml^{-1}. In the past these cultures were grown-up in the dairy from stock cultures or from freeze-dried preparations bought in from commercial suppliers. Nowadays frozen, concentrated cultures that are added directly to the cheese vat are increasingly used because of their ease of handling and the greater security they offer the cheesemaker. This applies particularly to the risk of bacteriophage inhibition of the fermentation which has been a major preoccupation of the cheesemaker since it was first identified in New Zealand in the 1930s. Problems of phage infection are not confined to cheesemaking but have also been encountered in the production of yoghurt and fermented meats.

A bacteriophage is a bacterial virus which in its virulent state infects the bacterial cell, multiplies within it, eventually causing the cell to burst (lysis). When this occurs during a cheese fermentation, acidification slows or even stops causing financial losses to the producer as well as an increased risk that pathogens might grow. An important source of phage in cheesemaking is thought to be the starter culture organisms themselves which carry within them lysogenic phages that can be induced into a virulent state. Problems occur particularly when starters contain a single strain or only a few strains and the same culture is reused over an extended period. During this time, phages specific to that organism build up in the plant and can be isolated from the whey and from environmental sources such as drains and the atmosphere, increasing the chance of fermentation failure. In the past, control of this problem has been based on the observation of rigorous hygiene in the dairy, the rotation of starter cultures with differing phage susceptibilities and propagation of starters in phage-inhibitory media which contain phosphate salts to chelate Ca^{2+} and Mg^{2+} required for successful phage adsorption to the bacterial cell. LAB possess their own resistance mechanisms to phage infection which include restriction/modification of non-host DNA, inhibition of phage adsorption by alteration or masking of specific receptors on the cell surface, and reduction of burst size (the number of phages released per infected cell). Most of these mechanisms appear to be plasmid encoded and this has opened the way for new strategies for phage control so that transconjugants with enhanced phage resistance are now available.

A time course for the production of Cheddar cheese showing pH changes and the timing of different process stages is shown in Figure 9.8. A good starter should produce around 0.2% acidity within an hour's incubation. It will multiply up to around 10^8–10^9 cfu g^{-1} in the curd producing an acidity of 0.6–0.7% before its growth is stopped by salting.

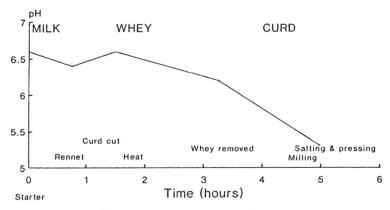

Figure 9.8   *pH changes during Cheddar cheese manufacture*

After about 45 min rennet is added. The time of renneting and the amount added are other important variables in cheesemaking which differ with cheese type. Rennet is a preparation from the fourth stomach or abomasom of suckling calves, lambs or goats. Its most important component is the proteolytic enzyme rennin or chymosin which cleaves *k*-casein, the protein responsible for the stability of the casein micelle, between phenylalanine 105 and methionine 106. This releases a 64 amino acid macropeptide into the whey leaving the hydrophobic *para-k*-casein attached to the micelle. Loss of the macro-peptide leads to the formation of cross-links between the micelles to form a network entrapping moisture and fat globules.

Authentic chymosin is produced as a slaughterhouse by-product but microbial rennets are available, produced from fungi such as *Mucor miehei*, *Mucor pusillus* and *Endothia parasitica*. These lack the specificity of animal rennet and have been associated with the production of bitter peptides in the cheese. Now however the genes for chymosin have been cloned into a number of organisms and nature-identical chymosin is available commercially, produced using the bacterium *E. coli* and yeasts.

After 30–45 min, coagulation of the milk is complete and the process of whey expulsion is started by cutting the curds into approximately 1 cm cubes. Whey expulsion is further assisted by the process known as scalding when the curds, heated to 38–42 °C, shrink and become firmer. The starter organisms are not inhibited by such temperatures and continue to produce acid which aids curd shrinkage. Cheeses produced using thermophilic starters can be scalded at higher temperatures without arresting acid development. When the acidity has reached the desired level (generally of the order of 0.25%), the whey is run off from the cheese vat.

It is at this stage that the process known as cheddaring occurs. The curd is formed into blocks which are piled up to compress and fuse the curds, expelling more whey. Nowadays, the traditional manual process is mechanized in a cheddaring tower.

At the end of cheddaring, the curd has a characteristic fibrous appearance resembling cooked chicken breast. The blocks of curd are then milled into small chips. This facilitates the even distribution of salt which, in Cheddar, is added at a level of between 1.5 and 2% w/w. The salted curd is formed into blocks which are then pressed to expel trapped air and whey.

Finally the cheese is ripened or matured at 10 °C to allow flavour development. During this stage, which can last up to five months to produce a mild Cheddar, the microflora is dominated by non-starter lactobacilli and a complex combination of bacterial and enzymic reactions give the cheese its characteristic flavour. In particular, proteases and peptidases from the starter culture continue to act, even though the organism can no longer grow. With other proteases from the rennet, they release free amino acids (principally glutamic acid and leucine in Cheddar) and peptides which contribute to the cheese flavour. In some cases this can give rise to a flavour defect: casein proteins contain a high proportion of hydrophobic amino acid residues such as leucine, proline and phenylalanine and if they are degraded to produce peptides rich in hydrophobic residues, the cheese will have a bitter taste.

The lipolytic and proteolytic activities of moulds play an important role in the maturation of some cheeses. In blue cheeses such as Stilton, *Penicillium roquefortii* grows throughout the cheese. It can grow at reduced oxygen tensions, but aeration is improved by not pressing the curds and by piercing the blocks of curd with needles. *P. camembertii* is associated with surface-ripened soft cheeses such as Camembert and Brie.

The keeping qualities of cheese vary with the type but are always much superior to those of milk. This is principally the result of the reduced pH (around 5.0 in Cheddar), the low water activity produced by whey removal and the dissolution of salt in the remaining moisture. Under these conditions yeasts and moulds are the main organisms of concern. The latter are effectively controlled by traditional procedures to exclude air such as waxing or by modern refinements such as vacuum packing.

9.7 FERMENTED VEGETABLES

9.7.1 Sauerkraut and Kimchi

Most horticultural products can be preserved by a lactic acid fermentation. In the West the most important commercially are cabbage, cucumbers and olives, although smaller amounts of others such as carrots, cauliflower, celery, okra, onions, sweet and hot peppers, and green tomatoes are also fermented. In Korea fermented vegetables known as kimchi are an almost ubiquitous accompaniment to meals. More than 65 different types of kimchi have been identified on the basis of differences in raw materials and processing. Cabbages and radishes are the main substrates but garlic, peppers, onions and ginger are often also used. Surveys have shown its importance in the Korean diet, variously reporting kimchi to comprise 12.5% of the total daily food intake or a daily adult consumption of 50–100 g in summer increasing to 150–200 g in winter.

Sauerkraut production is thought to have been brought to Europe from China by the Tartars. Like a number of other traditional fermentations, the commercial process is technologically simple (Figure 9.9), but involves some interesting and complex chemistry and microbiology.

Usually where sauerkraut is produced commercially special cabbage cultivars are grown. These tend to have a higher solids content than normal and so minimize production of liquid waste during processing.

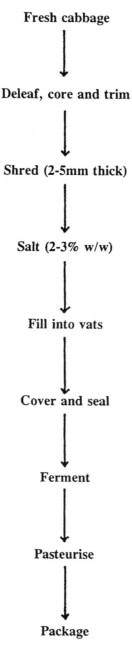

Figure 9.9 *Sauerkraut production*

The outer leaves are removed mechanically and the cabbages decored before cutting into shreds about 1 mm thick. The shredded cabbage is then salted and packed into vats for the fermentation stage.

The level of salting is critical to obtaining a satisfactory product, it must be within the range 2–3% w/w and is normally about 2.25%. Too little salt (<2%) and the product softens unacceptably, too much salt (>3%) and the correct microbial sequence is not obtained. The salt serves a number of purposes:

(i) it extracts moisture from the shredded cabbage by osmosis to form the brine in which the fermentation will take place;

(ii) it helps to inhibit some of the natural microflora of the cabbage such as pseudomonads which would otherwise cause spoilage and helps to select for the lactic acid bacteria;

(iii) it helps maintain the crisp texture of the cabbage by withdrawing water and inhibiting endogenous pectolytic enzymes which cause the product to soften;

(iv) finally, salt contributes to the flavour of the product.

Traditionally, fermentation vats have been made of wood but nowadays are more often of concrete with a synthetic polymer lining to protect from attack by the acid brine. The tanks are sealed by covering the salted cabbage with plastic sheeting. They are then filled with brine to press the sheeting on to the cabbage expelling the entrapped air.

Although commercial starter cultures for sauerkraut fermentation are available, they are used less often than in other food fermentations. The time course of a typical sauerkraut fermentation is shown in Figure 9.10 and shows how strongly selective the process is. At the start, lactic acid bacteria (LAB) comprise only about 1% of the total microflora, but many of the non-lactics fail to grow and two days later LAB account for more than 90% of the total microflora. During this time, they produce sufficient acid to decrease the pH to below 4 further inhibiting the competing microflora. Underlying this overall dominance by LAB is a natural succession of different species which contribute to the characteristic flavour of sauerkraut. The fermentation is initiated by *Leuconostoc mesenteroides* which is among the less acid- and salt-tolerant LAB but grows fastest during these early stages. As a heterofermenter it produces CO_2 which replaces entrapped air and helps establish anaerobic conditions within the product and prevent the oxidation of vitamin C and loss of colour. Since fructose is present as an alternative electron acceptor, it also produces appreciable amounts of acetic (ethanoic) acid from acetyl-CoA which is a major contributor to sauerkraut flavour. Reduction of fructose leads to the accumulation of mannitol. As the pH drops due to acid production in a weakly buffered medium so the *Leuconostoc* is inhibited and replaced, first by heterofermentative lactobacilli, and then by more acid-tolerant homofermentative lactobacilli such as *Lactobacillus plantarum*. Acid accumulation continues in the form of lactic acid although the pH stabilizes somewhere around 3.8 (the pK_a of lactic acid). At the end of fermentation which can last from 4–8

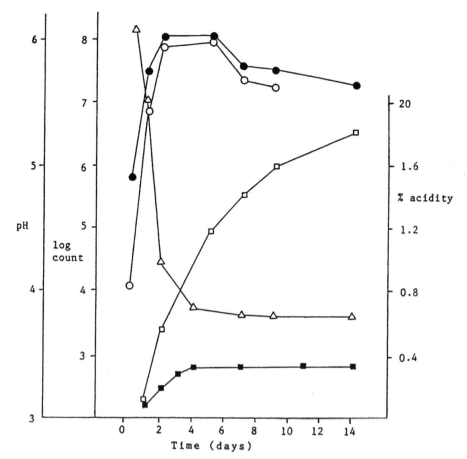

Figure 9.10 *Chemical and microbiological changes during sauerkraut production.* ●
Total bacterial count; ○ *lactic acid bacterio;* △ *pH;* □ *total titratable*
acidity (principally lactic plus acetic); ■ *volatile acidity (principally*
acetic);
(Adapted from Stamer, 1974.)

weeks the total acidity of the product is 1.7–2.3%, expressed as lactic
acid, with a ratio of volatile to nonvolatile acid of around 1 to 4.

Defects of sauerkraut arise mostly as a result of yeast and mould
growth. These can produce off-odours, loss of acidity, a slimy, softened
product as a result of pectolytic activity, or a pink discolouration due to
the growth of the yeast *Rhodotorula*. In the early stages of fermentation,
Leuconostoc mesenteroides fermenting sucrose will preferentially utilize
fructose, polymerizing the glucose moieties to produce a dextran slime.
This is however transient and is later degraded and utilized by other LAB.

In some brined and fermented vegetables nutrients are not particularly
well conserved. These tend to employ high salt levels which draw
nutrients and moisture from the product into a high-strength brine
which is often discarded and replaced before consumption. This is not

the case with sauerkraut which uses a low-salt brine and is not desalted before use. As a result several vitamins are partially conserved, particularly ascorbic acid, vitamin C. Sauerkraut was used extensively as an anti-scorbutic (for the prevention of scurvy) in the Dutch navy in the 18th century and was also highly regarded in this respect by Captain Cook who ordered servings of a pound per man twice weekly during his voyage of 1772. Some losses of vitamin C will occur during processing, a 50% reduction was observed in the first five weeks of kimchi fermentation, but nutritional labels on commercial sauerkraut in the United States usually show an ascorbic acid content of 50% of the Recommended Daily Allowance per 100 g serving.

Kimchi is similar to sauerkraut in some respects since cabbage is a common ingredient and the level of salt used is low (<3%). It differs principally in having a shorter fermentation time; the best taste is claimed after 3 days at 20 °C when the acidity is 0.6% and the pH around 4.2. Consequently *Leuconostoc mesenteroides* is the principal organism responsible for the fermentation and dominance of *Lactobacillus plantarum* is regarded as a defect which results in an excessively sour product.

9.7.2 Olives

Olives are native to the eastern Mediterranean region where they have been cultivated since at least 3000 BC. Today 98% of the world's hectarage of olives is in the Mediterranean region, most of this going to the manufacture of olive oil. Substantial quantities (> 600 000 tonnes annually) are also processed into table olives; some are preserved by a canning process similar to other foods but the production of most types includes a period of storage in brine during which a fermentation occurs contributing to the product's stability. Pickled olives in their various forms have a complex taste which often requires considerable application to acquire. In colder climes their consumption has a certain cachet summed up by the 19th century poet and philosopher Ralph Waldo Emerson who likened them to life at sea: exotic and distasteful.

In the production of Spanish-style green olives, which account for 38% of world production, the unripe fruits are first treated with lye (1.0– 2.6% sodium hydroxide solution) to hydrolyse the glucoside oleuropein which imparts a bitter flavour and also inhibits lactic acid bacteria (see Section 3.2.4). This lasts for up to 10 h during which the lye penetrates flesh between a half and three quarters of the way to the stone. The lye is then washed off with water over several hours and the fruits placed into a brine. Initially this contains 5–6% salt but the level is increased in strength during the course of the fermentation up to around 8%. Because some of the natural sugars in the olives will have been removed during the lye treatment and washing, fermentable sugar may be added to the brine.

Complex sequences of bacteria have been reported by different investigators but the most important species appears to be *Lactobacillus plantarum*. Several other LAB have been reported, including an early phase of growth by *Leuconostoc mesenteroides*. This, in particular, will depend upon the salt level used since *Leuc. mesenteroides* is not markedly salt tolerant. Essentially though, the decreasing pH, increasing acidity and the salt combine to eliminate the natural microflora dominated by Gram-negatives and replace it with one composed of lactic acid bacteria and some yeasts. The fermentation process lasts for several weeks and culminates in a product with a pH of 3.6–4.2 containing around 1% lactic acid. Starter cultures are available but rarely used at present, the most important measure taken to control the fermentation is to ensure that air is excluded from the fermenting product to prevent the growth of oxidative moulds and yeasts.

The traditional Greek-style product, natural black olives in brine, accounts for 31% of world production. Processing starts with ripe olives which are placed in a higher strength brine than Spanish-style olives, usually containing up to 10% w/v salt. Fermentation is very slow because the absence of a lye treatment means that oleuropein is still present and that nutrients diffuse slowly through the tough fruit skin. The microflora is usually dominated by yeasts of which a large number of different species have been isolated and identified including members of the genera *Saccharomyces*, *Hansenula*, *Candida*, *Torulopsis*, *Debaryomyces*, *Pichia*, *Kluyveromyces*, and *Cryptococcus*. Lactic acid bacteria may be significant if the salt content is low (<6–7%) but are generally a minor component. As a result, there is less acid production than in low-salt vegetable fermentations and the final product generally has a pH of 4.5–4.8 and a total acidity of 0.1–0.6% expressed as lactic acid. This is not sufficient to confer reasonable stability on the product so the salt content is usually increased to above 10% for storage.

9.7.3   Cucumbers

Lactic fermentation following pickling in a brine was once the only method for successful preservation of cucumbers. Since the 1940s, 'fresh pack' techniques have evolved which do not require a fermentation to confer stability. The first of these is based on direct acidification with vinegar or acetic acid followed by pasteurization while more recently direct acidification coupled with refrigerated storage has become increasingly popular. Today in the United States, where more than half a million tonnes of cucumbers are preserved each year, only about 40% are preserved by fermentation, approximately equal volumes are pasteurized and the remainder are preserved by refrigeration. Despite this recent trend, fermentation does have a number of advantages over other methods.

(1) Fermented cucumbers have flavour and texture characteristics not possessed by the other products.
(2) Bulk fermentation techniques facilitate quick and easy processing in busy harvest seasons.
(3) Under these conditions, products can be stored in bulk until they are required for further processing, so that year-round working is possible.
(4) Fermentation is more economical with energy than techniques which require pasteurization or an efficient cold chain.

Cucumber fermentations can be divided into essentially two different types: high-salt, or salt stock, and low-salt fermentations. Salt stock cucumbers are fermented in a brine containing 5–8% salt until they are stabilized by conversion of all the fermentable sugars to organic acids and other products. *Lb. brevis*, *Lb. plantarum* and *Pediococcus pentosaceus* are most commonly isolated. At these levels of salt *Leuconostoc mesenteroides* does not play the same crucial role as in sauerkraut or kimchi production and at 8% salt it is often not even detected. During the first phase of the fermentation which lasts for 2–3 days the microflora contains a large diversity of bacteria, yeasts and moulds. The environment is selective for LAB and yeasts which increase while other organisms decrease. The fermentation process is not restricted to the surrounding brine but also occurs within the cucumbers as a result of organisms entering through stomata. Sometimes this can lead to defects in the product known as 'bloaters'. Carbon dioxide accumulates within the fruit and is unable to diffuse out, some of this gas production arises from endogenous respiration of the tissues but much is the result of microbial action such as the malo-lactic fermentation and the heterofermentation of sugars. In a controlled fermentation process which has been developed, measures taken to control this problem include the use of strongly homofermentative starter cultures containing *Lactobacillus plantarum* or *Pediococcus pentosaceus* and intermittent purging of CO_2 from the system by bubbling nitrogen through the fermentation.

Genuine dill pickles are fermented in a lower salt brine (3–5%) in the presence of dill (an umbellifer, *Anethum graveolens*) and spices. The fermentation resembles sauerkraut production in the sequence of lactic acid bacteria that develops though it is usually conducted at a slightly higher temperature, 20–26 °C compared with 18 °C for sauerkraut. The full curing process can take up to 8 weeks although active fermentation usually lasts for only 3–4 weeks. The product brine has a pH of 3.2–3.6

and contains 0.7–1.2% acidity (as lactic acid) but will include appreciable amounts of acetic (ethanoic) acid.

9.8 FERMENTED MEATS

Fermented sausages are sometimes claimed to have originated in the Mediterranean region, although traditional products in China and Southeast Asia suggest that they probably developed independently in several locations. Like cheesemaking, meat fermentation is a method for improving the keeping qualities of an otherwise highly perishable commodity. Key features in this are the combination of lactic fermentation with salting and drying which, in many cases, produces a product which is shelf-stable at ambient temperatures.

A further similarity to cheese is the bewildering variety of different types, 330 produced in Germany alone. In the United States fermented sausages are divided into two categories: dry, which have a moisture content of 35% or less, and semi-dry typically containing about 50% moisture. Spreadable fermented sausages, produced in Germany, such as Teewurst, and Mettwurst are not dried during production and in this respect are similar to the Thai product *nam*.

The ingredients of a European-style fermented sausage may comprise:

lean meat, 55–70%
fat, 25–40%
curing salts, 3%
fermentable carbohydrate, 0.4–2.0%
spices and flavouring, 0.5%
starter, acidulant, ascorbic acid, *etc.* 0.5%

Pork is most commonly used in southern Europe but elsewhere beef, mutton and turkey meat are also used. The meat should always be of high quality since the products are usually consumed without cooking and so are essentially a raw-meat product. Unlike fermented milk products, it is not possible to heat treat the meat before processing as this would destroy the sausage's textural characteristics, but some are given a final pasteurization to ensure safety.

The curing salts added are a similar mixture of sodium chloride and sodium nitrate and/or nitrite to that used in the production of cured meats such as ham and bacon. Here too, they contribute to the taste, colour, safety, stability and texture of the product.

Spices are added primarily for reasons of flavour but are known to have potentially important roles in retarding microbial spoilage and promoting lactic fermentation. The antimicrobial effect of spice components, which has already been discussed in Section 3.2.4, could help inhibit the normal spoilage microflora of the meat. Spices have also been shown to stimulate the growth of lactic acid bacteria. This is a result of their manganese content, spice extracts or spices low in manganese do not have this stimulatory effect. Most aerobes have micromolar quantities of the enzyme superoxide dismutase (SOD) to scavenge the toxic superoxide anion radical produced by a one-electron reduction of molecular oxygen (see Section 3.2.3). Aerotolerant lactic acid bacteria do not have SOD but have developed an alternative protective mechanism based on the accumulation of millimolar concentrations of manganous (Mn^{2+}) ion.

$$2H^+ + O_2^{-\bullet} + Mn^{2+} \rightarrow H_2O_2 + Mn^{3+} \qquad (9.1)$$

The Mn^{2+} is regenerated by a subsequent reduction step. Increasing the manganese content of the medium can stimulate LAB growth.

Other ingredients which may be included are glucono-δ-lactone which improves acidulation by slowly hydrolysing to produce gluconic acid, ascorbic acid to improve colour production and stability, and glucose to supplement the available fermentable sugar in the starting mix.

Ingredients are blended together in a bowlchopper at low temperature. When the ingredients have been blended together they are packed into casings of the appropriate diameter. Traditionally collagen from the gastrointestinal tract of animals has been used but nowadays fibrous cellulose and regenerated collagen produced from animal hides are more common. The packing material must have certain properties: it must adhere to the meat mix and must shrink with it during processing and must be permeable to moisture and smoke.

Fermented sausages are still often made by natural fermentations in which the selectivity of the starting mix determines which components of the heterogeneous microflora dominate. Starter cultures are however being increasingly used for the greater assurance of a satisfactory fermentation they provide. The principal components of commercial starters are lactic acid bacteria and nitrate-reducing bacteria. Some will also include yeasts and moulds such as *Debaryomyces hansenii* and *Candida famata*, and moulds, usually *Penicillium* spp. such as *P. nalgiovense*. LAB included in early starters were mainly *Lactobacillus plantarum*, *Pediococcus acidilactici* and *Ped. pentosaceus*, not necessarily those most important in the natural fermentation. Surveys of naturally fermented products demonstrated that the dominant LAB were psychrotrophic, facultatively heterofermentative lactobacilli (see page 317) that were slightly less acid tolerant than usual (minimum pH 3.9 compared with 3.7–3.8). Most of these are now assigned to the species *Lactobacillus sake* and *Lactobacillus curvatus* and strains of these have been incorporated into commercial starters.

Members of the Micrococcaceae such as *Micrococcus varians* and *Staphylococcus carnosus* are important with respect to the reduction of nitrate to nitrite although this activity has also been demonstrated in some lactobacilli. Their presence would not be required in nitrite-cured products.

Sausage fermentations are conducted at temperatures ranging from 15 °C up to in excess of 40 °C, depending on the starter used, and last for 20–60 h. The relative humidity is also controlled to ensure that a slow drying of the product commences. Acid production and the decrease of the pH to below 5.2 promote the coagulation of meat proteins and this aids moisture expulsion and development of the desired texture and flavour. It also contributes to the microbiological stability and safety of the product.

North American and northern European sausages are often smoked. This confers a characteristic flavour but phenolic components of the wood smoke also have important anti-oxidative and antimicrobial properties which improve shelf-life. Fungal growth may occur on the surface of unsmoked sausages providing a particular character to these products as a result of fungal lipolytic, proteolytic and antioxidative activity.

In the final drying stage which can last up to 6 weeks, the moisture content is reduced further by storage at low temperatures, 7–15 °C, and at low relative humidity (65–85%).

The combination of antimicrobial hurdles or barriers introduced during sausage fermentation is normally sufficient to ensure product safety. *Staphylococcus aureus* with its ability to tolerate reduced a_w and pH and grow anaerobically would seem well suited to growth in these products. Occasional outbreaks of *Staph. aureus* food poisoning have been reported from the United States where higher fermentation temparature are used. However, studies suggest that *Staph. aureus* does not compete well with the LAB present, particularly if the latter have a large numerical superiority as a result of starter addition. The risk is also reduced since enterotoxin production appears to be more susceptible than growth to inhibition by adverse conditions (see Section 7.14.2).

Numbers of *Salmonella* and other Enterobacteriaceae have been shown to fall throughout fermentation and drying. At present though, our knowledge of such processes is insufficient to allow us to predict this lethal effect reliably. It is therefore, most important that only good quality raw materials are used so that undue reliance is not placed upon these factors.

Outbreaks of Verotoxin producing *E. coli* in the United States associated with fermented meats highlighted this problem and prompted the US Food Safety and Inspection Service to recommend that the procedures used in the production of ready-to-eat fermented products should achieve a $5 \log_{10} \mathrm{cfu\,g}^{-1}$ reduction in pathogen numbers.

One way of achieving this reliably would be to introduce a heating step. Following an outbreak of salmonellosis in the UK associated with a salami stick product imported from Germany, the production process was changed to incorporate a final pasteurization step without adverse effects on sensory quality.

Nam, the Thai fermented sausage, differs in several respects from European fermented sausages. It is a low-fat product which is subjected to a short fermentation and is not dried. It is also wrapped in water-impermeable plastic material or, traditionally, banana leaves. As the fermentation proceeds and the pH drops the moisture is expelled but is trapped within the packaging giving the consumer an indication of the age of the product. It is not always stored chilled and its largely anecdotal association with food poisoning has prompted test marketing of irradiated *nam* in some areas of Thailand.

9.9 FERMENTED FISH

The term fermented fish is applied to two groups of product, mostly confined to East and Southeast Asia: the more widely known fish/salt formulations such as fish sauces and pastes, and fish/salt/carbohydrate blends. Strictly speaking, only in the latter case is the description 'fermented' fully justified. Microbial action in the production of fish sauces and pastes is slight if not insignificant and the term is being used in its looser, non-microbiological, sense to apply to any process where an organic material undergoes extensive transformation.

In many areas where they are produced, fish sauces and pastes are the main flavour principle in the local cuisine and provide a valuable balanced source of amino acids. The names of some fish sauces and pastes and their countries of origin are given in Table 9.8.

Fish sauces and pastes are usually made from a variety of small fish which are packed into tanks or jars with salt usually at a ratio of around three parts fish to one part salt. This is more than sufficient to saturate the aqueous phase, to produce an a_w below 0.75 and arrest the normal pattern of spoilage. The only organisms likely to be able to grow under such conditions are anaerobic extreme halophiles. Although there have been recent reports of isolations of organisms such as the proteolytic *Halobacterium salinarium* from fish sauce, their importance remains to be established since earlier work has shown that acceptable fish sauce could be made using fish sterilized by irradiation.

The production process can take up to 18 months or more, during which the fish autolyse, largely through the action of enzymes in the gut and head of the uneviscerated fish, to produce a brown salty liquid rich in amino acids, soluble peptides and nucleotides. Products in which auto-lysis is less extensive are described as fish pastes.

Table 9.8 *Fish sauces and pastes and their countries of origin*

	Name	
Country	Sauce	Paste
	Amber/brown liquid, salty taste, cheese-like aroma	Red/brown salty paste
Burma	*ngapi*	*nga-ngapi*
Indonesia	*ketjap-ikan*	*trassi-ikan*
		trassi-udang (shrimps)
Kampuchea	*nuoc-mam*	*prahoc*
	nuoc-mam-gau-ca (livers only)	*mam-ruoac* (shrimps)
Laos	*nam-pla* (pa)	*padec*
Malaysia	*budu*	*belachan* (shrimps)
Philippines	*patis*	*bagoong*
Thailand	*nam-pla*	*kapi*
Vietnam	*nuoc-mam*	*man-ca*
		man-tom (shrimps)

Table 9.9 *Fermented fish/salt/carbohydrate products*

Country	Products
Japan	*I-shushi, e.g. ayu-sushi, funa-suchi tai-suchi,*
Kampuchea	*phaak, mam-chao*
Korea	*sikhae*
Laos	*som-kay-pa-eun, som-pa, mam-pa-kor, pa-chao, pa-khem som-pa-keng*
Malaysia	*pekasam, cencalok*
Philippines	*burong-isda, e.g. burong-ayungi, burong-dalag, burong-bangus*
Thailand	*pla-ra, pla-som, pla-chao, som-fak*

Authentic lactic-fermented fish products have to include as an ingredient an exogenous source of fermentable carbohydrate. Considerable variation in recipes has been noted but production is governed by two general principles: the higher the salt content of the product, the longer the production process takes but the better the product's keeping qualities; and the higher the level of added carbohydrate, the faster the fermentation and the more acidic the flavour.

Fish/salt/carbohydrate products (Table 9.9) are generally much less popular than the fish sauces and pastes and are produced on a smaller scale. Their production also tends to be more common away from the coast and to use freshwater fish. Though superficially their production appears similar to that of fermented meat sausages, they are quite distinctive.

In products such as *Burong-isda* (Philippines), *Pla-jao*, and *Pla-som* (Thailand) and *I-sushi* (Japan), cleaned fish flesh is dry salted with about 10–20% salt and left for a period of up to a day. The flesh is then usually removed from the brine that develops and may be subjected to further moisture reduction by sun-drying for a short period. Lactic fermentation is then initiated by addition of carbohydrate. This is usually in the form of rice although traditional saccharifying agents (*koji*, Japan; *look-pang*, Thailand; *ang-kak*, Philippines) employing mould enzymes may be added. These accelerate the fermentation, since most LAB are not amylolytic, and also increases the total acid produced. For example, *Burong-isda* containing *ang-kak* has a lower pH (3.0–3.9) than that produced with rice alone (4.1–4.5). Garlic is often added along with the rice as a flavouring ingredient and this may play a similar role in directing the fermentation as spices do in fermented sausage production. Garlic is also a source of the fermentable carbohydrate inulin. The product is normally ready for consumption after about two weeks of fermentation when the microflora is dominated by yeasts and LAB which are present at levels around $10^7\,\mathrm{cfu\,g^{-1}}$ and $10^8\,\mathrm{cfu\,g^{-1}}$ respectively.

With the exception of *I-sushi*, these products are usually cooked before consumption and this along with the low pH generally guarantees safety. However, the small, very often domestic-scale, production can lead to extreme variations in a product's character and failure to obtain a satisfactory rapid fermentation in *I-sushi* has led to outbreaks of botulism in Japan caused by *C. botulinum* type E.

9.10   BEER

The popularity of products resulting from the conversion of sugars into ethanol by yeasts is almost universal and there is hardly a culture without its own indigenous alcoholic beverage. All that is required is a material that will furnish sufficient fermentable carbohydrate; a condition fulfilled by honey, cereals, root crops, palm saps and many fruits, pre-eminently grapes, but also, apples, pears, plums and others. The ethanol concentration achieved by fermentation is limited by the sugar content of the raw material and also by the ethanol tolerance of the yeast which is normally around 14% v/v. *Sake*, Section 9.12.2 below, is something of an exception. Potency can be increased by distillation of a fermented wash to produce spirits such as whisky, vodka, brandy, calvados and arrack, and ethanol partially purified by distillation can also be added back to a fermented product to give fortified wines such as port, sherry and madeira. Here, we will concentrate on a single product which has spread throughout the world and is now produced more widely than any other alcoholic drink: European-style beer.

Brewing is thought to have originated in Mesopotamia where it is said that as much as 40% of total cereal production was used for this purpose. Because of the relative complexity of the process, it is likely that beer was a later discovery than wine. The Romans were disinterested and after tasting British ale in the 4th century the Emperor Julian was compelled to pen a little poem:

Who made you and from what
By the true Bacchus I know you not
He smells of nectar
But you smell of goat.

Clearly the unhopped ale of the time was not to his taste, but even today beer enjoys an inferior reputation to that of wine.

Barley is the principal cereal used in the production of beer, although other cereals are occasionally used and wheat beers such as Berliner Weisse and the Gueuze–Lambic beers of Belgium are notable exceptions. Africa has a number of traditional beers produced from local cereals such as sorghum or millet and some of these are produced on a substantial industrial scale. These however are the result of a mixed lactic/ethanolic fermentation and bear little resemblance to European-style beers.

One reason for barley's pre-eminence is that the grain retains the husk which affords protection during storage and transport and also acts as an aid to filtration during wort separation. The gelatinization temperature of malted barley starch is also low relative to that of other cereals (52–59 °C) and this enables the starch to be gelatinized (solubilized), prior to enzymic digestion, at temperatures which will not inactivate the starch degrading enzyme α-amylase. A further advantage is the presence in barley of substantial quantities of a second enzyme, β-amylase, which is essential for the rapid conversion of starch and dextrins to maltose.

Since the brewing yeast, *Saccharomyces cerevisiae*, is unable to ferment starch, the first stage in the production of any alcoholic beverage from starchy materials is conversion of the starch into fermentable sugars. Human ingenuity has come up with a number of ways of doing this. In the Oriental Technique, exemplified by products such as *sake*, mould enzyme preparations like *koji* are used, whereas the prevalent Western technique uses endogenous starch-degrading enzymes produced in the grain through the process of malting. A third technique used in some native cultures in South America is to use salivary amylase by chewing the substrate so that it becomes coated with saliva and then spitting it out to saccharify and ferment. This approach is not amenable to industrialization and is not, as far as we are aware, the basis of any large-scale commercial production of alcoholic beverages.

In malting, the grain is moistened by steeping in water and is then spread on to a malting floor and allowed to germinate. During germination, hydrolytic enzymes, produced in the aleurone layer surrounding the grain endosperm, attack the endosperm, mobilizing the nutrient and energy reserves it contains for the growing barley plant. To encouraged this, maltsters sometimes add gibberellins, plant growth hormones which are the natural regulators of this process.

The development from a seed to a plant is arrested by kilning which reduces the moisture content of the malt to 3–5%. During kilning, some non-enzymic browning reactions occur between amino acids and sugars in the malt and these contribute to the final beer colour. Darker beers tend to include malts that have been kilned at higher temperatures to promote browning reactions.

Nowadays malts are usually bought in by brewers as one of their raw materials and the brewing process proper starts with its conversion into a liquid medium (wort) capable of supporting yeast growth: a step known as mashing (Figure 9.11).

The malt is ground to reduce the particle size and increase the rate of enzymic digestion and is then mixed with hot water. Water, known in brewers' parlance as liquor, is an important ingredient in brewing and the quality of the local water was one of the reasons for the development of traditional UK brewing centres such as Burton on Trent, London and Edinburgh. In particular, calcium content has a significant impact on the brewing process because calcium ions precipitate out as calcium phosphate during mashing. This decreases the wort pH from 6.0 to 5.4, nearer the optimum for a number of malt enzymes, and thus increases the yield of fermentable extract. Starchy adjuncts may be added during mashing to boost the fermentable sugar content of the wort.

There are two traditional systems of mashing: the British technique of infusion mashing where the mash is held in a single vessel at a constant temperature of around 65 °C, and the continental decoction system where the mash is heated through a range of temperatures by removing a portion, heating it, then adding it back. Nowadays a number of variations on these techniques are used so that the differences are less distinct.

In mashing, a number of enzymic activities contribute to the production of the clear liquid medium known as sweet wort. For instance, it requires two enzymes operating in concert to break down starch into maltose, a disaccharide of glucose fermentable by the brewing yeast. Barley starch is composed of two fractions: amylose (20–25%), a linear polymer of α-1,4-linked glucose units, and amylopectin (75–80%), a branched polymer containing linear chains of α-1,4-linked glucose units with branches introduced by occasional α-1,6-linkages. Alpha amylase hydrolyses α-1,4-linkages to produce a mixture of lower molecular weight dextrins while the exoenzyme, β-amylase, attacks dextrins at their non-reducing end, snipping off maltose units. Limit dextrins containing the α-1,6-linkages are left in the wort largely untouched unless the non-malt enzyme amyloglucosidase is added to the mash.

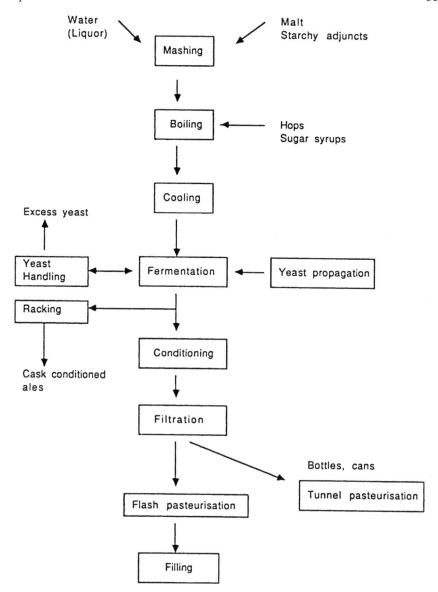

Figure 9.11 *The brewing process*

Proteinases solubilize malt proteins and supply yeast nutrients so that about 35–40% of malt protein is solubilized during mashing compared with 90–95% of the malt starch. Phosphatases release inorganic phosphate, which is important for yeast nutrition and for contributing to the buffering capacity of the wort. The activity of β-glucanases can also be useful in breaking down β-glucans which can cause subsequent handling problems with the beer.

After mashing, sweet wort is boiled. This stops the degradative processes by inactivating the malt enzymes. It also pasteurizes the wort,
completes ionic interactions such as calcium phosphate precipitation,
denatures and precipitates proteins and tannins which separate as a
material known as hot break or trub and helps dissolve any sugars which
may be added at this stage as an adjunct.

Hops are also added during boiling. These are the cones or strobili of
the plant *Humulus lupulus* whose principal purpose is the bittering of the
wort. The hop resin contains α-acids such as humulone and cohumulone
which are only partially soluble in wort. During boiling they isomerize to
isohumulones which are more soluble and more bitter than α-acids
(Figure 9.12). Although hop resins have some antibacterial action, they
play little part in assuring the bacteriological stability of beer as spoilage
bacteria such as lactobacilli rapidly acquire a tolerance to them.

Wort boiling lasts for 1–2 h during which 5–15% of the volume is
evaporated. Hop residues are then strained off, hot trub is removed in a
whirlpool separator, and the hopped wort cooled to the fermentation
temperature.

The yeasts used to brew ales and lagers are strains of *Saccharomyces
cerevisiae*, known as *S. cerevisiae* var. *cerevisiae* and *S. cerevisiae* var.
carlsbergensis (uvarum) respectively. The distinctions between the yeasts
used in ale and lager brewing are slight. Traditionally, ale yeasts were
regarded as top fermenters which formed a frothy yeast head on the
surface of brewing beer and was skimmed off to provide yeasts for
pitching (inoculating) subsequent batches, while lager yeasts were bottom fermenters which formed little surface head and were recovered
from the bottom of the fermenter. Nowadays this is a less useful
distinction as many ales are brewed by bottom fermentation.

The cardinal temperatures of the two organisms differ and this is
reflected in the different temperatures used for lager fermentations (8–
12 °C) and for ale fermentations (12–18 °C). They can also be distinguished by the ability of *S. cerevisiae* var. *carlsbergensis* to ferment the
disaccharide melibiose, although this is of no practical import since the
sugar does not occur in wort.

During fermentation the yeast converts fermentable carbohydrate to
ethanol via the EMP pathway. Although this is an anaerobic process, a
vigorous fermentation is often helped by aeration of the wort before
pitching with yeast. This supplies oxygen, necessary for the synthesis of
unsaturated fatty acid and sterol components of the yeast cell membrane,
and may sometimes be repeated later in the fermentation.

A time course of a typical ale fermentation is shown in Figure 9.13. After
an initial vigorous phase during which there is active yeast growth, ethanol
production and a drop in pH as nitrogen is removed from the wort, there is
a second phase of slower ethanol production in the absence of further yeast

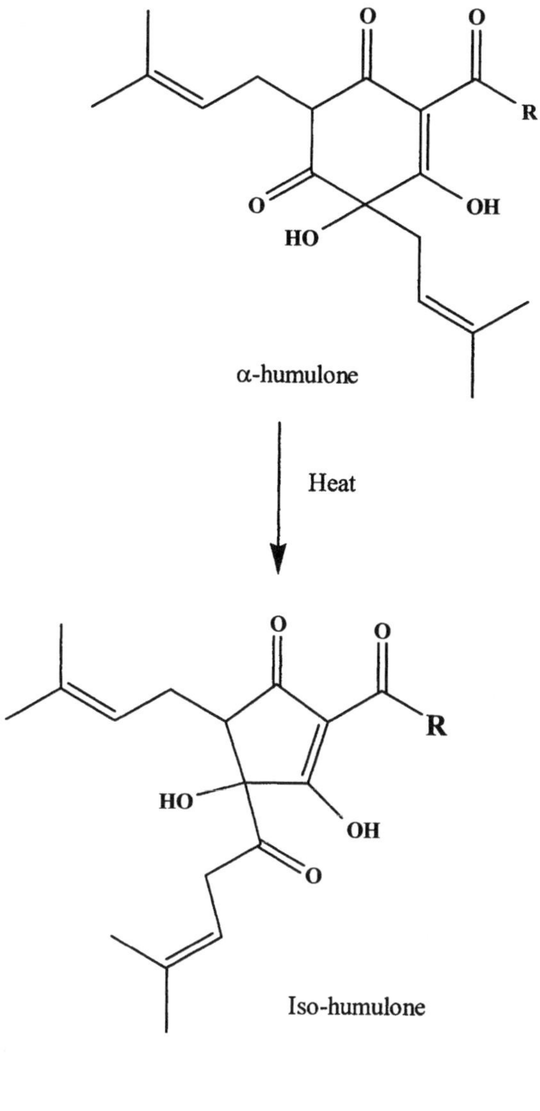

α-humulone

Heat

Iso-humulone

R = -CH₂.CH(CH₃)₂   humulone

R = -CH(CH₃)₂   cohumulone

R = -CH(CH₃).CH₂.CH₃   adhumulone

Figure 9.12 *Isomerization of hop α- and β-acids*

growth. Overall the yeast population increases about six-fold during fermentation. This yeast can be recycled, usually after an acid wash to control bacterial contamination, but eventually its performance drops as viability declines and it is used in animal feed and the manufacture of yeast extract.

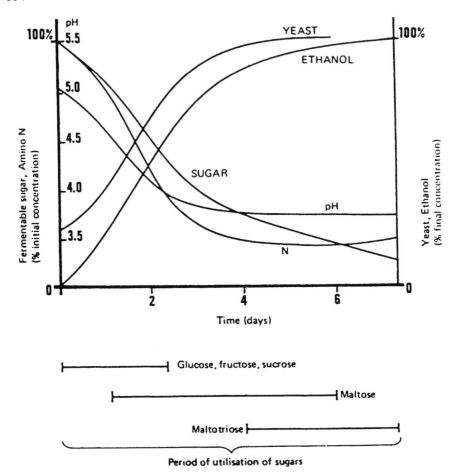

Figure 9.13 *Changes during the beer fermentation*
(Reproduced courtesy of the Institute of Brewing)

Lager fermentations take longer due to the lower temperature. The name lager originates from the German word for store and describes the period of secondary fermentation (storage) at low temperature which these products undergo to improve yeast settling, clarification and CO_2 dissolution. In the past, this could last for several months, but with modern techniques such as centrifugation and artificial carbonation it is less protracted and is now complete within one to two weeks.

Depending on the product, beer from the fermenter can be subjected to a variety of downstream processes. It may be run to casks where priming sugars are added to stimulate the secondary fermentation necessary in cask-conditioned beers, or it may be filtered prior to pasteurization and kegging. Bottled and canned beers usually undergo a combination of filtration and centrifugation before packing and pasteurization.

Table 9.10 *Possible taints in beer*

Taint	Associated Flavours	Possible Cause
Acetaldehyde	Apples, paint, grassy	Bacterial spoilage: Acetic acid bacteria/*Zymomonas*
Sulphur	Bad eggs, drains	Formed during fermentation, wild yeasts/*Zymomonas*
Cloves	Herbal phenolic	Bacterial spoilage/wild yeasts
Musty/fungal	Mouldy, stale, hessian	Mould or bacteria: Usually in water
Fruity	Estery, pineapples, solvent, bananas, pear drops	Wild yeast/*Brettanomyes*: Wort bacteria/*Enterobacter agglomerans*
Ethyl Acetate	Solvent-like	Wild yeast/*Hansenula anomala*
Diacetyl	Toffee, butterscotch, honey	Formed during fermentation/ lactic acid bacterial spoilage: *Lactobacillus*/*Pediococcus*
DMS	Sweetcorn, jammy	Bacterial spoilage: *Hafnia*/ *Obesumbacterium*. Wort bacteria
TCP Acetic Acid	Medicinal, antiseptic Sour, vinegar	Bacterial spoilage: Wort bacteria Bacterial spoilage: *Acetobacter*/*Gluconobacter*
Acidic	Sourness and creaminess Sourness and apples	Bacterial spoilage: *Lactobacillus Acetobacter*

Courtesy L. Hargreaves

Brewing is quite a robust process microbiologically due to a combination of factors such as low nutrient status, ethanol content and low pH and there is a limited range of micro-organisms of concern to the brewery microbiologist. Members of the Enterobacteriaceae such as *Obesumbacterium proteus*, *Klebsiella* and *Enterobacter* species are sensitive to the low pH and ethanol of beer but can grow in wort producing off-odours like dimethyl sulfide which can persist through to the final product. They also contribute to the production of nitrosamines by reducing wort nitrate to nitrite. *Obesumbacterium proteus* is particularly associated with top-fermenting yeasts but can be controlled by acid washing.

Acetic acid bacteria of the genera *Acetobacter* and *Gluconobacter* can be found throughout the brewery. As obligate aerobes they are particularly associated with cask-conditioned beer where they cause spoilage as a result of turbidity, ropiness and the oxidation of ethanol to ethanoic (acetic) acid. *Zymomonas mobilis* is an anaerobic, Gram negative rod which can ferment sugars to ethanol. It causes more problems in ale brewing where it grows in the primed beer producing turbidity and off-flavours.

Lactic acid bacteria of the genera *Lactobacillus* and *Pediococcus* can grow widely in the brewery environment and in beer where they produce acid, diacetyl, which gives beer a sweet butterscotch flavour, and polymeric material known as rope. Yeasts that are not used for the fermentation can cause hazes and off-odours and are known as wild yeasts. Normally these are described as being *Saccharomyces* and non-*Saccharomyces*. *Saccharomyces* wild yeasts can be detected using a medium containing copper sulfate to inhibit the brewing yeast. Non-*Saccharomyces* yeasts such as *Pichia, Hansenula, Brettanomyces* and others can be detected with a medium containing lysine as the sole nitrogen source which *Saccharomyces* cannot utilize.

Some of the possible taints in beer and their causes are presented in Table 9.10.

9.11 VINEGAR

Vinegar is the product of a two-stage fermentation. In the first stage, yeasts convert sugars into ethanol anaerobically, while in the second ethanol is oxidized to acetic (ethanoic) acid aerobically by bacteria of the genera *Acetobacter* and *Gluconobacter*. This second process is a common mechanism of spoilage in alcoholic beverages and the discovery of vinegar was doubtless due to the observation that this product of spoilage could be put to some good use as a flavouring and preservative. The name vinegar is in fact derived from the French *vin aigre* for 'sour wine' and even today the most popular types of vinegar in a region usually reflect the local alcoholic beverage; for example, malt vinegar in the UK, wine vinegar in France, and rice vinegar in Japan.

In vinegar brewing, the alcoholic substrate, known as vinegar stock, is produced using the same or very similar processes to those used in alcoholic beverage production. Where differences occur they stem largely from the vinegar brewer's relative disinterest in the flavour of the intermediate and his concern to maximize conversion of sugar into ethanol. In the production of malt vinegar for example, hops are not used and the wort is not boiled so the activity of starch-degrading enzymes continues into the fermentation. Here we will concentrate on describing the second stage in the process, acetification.

Acetification, the oxidation of ethanol to acetic acid is performed by members of the genera *Acetobacter* and *Gluconobacter*. These are Gram-negative, catalase-positive, oxidase-negative, strictly aerobic bacteria. *Acetobacter* spp. are the better acid producers and are more common in commercial vinegar production, but their ability to oxidize acetic acid to carbon dioxide and water, a property which distinguishes them from *Gluconobacter*, can cause problems in some circumstances when the vinegar brewer will see his key component disappearing into the air as CO_2. Fortunately over-oxidation, as it is known, is repressed by ethanol and can be controlled by careful monitoring to ensure that ethanol is not completely exhausted during acetification. Most acetifications are run on a semi-continuous basis; when acetification is nearly complete and acetic acid levels are typically around 10–14% w/v, a proportion of the fermenter's contents is removed and replaced with an equal volume of fresh alcoholic vinegar stock. Since a substantial amount of finished vinegar is retained in the fermenter, this conserves the culture and means that a relatively high level of acidity is maintained throughout the fermentation, protecting against contamination. It also protects against over-oxidation as it has been found that *Acetobacter europaeus*, a species commonly found in commercial vinegar fermenters, will not over-oxidize when the acetic acid concentration is more than 6%.

Many of the acetic acid bacteria associated with commercial acetification are difficult to culture on conventional solidified media, although some success has been enjoyed using a double-layer medium which provides colonies growing on the surface with a constant supply of ethanol and moisture from a lower, semi-solid layer. As a result, vinegar fermentations are usually initiated with seed or mother vinegar, an undefined culture obtained from previous fermentations. Depending on the type of acetification, the culture can be quite heterogeneous and *A. europaeus*, *A. hansenii*, *A. acidophilum*, *A. polyoxogenes*, and *A. pasteurianus* have all been isolated from high-acidity fermentations.

Oxidation of ethanol to acetic acid is the relatively simple pathway by which acetic acid bacteria derive their energy. It occurs in two steps mediated by an alcohol dehydrogenase and an aldehyde dehydrogenase (Figure 9.14). Both enzymes are associated with the cytoplasmic membrane and have pyrroloquinoline quinone (PQQ) as a coenzyme. PQQ acts as a hydrogen acceptor which then reduces a cytochrome *via* an intermediate quinone. The consequent electron transport establishes a proton motive force across the membrane which can be used to synthesize ATP.

Overall, acetification can be represented chemically by:

$$C_2H_5OH + O_2 = CH_3COOH + H_2O \tag{9.2}$$

From the stoichiometry of the equation it can be calculated that 1 litre of ethanol should yield 1.036 kg of acetic acid and 0.313 kg of water. This leads to the approximate relationship that 1% v/v ethanol will give 1% w/v acetic acid, and this is used to predict the eventual acidity of a vinegar and to calculate fermentation efficiency. It implies that, in the absence of over-oxidation, evaporative losses and conversion to biomass, the sum of the concentration of ethanol (%v/v) and the concentration of acetic acid (%w/v), known as the total concentration or GK (German: Gesammte Konzentration) should remain constant throughout acetification. The GK yield is the GK of the final vinegar expressed as a percentage of the GK at the start of acetification.

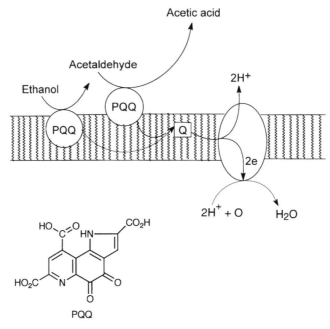

Figure 9.14 *Oxidation of ethanol by acetic acid bacteria*

There are a number of techniques for acetification which differ in the means by which the three interacting components, ethanol, bacteria and oxygen, are brought together. Surface culture techniques, where the bacteria form a surface film at the interface between the acetifying medium and air, are the simplest but can be applied with varying levels of sophistication. In the Orleans process, vinegar stock in partially filled casks drilled with air holes (Figure 9.15) is left to acetify until the acidity reaches the appropriate level determined by the initial GK value. At this point a proportion, typically one-third to two-thirds, is drawn off through the tap, replaced with fresh stock and the process restarted. The vinegar stock is usually added *via* a pipe passing through the top of the barrel and resting on the bottom. In this way the surface film of bacteria is not disturbed and the delays and losses that result from having to reform the film are avoided. Usually the time taken to complete one acetification cycle is of the order of 14 days.

Only a small proportion of the world's vinegar is produced by surface culture today, although it is claimed to produce the finest quality vinegar. More elaborate surface culture techniques based on series of trays have been described but these have received only very limited application.

The quick vinegar process derives its name from the faster rates of acetification achieved by increasing the area of active bacterial film and improving oxygen transfer to the acetifying stock. The acetic acid bacteria grow as a surface film on an inert support material packed into

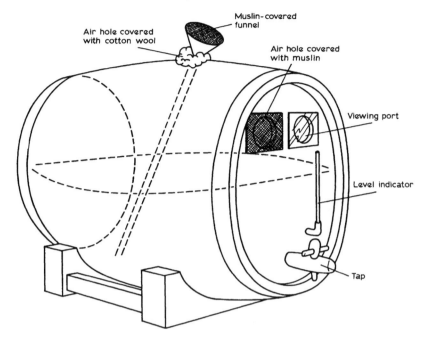

Figure 9.15 *The Orleans process of vinegar manufacture*

a false-bottomed vat. The acetifying stock is sprayed on to the surface of the packing material and trickles down against a counter-current of air which is either pumped through the bed or drawn up by the heat of reaction within it. The packing material normally consists of some lignocellulosic material such as birch twigs, vine twigs, rattan, wood wool, or sugarcane bagasse, although other materials such as coke have also been used. The vinegar stock is collected in a sump at the bottom of the vat and recirculated until the desired level of acidity is reached. The faster rate of reaction achieved means that the wash heats up during passage through the bed and, depending on the size of the fermenter, some cooling may be required.

The process is operated semicontinuously to maintain a high level of acidity throughout, and most of the biomass is retained within the packed bed. A well operated quick vinegar process fitted with temperature control and forced aeration can usually acetify a vinegar stock with a GK of 10 and an initial ethanol content of 3% in 4–5 days.

The fastest rates of acetification are achieved using submerged acetification in which acetic acid bacteria grow suspended in a medium which is oxygenated by sparging with air. The most commercially successful technique to have been developed is the Frings Acetator (Figure 9.16) which uses a patented self-priming aerator to achieve very efficient oxygen transfer.

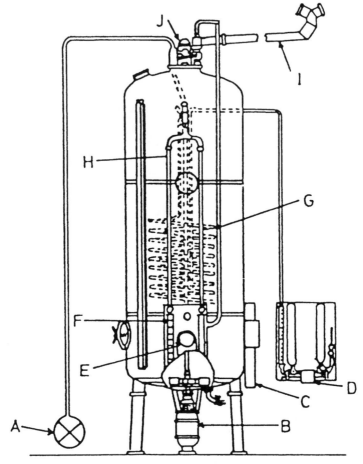

Figure 9.16 *The Frings Acetator. A, charging pump; B, aerator and motor; C, alkograph;*
D, cooling water value; E, thermostat controlling D; F, rotameter; G, cooling
coil; H, air line; I, air exhaust line; J, defoamer
Reproduced by kind permission of Heinrich Frings GmbH, Bonn

Submerged culture is very efficient and rapid, a semicontinuous run normally takes 24–48 h. It does however require far more careful control than simpler processes. The acetic acid bacteria are very susceptible to interruptions to the air supply, indicating that, in order to survive suspended in a medium with a pH of 2.5 and 10–14% acidity, the bacteria need a constant supply of energy from respiration. A stoppage of only one minute in a stock with a GK of 11.35 is enough to completely arrest acetification which will not resume when aeration is resumed.

Another possible cause of fermentation failure in submerged acetification is phage infection. The presence of bacteriophage particles has been demonstrated in disturbed vinegar fermentations both in submerged acetifiers and the quick vinegar process. The performance of quick vinegar generators appears to be less affected as their acetification rate may slow but rarely stops. This is probably due to the greater heterogeneity of the culture present which allows organisms of different phage susceptibility to take over in the event of phage attack.

Where legal definitions of vinegar exist, it is specified as a fermentation product. 'Artificial vinegars' made by diluting and colouring acetic acid are thus excluded and, in the UK, have to be known rather laboriously as 'non-brewed condiment'. Although vinegar can be made up to 14% acidity, it is usually diluted down to an appropriate strength for bottling. The minimum acetic acid content is usually prescribed to be something between 4 and 6% w/v, but higher strength vinegars are available for pickling.

Though most often thought of in terms of its use as a condiment, vinegar is an important food ingredient. It is used as a preservative and flavouring agent in a large and expanding range of products such as mayonnaise, ketchups, sauces and pickles. In the United States only about 30% of the vinegar produced is sold as table vinegar, the rest being used in food processing.

The antimicrobial action of organic acids such as acetic acid has already been discussed (see Section 3.2.2) and the use of vinegar in a formulated product usually restricts the spoilage microflora to yeasts, moulds and lactobacilli. Vinegar preserves were one of the earliest areas where predictive models were developed (see Chapter 3). Work at what became the Leatherhead Food Research Association indicated that to achieve satisfactory preservation of a pickle or sauce a minimum of 3.6% acetic acid, calculated as a percentage of the volatile constituents, is necessary. That is to say:

$$\% \text{ acetic acid on whole product} = 0.036 \times \% \text{ volatile constituents} \qquad (9.3)$$

A different formula has been described specifically for sweet cucumber pickles:

$$\% \text{ acetic acid on whole product} = (80 - S)/20 \qquad (9.4)$$

where S is the % sucrose on the whole product.

More elaborate formulae have been produced which apply to emulsified and non-emulsified sauces. These are based largely on work conducted at the laboratories of Unilever in the Netherlands and are known as the CIMSCEE code, after the French acronym of the European Sauces Trade Association. The code consists of two formulae; one to determine the potential for spoilage by acetic acid tolerant yeasts, moulds and LAB, and another derived from inactivation rates of salmonella to assess microbiological safety. Each contains terms for salt, sugar, and acetic acid content and pH. If, when the relevant values are substituted in the formulae, the result is higher than a specified value then this indicates that the product would be microbiologically stable or safe depending on the formula used.

The formula for safety is:

$$15.75(1 - \alpha)(\% \text{ total acetic acid}) + 3.08(\% \text{ salt}) + (\% \text{ hexose})$$
$$+ 0.5(\% \text{ disaccharide}) + 40(4.0 - \text{pH}) = \sum_s \qquad (9.5)$$

Where $(1 - \alpha)$ is the proportion of undissociated acetic acid, α is the proportion dissociated, given by the expression:

$$pH = pK + \log[\alpha/(1 - \alpha)] \qquad (9.6)$$

where $pK = 4.76$.

Any sauce based on acetic acid with $\sum_s > 63$ is regarded as intrinsically safe, since viable numbers of *E. coli* in it will decline by more than 3 log cycles in less than 72 h at 20 °C. The formula for stability is:

$$15.75(1 - \alpha)(\% \text{ total acetic acid}) + 3.08(\% \text{ salt}) + (\% \text{ hexose})$$
$$+ 0.5(\% \text{ disaccharide}) = \sum \qquad (9.7)$$

If $\sum > 63$, any sauce with this formulation should be microbiologically stable without refrigeration, even after opening.

The mould *Moniliella acetoabutans*, the micro-organism most resistant to acetic acid, would still grow in such products but can be controlled by good hygienic practices and pasteurization of ingredients containing vinegar. Experience in the pickle and sauce industry indicates that spoilage by *Moniliella* is rare.

Products meeting the CIMSCEE requirement for stability would have a relatively strong taste of acid or salt. If levels of these ingredients are reduced to produce a milder taste then some supplementary preservative measure would be necessary such as sorbate, a final pasteurization step or refrigerated storage.

9.12 MOULD FERMENTATIONS

Mould fermentations are an aspect of fermented foods that, up until now, we have mentioned only briefly. Failure to remedy this neglect would give a seriously unbalanced view for, in the East particularly, moulds play a key role in a number of food fermentations.

9.12.1 Tempeh

Tempeh is a traditional mould-fermented food in Indonesia, though it has also attracted interest in the Netherlands and United States. The most

The Tempeh Fermentation

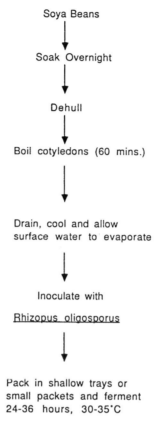

Soya Beans

↓

Soak Overnight

↓

Dehull

↓

Boil cotyledons (60 mins.)

↓

Drain, cool and allow
surface water to evaporate

↓

Inoculate with

Rhizopus oligosporus

↓

Pack in shallow trays or
small packets and ferment
24-36 hours, 30-35°C

Figure 9.17 *Tempeh production*

popular type of tempeh is produced from soya beans and is also known as tempeh kedele. The process of tempeh production is outlined in Figure 9.17. Whole clean soya beans are soaked overnight in water to hydrate the beans. A bacterial fermentation occurs during this stage decreasing the pH to 4.5–5.3. The hydrated beans are dehulled and the moist cotyledons cooked; a process which pasteurizes the substrate, destroys the trypsin inhibitor and lectins contained in the bean and releases some of the nutrients required for fungal growth. After cooking, the beans are drained and may be pressed lightly to remove excess moisture before spreading into shallow bamboo trays and allowing to cool. Starter culture is added either by mixing some tempeh in with the cooked soya beans prior to packing in the trays or by sprinkling a spore inoculum, prepared by extended incubation of a piece of tempeh, on to the beans.

The fermentation is invariably a mixed culture of moulds, yeasts and bacteria but the most important component appears to be *Rhizopus oligosporus*, although other *Rhizopus* and *Mucor* species are often isolated. Over two days incubation at ambient temperature (30–35 °C), the mycelium develops throughout the mass of beans knitting it together. During fermentation the pH rises to around 7, fungal proteases increase the free amino acid content of the product and lipases hydrolyse over a third of the neutral fat present to free fatty acids.

Unlike many fermented foods, tempeh production is not a means of improving the shelf-life of its raw material which is in any case inherently quite stable. Tempeh contains antioxidants which retard the development of rancidity but will keep for only one to two days as sporulation of the mould discolours the product and a rich ammoniacal odour develops as proteolysis proceeds.

Tempeh production does however improve the acceptability of an otherwise rather unappealing food. Fresh tempeh has a pleasant nutty odour and flavour and can be consumed in a variety of ways, usually after frying in oil.

In addition to improving acceptability, fermentation also improves the nutritional quality of soya beans. In part this stems from the reduction or removal of various anti-nutritional factors at different stages in the processing. Destruction of the trypsin inhibitor and lectins during cooking of the beans has already been mentioned and levels of phytic acid, which can interfere with mineral nutrition, are also reduced by about a third in the course of processing. The notorious ability of beans to produce flatulence is also regarded as an anti-nutritional property and flatulence-inducing oligosaccharides such as stachyose and raffinose are partially leached out of the beans during the soaking stage.

Despite the extensive proteolytic changes which occur during fermentation, studies have failed to show that the protein in tempeh is more easily digested. With the exception of thiamine which decreases, other vitamins increase to varying degrees during fermentation. Vitamin B_{12}, the anti-pernicious anaemia factor, shows the most marked increase and this is associated with the growth of the bacterium *Klebsiella pneumoniae* during fermentation. The usual source of this vitamin in the diet is animal products and it has been suggested that tempeh could be an important source of B_{12} for people subsisting on a largely vegetarian diet.

Tempeh can be made from a number of different plant materials including other legumes, cereals and agricultural by-products. One variety that has achieved some notoriety is tempeh bongkrek which is made in central Java using the presscake remaining after extraction of coconut oil. Tempeh bongkrek has been associated with occasional serious outbreaks of food poisoning due to the bacterium *Burkholderia cocovenenans* growing in the product and elaborating the toxins bongkrekic acid and

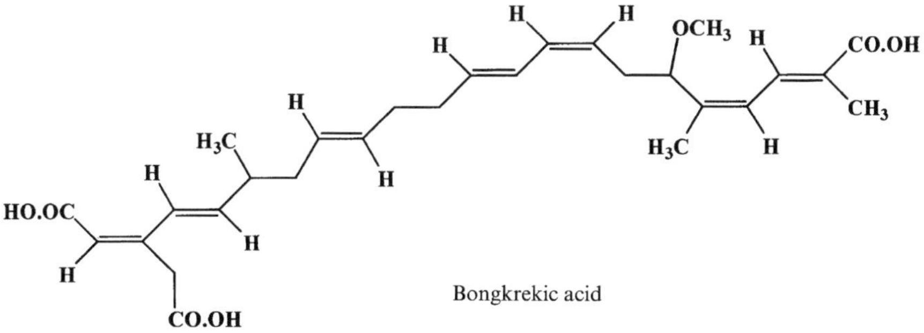

Bongkrekic acid

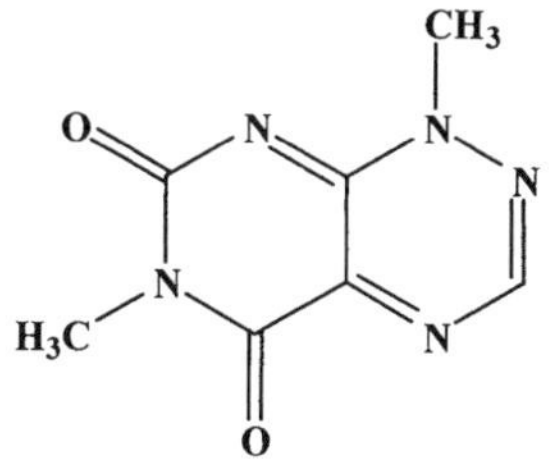

Toxoflavin

Figure 9.18 *Toxoflavin and bongkrekic acid*

toxoflavin (Figure 9.18). Since 1951, at least 1000 people are known to have died as a result of this intoxication and in 1988 the Indonesian Government prohibited the production of tempeh bongkrek.

Two factors are thought to give rise to this problem. Reduction or omission of an initial soaking of the presscake may fail to give a lactic fermentation sufficient to reduce the pH below 6, a level at which the bacterium cannot grow. Also, the fungal inoculum may be too small since it has been shown that *B. cocovenenans* cannot grow if *Rhizopus oligosporus* has more than a tenfold numerical superiority (estimated by plate counts).

Ontjom is a tempeh-like product produced in Indonesia from peanut presscake which normally has a fruity/mincemeat character. It can be produced using the tempeh mould but *Neurospora intermedia* is also often used. This mould has strong α-galactosidase activity which can further contribute to the reduction of flatulence-inducing oligosaccharides.

9.12.2 Soy Sauce and Rice Wine

Though they are markedly different in character, rice wine and soy sauce share sufficient common features in their production to warrant discussing them together. Both are representatives of products which involve mould activity in a two-stage fermentation process. The mould starter used is often known as *koji*, a Japanese term derived from the Chinese character for mouldy grains. In the *koji* stage, aerobic conditions allow moulds to grow on the substrate producing a range of hydrolytic

Soy Sauce Production

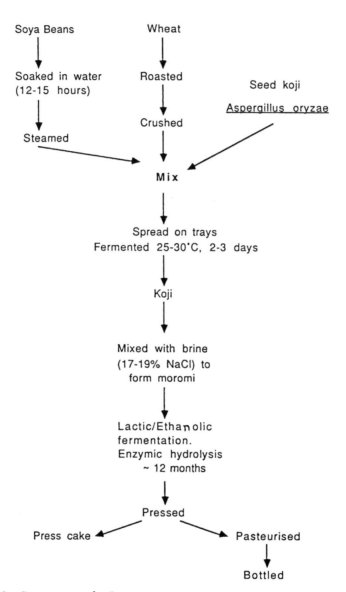

Figure 9.19   *Soy sauce production*

enzymes necessary for utilizing the macromolecular material present. In the case of soy sauce (Figure 9.19), soaked and cooked soya beans are mixed with roasted cracked wheat in about equal proportions and inoculated with *tane koji* or seed *koji*. This has been previously grown-up on a similar mixture of substrates and contains a mixture of strains of

Aspergillus oryzae. The moulds are then allowed to grow throughout the mass of material spread as layers about 5 cm deep for 2–3 days at 25–30 °C.

In the second, mash or *moromi* stage, conditions are made anaerobic so no further mould growth can occur. In soy sauce production, this is achieved by mixing the *koji* with an approximately equal volume of brine to give a final salt concentration of 17–20%. Although the moulds can no longer grow, the activity of a whole battery of hydrolytic enzymes continues breaking down proteins, polysaccharides and nucleic acids to produce a liquid rich in soluble nutrients. Yeasts and lactic acid bacteria dominate the microflora producing a number of flavour components and converting roughly half of the soluble sugars to lactic acid and ethanol so that the final soy sauce normally has a pH of 4.5–4.9 and ethanol and lactic acid contents of 2–3% and 1% respectively. The halophilic lactic acid bacterium *Tetragenococcus halophilus* (formerly *Pediococcus halophilus*) and the yeasts *Zygosaccharomyces rouxii* and *Torulopsis* have been identified as being important in this stage.

The *moromi* stage can be quite protracted, lasting up to a year or more, at the end of which the mash is pressed to remove the solid residues which may then be mixed with brine to undergo a second fermentation and produce a lower grade product. The liquid is pasteurized and filtered, possibly after a period of maturation, and then bottled.

Rather similar steps are involved in the production of soya bean pastes known as *miso* in Japan and *chiang* in China. These include up to 40% of a grain such as rice or barley, use dry salt rather than brine, and employ a shorter fermentation so the product has the consistency of a paste rather than a liquid.

In the brewing of the Japanese rice wine, *sake*, a *koji* prepared on steamed rice is used. Although the mould used is the same species as in soy sauce production, *Aspergillus oryzae*, the strains used in *sake* production are particularly noted for their ability to produce amylolytic enzymes. In the *moromi stage*, water is added along with strains of the yeast *Saccharomyces cerevisiae* specially adapted to the *sake* fermentation. During this stage amylolytic enzymes from the mould continue to break down the starch in the rice to produce fermentable sugars which are then converted to ethanol by the yeast. The high alcohol content of around 20% v/v achieved in such fermentations is thought to be due to a combination of factors. Particularly important is the slow rate of fermentation which results from the relatively low fermentation temperature (13–18 °C) and the slow release of fermentable sugars. The high solids content in the *moromi* is also thought to help in keeping the

Table 9.11 *Amylolytic mould preparations*

koji	Japan
look pang	Thailand
ang-kak	Philippines
ragi-tapai	Malaysia/Indonesia
nuruk	Korea
peh-yueh	China
bakhar	N. India

yeast in suspension and active at such high alcohol concentrations. At the end of fermentation which typically lasts for three weeks, the product is settled, filtered and blended before being pasteurized and bottled.

A number of other rice-based mould starters are used in the countries of East and Southeast Asia to fulfil a similar role to *koji* (Table 9.11). They are used to produce sweetened rice products which can be consumed fresh or added to other products (see Section 9.9) or can be used as a base for the production of rice wine and rice vinegar. The microbiological composition of these generally differs from that of *koji* and comprises primarily *Rhizopus* and *Mucor* species and amylolytic yeasts.

9.12.3 Mycoprotein

Products such as *tempeh* and *koji* contain a significant amount of mould biomass and a reasonable extension of this type of approach would be to grow up mycelium itself as a source of food. Of the many investigations into the growth of moulds on readily available substrates one has successfully emerged as a commercial product. Mycoprotein, marketed as Quorn, is essentially the mycelium of *Fusarium venenatum* (formerly *F. graminearum*) grown in continuous culture in a medium containing glucose, ammonium salts and a few growth factors. Advantages associated with the use of a filamentous organism are that it can be harvested by filtration and washing and can be readily textured to give the product an acceptable mouth-feel.

To be acceptable as a food for human consumption it is necessary to reduce the level of RNA, which is nearly 10% based on mycelial dry weight, to below the levels likely to lead to kidney-stone formation or gout. This is achieved by a mild heat treatment prior to filtration which activates the mould's RNAases and leads to a dramatic reduction of RNA to about 1% which is acceptable. The product has a useful protein content of 44% and is high in 'fibre' because of the cell walls of the filamentous fungal structure.

9.13 CONCLUSION

Here we have described a limited selection of fermented foods which we believe illustrates their diversity and importance as well as some of their general microbiological features. There is a large and growing literature on this topic and details of others among the plethora of fermented foods produced can be found in some of the references recommended as further reading in Chapter12.

CHAPTER 10

Methods for the Microbiological Examination of Foods

10.1 INDICATOR ORGANISMS

It is frequently necessary to conduct a microbiological examination of food to determine its quality. This may be necessary to estimate its shelf-life, its suitability for human consumption or to confirm that it meets some established microbiological criterion (see Chapter 11). The total mesophilic plate count is widely used as a broad indication of microbiological quality, although it is unsuitable for this purpose in fermented foods which contain large numbers of organisms as a natural consequence of their preparation. In other foods, knowledge of the total count may be useful but it is often of more value to obtain an estimate of the numbers of a particular component of the total flora. Examples of this could be moulds in a cereal, psychrotrophic bacteria in a chilled food, anaerobes in a vacuum-packed food, or yeasts in a fruit beverage.

A quite different reason for a microbiological examination of a food may be to identify the cause of spoilage or the presence of a pathogen where a food has been implicated in foodborne illness. The methods for determining an estimate of the total mesophilic count are very different from those required for demonstrating the presence of a pathogen, or its isolation for further study.

The isolation of specific pathogens, which may be present in very low but significant numbers in the presence of larger numbers of other organisms, often requires quite elaborate procedures, some of which are outlined in Chapter 7. It may involve enrichment in media which encourage growth of the pathogen while repressing the growth of the accompanying flora, followed by isolation on selective diagnostic media, and finally the application of confirmatory tests (see, for example, Section 7.12.4).

Though microbiological criteria or the investigation of an outbreak of foodborne illness may often require the monitoring of certain products for specific pathogens, the difficulties associated with detecting low numbers of pathogens make it impracticable as a routine procedure to be applied without good cause.

An alternative to monitoring for specific pathogens is to look for an associated organism present in much larger numbers – an indicator organism. This is a concept developed originally for pathogens spread by the faecal–oral route in water and which has since been applied to foods, often rather uncritically. A good indicator organism should always be present when the pathogen may be present, it should be present in relatively large numbers to facilitate its detection, it should not proliferate in the environment being monitored and its survival should be similar to that of the pathogen for which it is to be used as an indicator.

Escherichia coli is a natural component of the human gut flora and its presence in the environment, or in foods, generally implies some history of contamination of faecal origin. In water microbiology in temperate climates *E. coli* meets these criteria very well and has proved a useful indicator organism of faecal pollution of water used for drinking or in the preparation of foods. There are, however, limitations to its use in foods where there appears to be little or no correlation between the presence of *E. coli* and pathogens such as *Salmonella* in meat, for example. Although *E. coli* cannot usually grow in water in temperate countries, it can grow in the richer environment provided by many foods.

Testing for *E. coli* can itself be relatively involved and a number of simpler alternatives are often used. These are less specific and therefore the relationship between indicator presence and faecal contamination becomes even more tenuous. Traditionally the group chosen has been designated the coliforms – those organisms capable of fermenting lactose in the presence of bile at 37 °C. This will include most strains of *E. coli* but also includes organisms such as *Citrobacter* and *Enterobacter* which are not predominantly of faecal origin (Table 10.1). The faecal coliforms, a more restricted group of organisms, are those coliforms which can grow at higher temperatures than normal, *i.e.* 44–44.5 °C and the methods developed for their detection were intended to provide rapid, reproducible methods for demonstrating the presence of *E. coli* without having to use time-consuming confirmatory tests for this species. However, the verocytotoxigenic strain O157:H7 (VTEC), which has caused so much concern during the late 1990s, does not grow well at 44 °C. Faecal coliforms contain a higher proportion of *E. coli* strains and the test can be made even more specific for *E. coli* type 1 by including a test for indole production from tryptophan to exclude other thermotolerant coliforms. Further specificity can be introduced by using a medium diagnostic for

Table 10.1 *Significance of genera of the Enterobacteriaceae in the monitoring of foods*

Genus	Predominantly faecal origin	Usually detected in 'coliform tests'	Typically enteropathogenic in humans
Citrobacter	No	Yes	No
Edwardsiella	Yes	No	No
Enterobacter	No	Yes	No
Erwinia	No	No	No
Escherichia	Yes	Yes	No
Hafnia	No	No	No
Klebsiella	No	Yes	No
Proteus	No	No	No
Salmonella	Yes	No	Yes
Serratia	No	No	No
Shigella	Yes	No	Yes
Yersinia	Yes	No	No

These comments are generalizations and there are exceptions to most of them. Adapted from 'Micro-organisms in Foods 1: Their Significance and Methods of Enumeration'. 2nd Edition ICMSF, University of Toronto Press, 1978

β-glucuronidase activity; an enzyme possessed by most, but not all, strains of *E. coli* and relatively uncommon in other bacteria.

One criticism of using coliforms and faecal coliforms is that their absence could give a false reassurance of safety when lactose-negative organisms predominate. The lactose-negative organisms include, not only *Salmonella* and *Shigella*, but also enteroinvasive strains of *E. coli* (EIEC) such as O124. For this reason, tests for the whole of the Enterobacteriaceae are increasingly being used. The Enterobacteriaceae includes even more genera of non-faecal origin than the coliforms, such as *Erwinia* and *Serratia* which are predominantly plant associated. For this reason Enterobacteriaceae counts are used more generally as an indicator of hygienic quality rather than of faecal contamination and therefore say more about general microbiological quality than possible health risks posed by the product. For instance, the presence of high numbers of Enterobacteriaceae in a pasteurized food would be cause for concern although it would not necessarily imply faecal contamination, and one would expect to find Enterobacteriaceae on fresh vegetables without the product necessarily being hazardous. The potential significance of genera of the Enterobacteriaceae is summarized in Table 10.1.

Some food microbiologists have tried to distinguish between 'indicator' organisms, which relate to general microbiological quality, and so-called 'index' organisms, which suggest that pathogens may be present. As will be apparent from the discussion above, this is not a simple distinction to make and the terminology has not been widely adopted.

10.2   DIRECT EXAMINATION

When examining foods, the possibility of detecting the presence of micro-organisms by looking at a sample directly under the microscope should not be missed. A small amount of material can be mounted and teased out in a drop of water on a slide, covered with a cover slip, and examined, first with a low magnification, and then with a ×45 objective. The condenser should be set to optimize contrast even though this may result in some loss of resolution. Alternatively dark-field illumination or phase-contrast microscopy may be used. It is usually relatively easy to see yeasts and moulds and with care and patience it is possible to see bacteria in such a preparation. The high refractive index of bacterial endospores makes them particularly easy to see with phase-contrast optics and, if the preparation is made as a hanging drop on the cover glass mounted over a cavity slide, it should also be possible to determine whether the bacteria are motile.

Since only a small sample of product is examined in this way, micro-organisms will not be seen unless present in quite large numbers, usually at least $10^6 \, ml^{-1}$. In the case of some liquid commodities, such as milk, yoghurt, soups and fruit juices, it may be possible to prepare and stain a heat-fixed smear. But the food constituents often interfere with the heat fixing and care is needed to prevent the smear being washed away during staining. It may be necessary to dilute the sample with a little water, although that will reduce the concentration of micro-organisms further. The great advantage of such techniques is their rapidity, although in their simplest forms they do not distinguish between live and dead cells. The Breed smear is a quantitative version in which the field of view of the microscope is calibrated and a known volume of sample is spread over a known area of the slide.

The direct epifluorescent filter technique or DEFT is a microscopy technique which has been applied to the enumeration of micro-organisms in a range of foods, although it was originally developed for estimating bacterial counts in raw milk. The technique was developed in response to the need for a rapid method for judging the hygienic quality of farm milks. It achieves a considerably increased sensitivity (10^3–10^4 bacteria ml^{-1}) over conventional microscopy techniques by concentrating bacteria from a significantly larger volume of sample by filtering it through a polycarbonate membrane filter. The retained bacteria are then stained on the membrane with acridine orange and counted directly under the epifluorescence microscope. It may be necessary to pretreat the sample to allow filtration thus, for example, milk can be pretreated with detergent and a protease enzyme. It is also essential to ensure that the bacteria are trapped in a single focal plane because of the limited depth of focus of the microscope at the magnifications required. This is achieved by using a polycarbonate membrane where relatively uniform pores are produced following neutron bombardment of a plastic film, rather than cellulose acetate filters which have tortuous pores where bacteria will be held at different levels.

Acridine orange is a metachromatic fluorochrome, fluorescing either green or orange depending on the nature of the molecules within the cell to which it is bound. When bound to double-stranded DNA it fluoresces green but when bound to single-stranded RNA it fluoresces orange, as long as there is an adequate concentration of dye to saturate all the binding sites. Generally it is assumed that those cells which fluoresce orange are viable while those that fluoresce green are nonviable. This is certainly not always true. The actual colour of an individual cell depends on many factors but, probably the most important is the concentration of acridine orange within the cell. In many micro-organisms the integrity of the cell membrane restricts the passage of the dye into the cell and it is often the case that viable micro-organisms will fluoresce green and dead micro-organisms, in which the membrane is more leaky, will fluoresce orange. Thus, although there are limitations to the use of acridine orange as a vital stain, the method has been adapted for the enumeration of micro-organisms in a range of food commodities including fresh meat and fish, meat and fish products, beverages and water samples. Although not commonly used in DEFT, there are alternative viability stains such as cyanoditolyl tetrazolium chloride (CTC) and fluorescein diacetate (FDA). These only cause cells to fluoresce if they retain a functioning electron transport chain (CTC) or esterase activity (FDA).

In a modification of the technique, specific groups of micro-organisms can be enumerated. The membrane filter is incubated on an appropriate medium containing optical brighteners and the microcolonies that develop on the membrane enumerated using the fluorescence microscope.

10.3 CULTURAL TECHNIQUES

Although there is clearly a place for the direct examination of a food for micro-organisms, a full microbiological examination usually requires that individual viable propagules are encouraged to multiply in liquid media or on the surface, or within the matrix, of a medium solidified with agar.

Agar is a polysaccharide with several remarkable properties which is produced by species of red algae. Although it is a complex and variable material, a major component of agar is agarose which is made of alternating units of 1,4-linked 3,6-anhydro-L-galactose (or L-galactose) and 1,3-linked D-galactose (or 6-O-methyl-D-galactose). The properties of agar which make it so useful to microbiologists include the ability to form a gel at low concentrations (1.5–2%) which does not significantly influence the water potential of the medium. Such a gel is stable to quite high temperatures and requires a boiling water bath, or autoclave temperatures, to 'melt' it. Once molten however, agar solutions remain liquid when cooled to relatively low temperatures (*ca.* 40 °C) making it possible to mix it with samples containing viable organisms before, or during, dispensing. A further convenient property of agar is its stability to microbial hydrolysis, despite being a polysaccharide. Only a relatively small group of micro-organisms are able to degrade agar, presumably due to the presence of the unusual L-form of galactose in the polymer.

A very wide range of media are available to the microbiologist and details of their formulation, and how they are used, may be found in a number of readily available books and manuals. A selection of some commonly used media is listed in Table 10.2.

The formulation of a medium will depend, not only on what group of organisms is being studied, but also on the overall purpose of the study; whether it be to encourage good growth of the widest possible range of organisms, to be selective or elective for a single species or limited group, to resuscitate damaged but viable propagules, or to provide diagnostic information.

General purpose media such as nutrient agar and plate count agar for bacteria, or malt extract agar and potato/dextrose agar for fungi, have evolved to provide adequate nutrition for the growth of non-fastidious, heterotrophic micro-organisms. They do not deliberately contain any inhibitory agents but they may nevertheless be selective because of the absence of specific nutrients required by more fastidious organisms.

Selective media contain one or more compounds which are inhibitory to the majority of organisms but significantly less so to the species, or group of species, which it is required to isolate. It must be noted that all selective media, because they are based on the presence of inhibitory reagents, will generally be inhibitory to some extent to the organisms to

Table 10.2 *A selection of media commonly used in food microbiology*

Medium	Use
Plate Count Agar	Aerobic mesophilic count
MacConkey Broth	MPN of coliforms in water
Brilliant Green/Lactose/Bile Broth	MPN of coliforms in food
Violet Red/Bile/Glucose Agar	Enumeration of Enterobacteriaceae
Crystal Violet/Azide/Blood Agar	Enumeration of faecal streptococci
Baird–Parker Agar	Enumeration of *Staphylococcus aureus*
Rappaport–Vassiliadis Broth	Selective enrichment of *Salmonella*
Thiosulfate/Bile/Citrate/Sucrose Agar	Isolation of vibrios
Dichloran/18% Glycerol Agar	Enumeration of moulds
Rose Bengal/Chloramphenicol Agar	Enumeration of moulds and yeasts
Cefixime/Tellurite/Sorbitol/MacConkey Agar	*E. coli* O157

be selected. If cells of the target organism have been subject to sublethal injury, then they may not be able to grow on the medium without a resuscitation step to allow them to repair.

Elective media on the other hand, are designed to encourage the more rapid growth of one species or group of micro-organisms so that they out-compete others even in the absence of inhibitory agents. Thus cooked-meat broth incubated at 43–45 °C allows rapid growth of *Clostridium perfringens* so that it may become the dominant organism after only 6–8 hours incubation.

The difference between selective and elective media must be seen from the viewpoint of the organism which it is desired to recover. By ensuring optimal growth in the elective medium for one organism, it is desirable that conditions are sub-optimal, or even inhibitory, to others. A problem in the use of elective media is that growth of the desirable species may change the medium in a manner which now encourages the growth of other species. On the other hand a selective medium, if well designed, should remain inhibitory to unwanted organisms even when the organisms required are growing.

Resuscitation media are designed to allow the recovery of propagules which are sub-lethally damaged by some previous condition such as heat treatment, refrigeration, drying or exposure to irradiation. Such damaged micro-organisms may not only be more sensitive to inhibitory agents present in selective media, but may be killed if exposed to conditions encouraging rapid growth of healthy cells. Typically resuscitation media are nutritionally weak and may contain compounds which will scavenge free radicals such as those which may be generated by the metabolism of oxygen.

Diagnostic media contain a reagent or reagents which provide a visual response to a particular reaction making it possible to recognize individual species or groups because of the presence of a specific metabolic pathway or even a single enzyme.

Many media used in practice combine selective reagents, elective components and diagnostic features. An interesting example is the Baird–Parker agar used for the presumptive isolation of *Staphylococcus aureus*. The selective agents are sodium tellurite and lithium chloride, the elective agents are sodium pyruvate and glycine and the diagnostic features are provided by the addition of egg yolk. The production of black colonies due to the reduction of tellurite is characteristic of *S. aureus* as well as several other organisms able to grow on this medium such as other species of *Staphylococcus*, *Micrococcus* and some species of *Bacillus*. The additional diagnostic feature shown by most strains of *S. aureus* is the presence of an opaque zone due to lecithinase activity surrounded by a halo of clearing due to proteolytic activity (see also Section 7.14.4).

10.4 ENUMERATION METHODS

10.4.1 Plate Counts

It has already been suggested that to count micro-organisms in a food sample by direct microscopy has a limited sensitivity because of the very small sample size in the field of view at the magnification needed to see micro-organisms, especially bacteria. In a normal routine laboratory the most sensitive method of detecting the presence of a viable bacterium is to allow it to amplify itself to form a visible colony. This forms the basis of the traditional pour plate, spread plate or Miles and Misra drop plate still widely used in microbiology laboratories. Table 10.3 compares the sample size examined and potential sensitivity of all these methods. In the pour plate method a sample (usually 1 ml) is pipetted directly into a sterile Petri dish and mixed with an appropriate volume of molten agar. Even if the molten agar is carefully tempered at $40-45\,°C$, the thermal shock to psychrotrophs may result in them not producing a visible colony. The spread-plate count avoids this problem and also ensures an aerobic environment but the sample size is usually limited to 0.1 ml.

In a thoroughly mixed suspension of particles such as micro-organisms, the numbers of propagules forming colonies on replicate plates is expected to have a Poisson distribution, a property of which is that the variance (standard deviation squared) is equal to the mean ($\bar{x}$), *i.e.*

$$\bar{x} = \text{var} = s^2 \tag{10.1}$$

A consequence of this is that the limiting precision of a colony count is dependent on the number of colonies counted. The 95% confidence limits (CL) can be estimated as approximately

$$2s = 2\sqrt{x} \tag{10.2}$$

Table 10.3 *A comparison of the sensitivity of methods of enumeration*

Method	Volume of sample (ml)	Count (cfu g $^{-1}$) corresponding to a single organism or colony seen[a]
Direct microscopy	5×10^{-6}	2×10^6
Miles and Misra	0.02	5×10^2
Spread plate	0.1	10^2
Pour plate	1.0	10
MPN	3×10.0 $+3\times1$ $+3\times0.1$	0.36

[a] Based on a 10^{-1} dilution of a sample obtained by, for example, stomaching 1g (or ml) of food with 9 ml of diluent

(if x is the count on a single plate and has to be our estimate of the mean). Thus for a plate with only 16 colonies, the 95% CL would be approximately ±50% (*i.e.* we would have 95% confidence that the count lies between 8 and 24). For a count of 30, it would be ±37% and for a count of 500 only ±9%. However, if the number of colonies on a plate was as high as 500, it would not only be difficult to count them accurately, but such a crowded plate is likely to result in many colony-forming units never forming a visible colony leading to an underestimate. Thus it is widely accepted that reasonably accurate results are obtained when plates contain between 30 and 300 colonies.

To obtain plates with this number of colonies it is often necessary to dilute a sample before enumeration. The most widely used dilution technique is the ten-fold dilution series. With a completely unfamiliar sample it is necessary to plate-out a number of dilutions to ensure that some plates are obtained with counts in the desired range, but with experience of a particular material plating only one dilution may be sufficient.

The diluent used must not cause any damage, such as osmotic shock, to the micro-organisms. Sterile distilled water is therefore unsuitable. A commonly used diluent, known as maximum recovery diluent, contains 0.1% peptone and 0.85% sodium chloride.

It is possible to increase the confidence in a plate count by plating a number of replicate samples from each dilution. From the results of a single dilution the count in the original sample can be calculated using Equation (10.3),

$$N = \frac{\bar{x}}{V.d} \tag{10.3}$$

where $N = \mathrm{cfu}\ \mathrm{g}^{-1}$; $\bar{x} = $ mean count per plate; $V = $ volume of sample plated; $d = $ dilution factor. Sometimes we can use the results from several sequential dilutions but it is a common experience that the apparent cfu g^{-1} increases the higher the dilution used in making the calculation. This may reflect the breaking up of clumps by the action of pipetting, and/or reduced competition on less crowded plates. The smaller numbers on plates at higher dilutions are associated with reduced levels of confidence but they can be used in the calculation of a weighted mean using Equation (10.4) based on a ten-fold dilution series:

$$N_w = \frac{(C_1 + C_2)}{(n_1 + n_2/10)} \cdot \frac{1}{V} \cdot \frac{1}{d_1} \tag{10.4}$$

where $N_W = $ weighted mean; $C_1 = $ the total count on n_1 replicates at dilution d_1; $C_2 = $ total counts on n_2 replicates of the next dilution. The use of this formula is best shown with a worked example:

Dilution	n	colony count/plate	total C	$\bar{x}$	cfu ml^{-1a}
10^{-4}	3	63, 74, 61	198	66	6.6×10^5
10^{-5}	3	5, 11, 9	25	8.3	8.3×10^5

^{a} Calculated from a single dilution

$$\text{Weighted mean} = \frac{(198 + 25)}{(3 + 0.3)} \cdot \frac{1}{0.1} \cdot \frac{1}{10^{-4}} = 6.8 \times 10^5$$

Traditional plate counts are expensive in Petri dishes and agar media, especially if adequate replication is carried out, and the Miles and Misra drop count and spiral plater have been developed to reduce costs. In the Miles–Misra technique materials are conserved by culturing a smaller volume of each dilution, usually 20 μl. This way a number of dilutions can be grown on a single plate by dividing it into sectors each representing a different dilution. The spiral plater employs a mechanical system which dispenses 50 μl of a liquid sample as a spiral track on the surface of an agar plate. The system is engineered so that most of the sample is deposited near the centre of the plate with a decreasing volume applied towards the edge. This produces an effect equivalent to a dilution of the sample by a factor of 10^3 on a single plate, thus producing a two-thirds' saving on materials as well as saving the time required in preparing and plating extra dilutions. After incubation, colonies are counted using a specially designed grid which relates plate area to the volume applied, thus enabling the count to be determined. The system is not suitable for all food samples though, as particulate material can block the hollow stylus through which the sample is applied.

The limit in sensitivity of the traditional plate count arises from the small volumes used and clearly, the sensitivity can be increased by increasing the volume size and the number of replicate counts or plates. It may be possible to filter a larger volume through a membrane, which retains the viable organisms, and then lay the membrane onto an appropriate medium.

In all of these methods of enumeration it is essential to appreciate the statistical background to sampling and to recognize that extrapolations from colony counts depend on several assumptions that may not be justified. Thus a colony may not be derived from a single micro-organism, but from a clump of micro-organisms, and the material being examined may not be homogeneous so that the subsample actually studied is not representative of the whole.

10.4.2 Most Probable Number Counts

An alternative method of enumerating low numbers of viable micro-organisms is that referred to as the Most Probable Number (MPN) method. The method is usually based on inoculating replicate tubes of an appropriate liquid medium (usually 3, 4 or 5) with three different sample sizes or dilutions of the material to be studied (*e.g.* 10 g, 1.0 g and 0.1 g). The medium used has to be designed to make it possible to decide whether growth or no growth has occurred and the number of positives at each sample size or dilution is determined after incubating the tubes. The MPN is obtained by referring to a table such as that shown in Table 10.4. There are computer programmes for generating MPN values from different designs of the experiment and these programmes can also provide confidence limits for the MPN and suggest what the likelihood of particular combinations of positive results should be.

A modern variation on the MPN theme is the use of the hydrophobic grid membrane filter (HGMF). A sample is filtered through the HGMF which is divided by a hydrophobic grid into a number (normally 1600) of small cells or growth compartments. After incubation of the filter on an appropriate medium, each of these cells is scored for growth or no-growth. This can be done either manually or automatically and the count in the original sample determined as equivalent to a single dilution MPN using 1600 tubes.

One application of the MPN, which allows one to calculate the maximum number of organisms in a batch of material, is based on the two-class attributes sampling plan (see Section 11.2.1). If a number of equal sized samples (n) is taken from a batch of material and all shown to be negative for a particular organism then the maximum percentage (d)

Table 10.4 *A selection of MPN values*[a]

Number of positive tubes	MPN	95% Confidence Limits
0 0 0	<0.30	
1 0 0	0.36	0.02 to 1.7
2 0 0	0.92	0.15 to 3.5
2 1 0	1.5	0.4 to 3.8
3 0 0	2.3	0.5 to 9.4
3 1 0	4.3	0.9 to 18.1
3 1 1	7.5	1.7 to 19.9
3 2 0	9.3	1.8 to 36
3 2 1	15	3.0 to 38
3 3 0	24	4.0 to 99
3 3 1	46	9.0 to 198
3 3 2	110	20.0 to 400
3 3 3	>110	

[a] Based on 3×1 g(ml) $+ 3\times0.1$ g(ml) $+ 3\times0.01$ g(ml) samples (expressed as organisms per 1 g)

of such samples containing at least one viable propagule at a required probability P is given by Equation (10.5).

$$d = 100[1 - \sqrt[n]{(1 - P)}] \qquad (10.5)$$

Thus, if 10 samples of 25 g from a batch of material are all found to be negative then, with 95% confidence, the maximum percentage of 25g samples containing at least one viable organism would be:

$$d = 100[1 - \sqrt[10]{(1 - 0.95)}] = 26\%$$

i.e. there would be less than 10 organisms kg^{-1} (1 kg $= 40 \times 25$ g; 26% of $40 = ca.$ 10).

The enumeration of micro-organisms assumes that there are distinctive propagules to be counted. This is acceptable for single-celled organisms such as the majority of bacteria or yeasts but in the case of filamentous fungi there may be a problem in interpreting the significance of numbers of colony-forming units. To assess the quantity of fungal biomass in a food commodity may require quite different techniques. One possibility is to make a chemical analysis for a constituent which is specifically associated with fungi, such as chitin, which is a constituent of the cell walls of zygomycetes, ascomycetes, basidiomycetes and deuteromycetes (but also present in insect exoskeleton), or ergosterol which is a major constituent of the membranes of these groups of fungi. Some moulds, such as species of *Aspergillus* and *Penicillium* associated with the spoilage of cereals, produce volatile metabolites such as methylfuran, 2-methylpropanol, 3-methylbutanol and oct-1-en-3-ol (this last compound having a strong mushroomy smell). It would be possible to detect and analyse these compounds in the head-space gases of storage facilities using gas–liquid chromatography (glc).

Equipment known as an artificial nose is now being developed in which a sample of head space gas from, for example, a grain silo is taken directly into a glc, the data read into a computer, and compared with a memory bank of previously recorded patterns. The pattern of peaks is often diagnostic for a particular mould species and it may even be possible to recognize patterns formed by mixtures of species.

10.5 ALTERNATIVE METHODS

Cultural methods are relatively labour intensive and require time for adequate growth to occur. Many food microbiologists also consider that the traditional enumeration methods are not only too slow but lead to an overdependence on the significance of numbers of colony-forming units. Food manufacturers require information about the microbiological quality of commodities and raw materials rapidly and it could be argued that an assessment of microbial activity is as important as a knowledge of numbers.

A number of methods have been developed which aim to give answers more quickly and hence are often referred to as 'Rapid Methods'.

10.5.1 Dye-reduction Tests

A group of tests which have been used for some time in the dairy industry depend on the response of a number of redox dyes to the presence of metabolically active micro-organisms. They are relatively simple and rapid to carry out at low cost. The redox dyes are able to take up electrons from an active biological system and this results in a change of colour. Usually the oxidized form is coloured and the reduced form colourless but the triphenyltetrazolium salts are an important exception. Figure 10.1 shows the structures of the oxidized and reduced forms of the three most widely used redox dyes, methylene blue, resazurin and triphenyltetrazolium chloride.

From 1937, and until relatively recently, the methylene blue test was a statutory test for grading the quality of milk in England and Wales. Changes in the technology of handling bulk milk, especially refrigeration have made this test less reliable and it is no longer a statutory test because results show little correlation with the numbers of psychrotrophic bacteria. Since the reduction of resazurin takes place in two stages, from blue to pink to colourless, there is a wider range of colour that can be scored using a comparator disc and the ten-minute resazurin test is still useful for assessing the quality of raw milk at the farm or dairy before it is bulked with other milk.

Triphenyltetrazolium salts and their derivatives are initially colourless and become intensely coloured, and usually insoluble, after reduction to formazans. Triphenyltetrazolium chloride itself is most widely used as a component of diagnostic and selective agar media on which some bacterial colonies will become dark red to maroon as formazan becomes precipitated within the colony. The crystals of the formazan produced from 2-(*p*-iodophenyl)-3-(*p*-nitrophenyl)-5-phenyltetrazolium chloride (INT) are so intensely coloured that they are readily seen in individual microbial cells under the microscope and their presence may be used to assess the viability of cells.

10.5.2 Electrical Methods

When micro-organisms grow, their activity changes the chemical composition of the growth medium and this may also lead to changes in its electrical properties. Measuring this effect has become the basis of one of the most widely used alternative techniques of microbiological analysis.

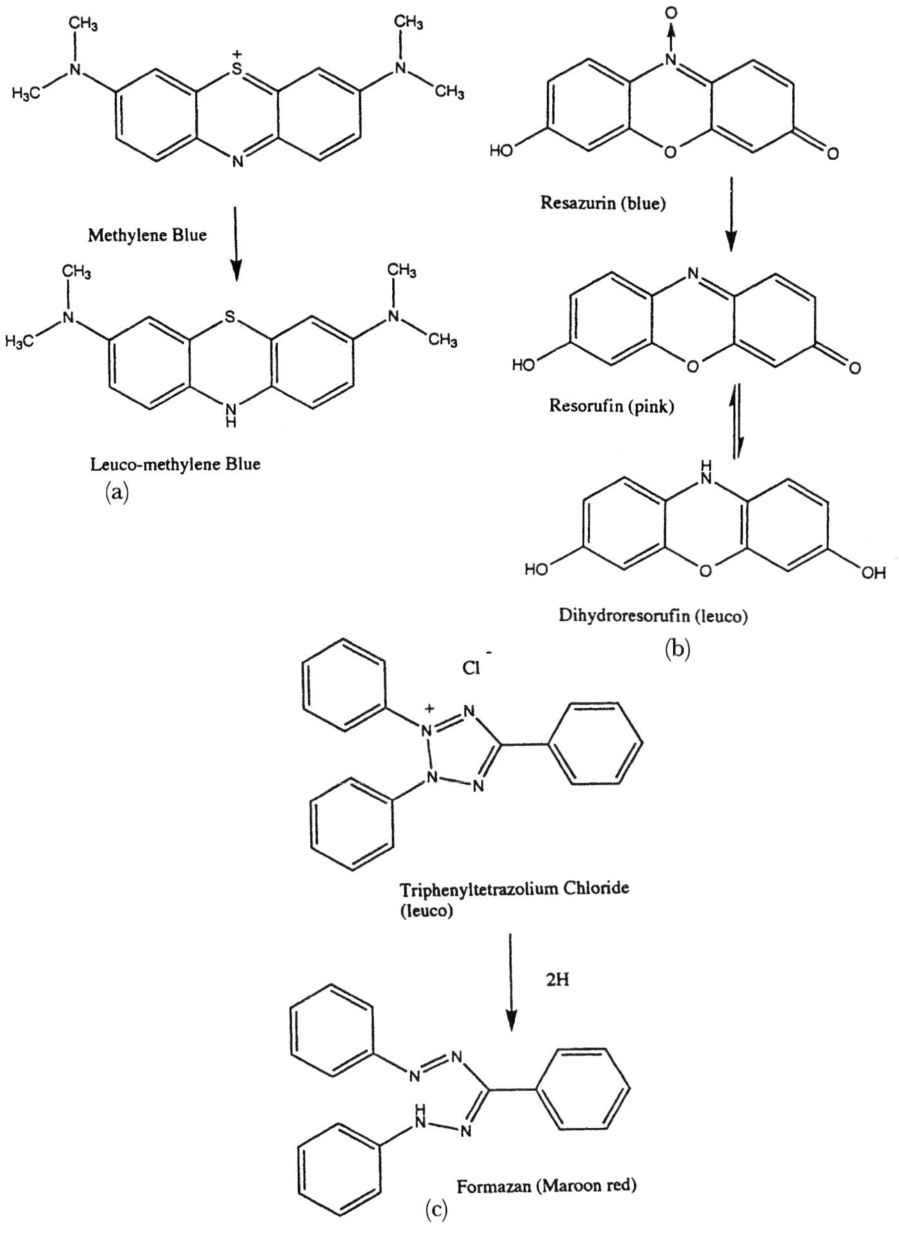

Figure 10.1 *Structures of redox dyes used in food microbiology. (a) Methylene blue. (b) resazurin. (c) triphenyltetrazolium chloride*

The electrical properties most frequently monitored are conductance (G), capacitance (C) and impedance (Z), the latter being influenced by both capacitance and resistance (R) as well as the frequency of the alternating current applied (f). The conductance is simply the reciprocal of resistance, *i.e.*

$$G = 1/R \qquad\qquad (10.6)$$

The relationship between impedance, resistance and capacitance is given by:

$$Z^2 = R^2 + (1/2\pi f\,C)^2 \qquad\qquad (10.7)$$

It is possible to take frequent measurements of the electrical properties of a growth medium by growing organisms in cells supplied with two metal electrodes. By saving the data obtained for subsequent analysis on a computer, large numbers of samples can be monitored at the same time. Central to the successful application of the method is the choice or design of a medium which will both support rapid growth of the micro-organisms to be monitored, and will change its electrical properties as a result of their growth.

With a suitable medium, the traces obtained resemble the bacterial growth curve, although this analogy can be misleading as the curves are not superimposable. In practice it requires quite a large number of bacteria to initiate a signal, usually about 10^6–10^7 cfu ml^{-1} in the cell. The time it takes to reach this number and produce a signal (referred to as the detection time) will therefore include any lag phase plus a period of exponential growth and will depend on both the initial number and the growth rate. Thus, for a particular organism/medium/ temperature combination, the detection time will be inversely related to the logarithm of the number and activity of the organisms in the original sample. Figure 10.2 shows typical traces of samples taken from a ten-fold

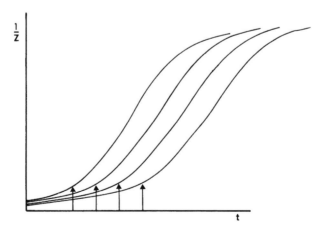

Figure 10.2   *Traces of 1/Z against time of ten-fold dilution series of Escherichia coli growing in brain/heart infusion broth at 37°C. The detection times are marked with arrows*

dilution series of a pure culture. In the case of *Escherichia coli* growing in brain/heart infusion broth, incubated at 37 °C, it is possible to detect the presence of one or two viable cells in five or six hours.

By obtaining detection times using samples where the microbial population is known, calibration curves relating detection time and microbial numbers can be drawn so that count data can be derived from detection times. One such example for *Salmonella* Enteritidis is presented in Figure 10.3. Some claim that the only value in converting detection times to counts is that the food microbiologist derives a sense of security from having data in a familiar form. Since electrical methods measure microbial activity directly, detection time may be a more appropriate measure of the potential to cause spoilage than a viable plate count.

In the food industry the potential for simultaneously testing many samples makes electrical methods a useful means for assessing the quality of raw materials and products. They have the additional advantage that the worse the microbiological quality, the shorter is the detection time, and the sooner the manufacturer knows that there may be a problem. In modern instruments, which can accommodate more than 500 samples, the results can be displayed using an unambiguous quality colour code of acceptable (green), marginally acceptable (orange), and unacceptable (red).

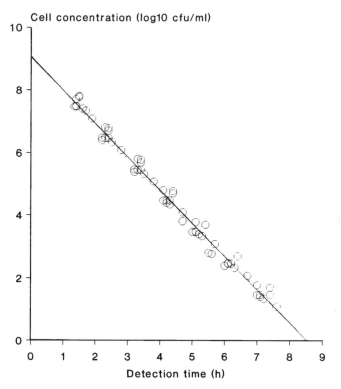

Figure 10.3 *A calibration curve for* Salmonella *Enteritidis using a Bactometer*

By carefully designing the medium to contain selective agents, and diagnostic compounds which will give a strong signal when they are metabolized, it is possible to use electrical methods to estimate the activity (and hence, by calibration, numbers) of specific groups of organisms. Thus the incorporation of trimethylamine oxide (TMAO) into a selective enrichment broth can be used as a pre-screening for the presence of *Salmonella*. The presence of the enzyme TMAO reductase converts this neutral molecule into the strongly charged trimethylammonium ion which has a considerable effect on the conductance of the medium. Absence of a detection time in control sample cells inoculated with a *Salmonella*-specific phage act as further confirmation of the presence of *Salmonella*.

10.5.3 ATP Determination

Adenosine triphosphate is found in all living cells and is the universal agent for the transfer of free energy from catabolic processes to anabolic processes. A number of quite different living organisms have evolved a mechanism for producing light by the activity of enzymes known as luciferases on substrates known as luciferins. These reactions require the presence of ATP and magnesium ions and produce one photon of light at the expense of the hydrolysis of one molecule of ATP through a series of intermediates. An ATP molecule facilitates the formation of an enzyme–substrate complex which is oxidized by molecular oxygen to an electronically excited state. The excited state of the molecule returns to the lower energy ground state with the release of a photon of light and dissociates to release the enzyme luciferase again. These reactions are summarized below:-

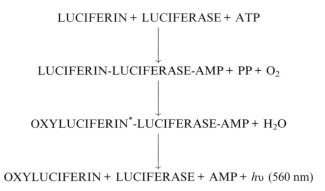

LUCIFERIN + LUCIFERASE + ATP

LUCIFERIN-LUCIFERASE-AMP + PP + O_2

OXYLUCIFERIN*-LUCIFERASE-AMP + H_2O

OXYLUCIFERIN + LUCIFERASE + AMP + $h\upsilon$ (560 nm)

Because instruments are now available which can accurately measure low levels of light emission and pure luciferin and luciferase from, for example fireflies, can be manufactured, the reaction can be used as a very sensitive assay for ATP. Sensitive instruments using photomultiplier tubes can detect as little as 10^2–10^3 fg (fg = femtogram = 10^{-15} g) which corresponds to as few as 10^2–10^3 bacterial cells.

Although the method gives very good results with pure cultures, when applied to foods it is essential to ensure that non-microbial ATP, which will be present in foods in considerably larger quantities than the microbial ATP, has been destroyed or that the micro-organisms have been separated from interfering food components. Non-microbial ATP can be selectively removed by treating with an ATPase after disrupting the somatic cells of animal and plant origin with a mild surfactant. The next stage is to destroy the ATPase activity and then extract the microbial ATP using a more powerful surfactant. The alternative, of removing microbial cells from the food before the ATP assay, can be achieved by centrifugation or filtration of liquid foods but is very much more difficult from suspensions of solid foods. Even when they are successfully separated there are problems arising from the different amounts of ATP in different microbial cells. Thus yeast cells may contain 100 times more ATP than bacterial cells and sub-lethally stressed micro-organisms may contain very low levels of ATP and yet be capable of recovering and growing on a food during long-term storage.

Immunomagnetic separation is a versatile method for removing specific groups of bacteria from a complex matrix of food and other organisms. It has been successfully applied prior to ATP determination as well as in a number of other contexts to extract and purify microorganisms. Antibodies specific to the target organism or lectins, plant proteins which recognize and bind to specific carbohydrate residues exposed on the outer surface of a micro-organism, are attached to magnetisable beads. The bacteria adhere to the beads *via* the antibody–antigen, or lectin–carbohydrate, reaction and can be removed from the food suspension by a powerful magnet acting through the walls of the container. After the food materials have been poured away and the cells washed they can be released into suspension for assay by removing the magnet.

The need for often complex sample preparation has meant that, rapid and sensitive though it is, ATP measurement is not widely used for routine monitoring of microbial contamination of foods. It is however being increasingly used to monitor hygiene in food processing plant. Instruments are available where a swab taken from equipment can be assayed directly for ATP giving a virtually immediate measure of surface contamination. In these cases it is not necessary to distinguish between microbial and non-microbial ATP since the presence of either at high levels would indicate poor hygiene. This speed and simplicity make ATP determination the most overtly microbiological test that can be applied for the routine-monitoring of critical control points as part of the HACCP technique of quality assurance (see Section 11.6).

10.6 RAPID METHODS FOR THE DETECTION OF SPECIFIC ORGANISMS AND TOXINS

10.6.1 Immunological Methods

Because of the potential specificity of immunoassays using polyclonal or monoclonal antibodies, there has been considerable effort devoted to developing their application in food microbiology. Commercial immunoassay kits are now available for detecting a variety of foodborne micro-organisms and their toxins, including mycotoxins.

Raising antibodies to specific surface antigens of micro-organisms, or to macromolecules such as staphylococcal or botulinum toxins, is relatively straightforward and can be achieved directly. Mycotoxins, however, belong to a class of molecules known as haptens which can bind to an appropriate antibody but are of relatively low molecular weight and are not themselves immunogenic. Haptens can be made immunogenic by binding them chemically to a carrier protein molecule, and antibodies have now been raised using this technique to a wide range of mycotoxins including the aflatoxins, trichothecenes, ochratoxin and fumonisins.

Although a number of different formats are used in immunoassays, their essential feature is the binding of antibody to antigen. A commonly used protocol is that of the sandwich ELISA (enzyme linked immunosorbent assay) in which a capture antibody is immobilized on a solid surface of say a microtitre plate well. The sample containing antigen is then added to the well, mixed and removed leaving any antigens present attached to the antibodies. These are then detected by adding second antibody which is coupled to an enzyme such as horseradish peroxidase or alkaline phosphatase. This antibody will also bind to the antigen producing an antibody sandwich. Binding is detected by addition of a chromogenic substrate for the enzyme attached to the second antibody and measuring the colour developed (Figure 10.4). Alternative detection systems are used, such as attachment of antibodies to latex and looking for agglutination in the presence of the antigen and fluorescence-labelled antibodies which can be used to detect target organisms using a fluorescence microscope or flow cytometry.

Commercial ELISAs are available for such organisms as *Salmonella* and *Listeria monocytogenes* but they still require the presence of at least 10^5–10^6 organisms. Detection of smaller numbers therefore depends on some form of enrichment or concentration by one of the separation methods briefly mentioned above, so that although the immunoassay itself may be rapid the whole analytical protocol may take almost as long as conventional procedures. Some advantage can be gained from the automation of the assay and a number of instruments are commercially available. There may also be some concern over the specificity of

(a)   (b)

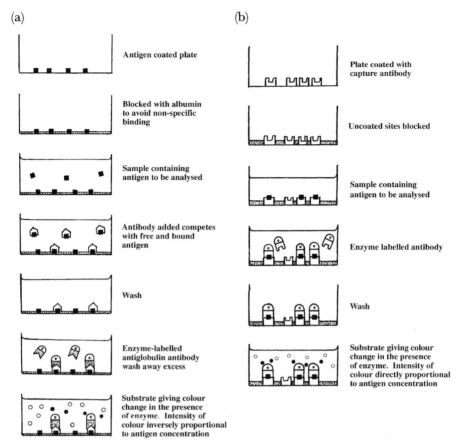

Figure 10.4   *Sandwich ELISAs. (a) Competitive sandwich ELISA, (b) direct sandwich ELISA*

immunoassays. While striving for antibodies that are sufficiently broad in their specificity to recognize all strains of the desired target organism, it is difficult to avoid the problem of cross reactivity with organisms other than those under investigation.

10.6.2   DNA/RNA Methodology

All biochemical, immunological and other characteristics used in the detection of micro-organisms are governed directly or indirectly by the base sequences encoded in the organism's genome. The specificity of this information can now be mobilized to provide methods capable of identifying genera, species or even strains within a species. Nucleic acid probes can be designed which recognize and bind (hybridize) to specified regions of either chromosomal or plasmid DNA or to RNA, and the region chosen to give the desired level of specificity. Thus, for example, ribosomal RNA contains both conserved and variable regions, the former being suitable for recognition at the genus level whereas the latter may be considerably more specific. Although RNA is a more labile molecule than DNA, there are many more copies of ribosomal RNA in a cell than genomic DNA which should make methods based on this molecule more sensitive.

The nucleic acids have to be released from the cells by some form of lysis and, in the case of double-stranded DNA, it has also to be denatured, usually by heat treatment, to the single-stranded form. The denatured nucleic acid is then adsorbed onto a membrane, fixed to it by heat or alkali treatment, and the membrane is treated with some form of blocking agent to prevent non-specific binding of the probe. After incubating with the labelled probe and washing off unadsorbed probe, the presence of the hybridization product is measured using the label attached to the probe. In the earliest stages of the development of this methodology probes were directly labelled with radioactive isotopes such as ^{32}P or ^{35}S and hybridization was detected by autoradiography. This is a very sensitive method but the routine use of radioactive compounds in a food-associated environment is not usually acceptable. Probes can be labelled with an enzyme and detected with a chromogenic substrate or they can be labelled with a small molecular weight hapten for which an enzyme-linked monoclonal antibody is available. Such probes are available for the enterotoxin gene of *Staphylococcus aureus*, the haemolysin gene and rRNA of *Listeria monocytogenes*, 23S rRNA of *Salmonella*, as well as several other systems. One interesting example is a ribosomal RNA probe to detect *Listeria monocytogenes* which uses a chemiluminescent label. The single-stranded DNA probe has a chemiluminescent molecule bound to it. When the probe binds to its RNA target, the chemiluminescent molecule is protected from degradation in a subsequent step so that successful hybridization is indicated by light emission measured in a luminometer (Figure 10.5).

Like the ELISA methods, nucleic acid methods also require some enrichment of the target to produce sufficient nucleic acid to reach the threshold of sensitivity of about 10^6 copies of the target sequence. They are particularly well suited for rapid confirmation of isolated colonies on an agar plate.

The polymerase chain reaction (PCR) provides a method for amplifying specific fragments of DNA, usually less than 3kb in length, and in principle could allow detection of a single copy of the target sequence. The method uses two short oligonucleotide primer sequences (usually about 20 nucleotides long) which will hybridize to opposite strands of heat-denatured DNA at either end of the region to be amplified (Figure 10.6). A DNA polymerase then catalyses extension of the primers to produce two double-stranded copies of the region of interest.

The whole process is then repeated a number of times. In each cycle, the reaction mixture is heated to 94–98 °C to separate the double

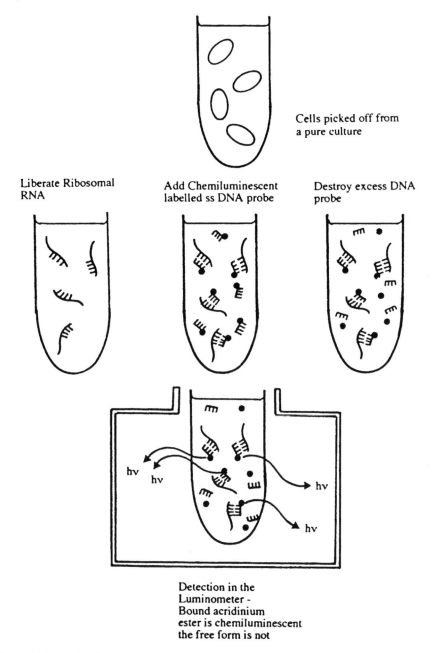

Cells picked off from a pure culture

Liberate Ribosomal RNA

Add Chemiluminescent labelled ss DNA probe

Destroy excess DNA probe

Detection in the Luminometer - Bound acridinium ester is chemiluminescent the free form is not

Figure 10.5 *Hybridization protection assay*

stranded DNA into single strands. The mix is then cooled to 37–65 °C to allow the primers to anneal to the single strands and then warmed to 72 °C to allow synthesis of their complementary strands. Thus as the cycles progress one double stranded segment of DNA will become two,

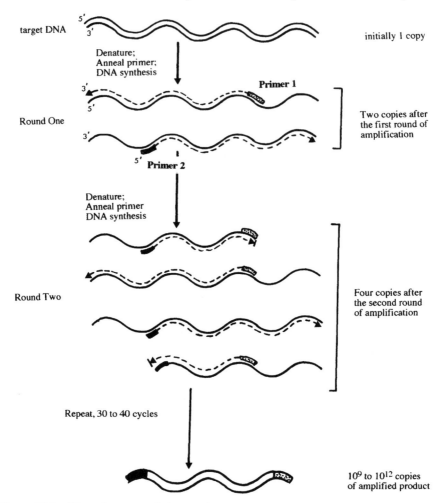

Figure 10.6   *The polymerase chain reaction*

four, eight, sixteen and so on so that after 20 cycles approximately 10^6 copies can be generated from the original.

Two key components essential to the successful application of PCR are precision automated thermal cycling equipment and a heat stable DNA polymerase which will survive the DNA denaturation step and catalyse subsequent extension of the primers. The latter, known as Taq polymerase, is obtained from the very thermophilic bacterium *Thermus aquaticus*.

There have been numerous embellishments of the basic PCR technique. These include:-

- Multiplex PCR which uses several primer pairs for the simultaneous amplification of a number of targets.
- Nested PCR which improves the sensitivity of the reaction by first amplifying a larger sequence using a pair of outer primers followed

by amplification of a shorter sequence within the amplicon using a second set of inner primers.

- Reverse transcriptase PCR which uses conventional PCR to amplify an RNA template after it has been transcribed into DNA. This can have higher sensitivity because of the multiple copies of RNA within the cell but is also essential for the detection of RNA viruses such as Norovirus.
- Real time PCR. Traditionally PCR products are detected by agarose gel electrophoresis on the basis of their size and/or their reaction with a complementary probe. An important development of the technique has been the advent of real time PCR where a fluorescent signal is produced. There are several techniques for doing this but the simplest uses a dye Sybr Green which fluoresces when it intercalates with double stranded DNA. Thus as the PCR reaction progresses more double stranded DNA is present and the more intense the fluorescence. The time taken to produce a detectable fluorescence will depend on the amount of target present initially, the more that is there the sooner the threshold level will be reached. This means that real time PCR can be both qualitative and quantitative.

The polymerase chain reaction can be inhibited by food components. This can be reduced by separation of the cells from the inhibitory food matrix by procedures such as immunomagnetic separation or by cultural enrichment to allow subsequent dilution of the sample and any inhibitory components. A cultural enrichment before PCR also helps overcome the objection that since PCR detects fragments of DNA these may not necessarily originate from a viable cell.

10.6.3 Subtyping

The ability to identify an organism to the species level is not sufficient to establish a firm link between food and clinical isolates or to identify whether a number of apparently unrelated cases have a common source. To do this it is necessary to have more highly discriminating methods that can distinguish between different strains of the same species.

Traditional subtyping procedures have been based on phenotypic characteristics. Thus biotyping employs particular biochemical activities for discrimination while phage typing and serotyping are based on the presence or absence of particular phage receptors or antigens on the cell surface. While these techniques have proved invaluable in particular circumstances they are not universally applicable due to factors such as the variability of gene expression. Genotypic subtyping methods however have broader applicability being based on an organism's underlying genetic make up rather than its phenotypic expression.

In restriction fragment length polymorphism (RFLP), a restriction enzyme is used to cut DNA into a number of fragments depending on the number and location of restriction sites present. Separation of these fragments by electrophoresis will reveal a fingerprint of that particular organism. Ribotyping, a particular type of RFLP where the probes used are specific for rRNA genes, has been automated and equipment is available commercially that can ribotype an organism in 8 hours.

A commonly encountered problem is that RFLP patterns are excessively complex with many poorly resolved bands. Use of restriction enzymes with fewer sites on the bacterial chromosome overcomes this problem but specialist electrophoretic techniques are needed to separate the large DNA fragments produced. This is pulsed field gel electrophoresis (PFGE) which uses an alternating electric field to tease apart the large DNA molecules.

10.7 LABORATORY ACCREDITATION

From what has already been said it should be clear that there can be a number of different ways of detecting the same organism in a food matrix. The choice of method used can be governed by several factors and the relative merits of different methods is a topic of constant investigation and debate. This can however lead to the situation where differences in a result reported by two laboratories simply reflect the different method used.

In addition to problems arising from intrinsic differences in the performance of different methods, the same method in different laboratories can be subject to variation introduced by factors such as differences in procedures, equipment and its calibration. Some possible examples would include autoclave temperature profile when sterilizing media, time and temperature of incubation, sources of medium components and, of course, competence and experience of laboratory personnel.

A number of approaches are adopted to avoid such potential problems. Several national and international bodies approve standard methods for conducting certain analyses and one of these should be adopted for routine work and strictly adhered to wherever possible. Testing laboratories also often participate in proficiency testing schemes where a central body distributes standard samples for analysis, often specifying the precise time this should be conducted and the method to be used. Results are reported back, collated and a report circulated to participating laboratories which can then judge their performance against that of others. Finally, laboratories can seek some form of third party, independent recognition. There are quality systems such as the Good Laboratory Practice scheme and standards such as (ISO 9000 series) which are concerned with the quality of management within the organization but which do not set a particular level of quality or competence to be achieved (see Section 11.7). There are also schemes of laboratory accreditation more concerned with the quality of performance in specific tests. In the UK this accreditation is usually sought through the UK Accreditation Service (UKAS) which accredits laboratories over a whole range of activities, not just microbiological testing. Most countries have their own equivalent organization such as NATA (Australia), DANAK (Denmark), ILAB (Ireland) and STERLAB (The Netherlands). The accrediting body inspects the laboratory and its procedures to ensure that tests are carried out consistently and correctly using approved methods with suitable quality control measures in place. Among the features investigated are the training and qualifications of staff, the suitability of equipment and procedures for its calibration and maintenance, participation in a proficiency testing scheme and the presence of full documentation prescribing the laboratory's operating procedures.

Obtaining laboratory accreditation can be a costly exercise but, if achieved, provides independent testimony to a laboratory's proficiency and will give increased confidence to potential customers.

CHAPTER 11

Controlling the Microbiological Quality of Foods

In Chapter 10 we surveyed the different methods used in the microbiological examination of food and how these give us information about the size and composition of a food's microflora. We will now go on to describe how data obtained from such tests can be used to make decisions on microbiological quality and how accurate these judgements are likely to be. It will become apparent that reliance on this approach alone is an ineffective means of controlling quality so, finally, we will examine how best to ensure the production of consistently good microbiological quality products.

11.1 QUALITY AND CRITERIA

We all feel we know what is meant by quality and the difference between good quality and poor quality. One dictionary defines quality as the 'degree of excellence' possessed by a product, that is to say how good it is at serving its purpose. In terms of the microbiology of foods, quality comprises three aspects:

(1) *Safety*. A food must not contain levels of a pathogen or its toxin likely to cause illness when the food is consumed.
(2) *Acceptability/shelf-life*. A food must not contain levels of micro-organisms sufficient to render it organoleptically spoiled in an unacceptably short time.
(3) *Consistency*. A food must be of consistent quality both with respect to safety and to shelf-life. The consumer will not accept products which display large batch-to-batch variations in shelf-life and is certainly not prepared to play Russian roulette with illness every time he or she eats a particular product.

Regulatory bodies and the food industry are the two groups most actively interested in determining and controlling the microbiological quality of foods. The regulatory authorities must do so to fulfil their statutory responsibility to protect the public from hazardous or inferior goods. The extent to which they intervene in food production and supply will depend of course upon the food laws of the country in which they operate. Commercial companies, both food producers and retailers, also have a major interest, since their association with products that are consistently good and safe will protect and enhance their good name and their market.

To distinguish food of acceptable quality from food of unacceptable quality requires the application of what are known as microbiological criteria. Three different types of microbiological criterion have been defined by The International Commission on Microbiological Specifications for Foods (ICMSF).

(1) A *microbiological standard* is a criterion specified in a law or regulation. It is a legal requirement that foods must meet and is enforceable by the appropriate regulatory agency.

(2) A *microbiological specification* is a criterion applied in commerce. It is a contractual condition of acceptance that is applied by a purchaser attempting to define the microbiological quality of a product or ingredient. Failure of the supplier to meet the specification will result in rejection of the batch or a lower price.

(3) A *microbiological guideline* is used to monitor the microbiological acceptability of a product or process. It differs from the standard and specification in that it is advisory rather than mandatory.

The ICMSF have also specified what should be included in a microbiological criterion as set out below:

(1) A statement of the food to which the criterion applies. Clearly foods differ in their origin, composition, and processing; will present different microbial habitats; and will therefore pose different spoilage and public health problems.

(2) A statement of the micro-organisms or toxins of concern. These may cover both spoilage and health aspects, but decisions on what to include must be realistic and based on a sound understanding of the microbial ecology of the food in question.

(3) Details of the analytical methods to be used to detect and quantify the micro-organisms/toxins. Preferred methods for standards or specifications would be those elaborated by international bodies, although less sensitive or less reproducible methods may be used for simplicity and speed in confirming compliance with guidelines.

(4) The number and size of samples to be taken from a batch of food or from a source of concern such as a point in a processing line.

(5) The microbiological limits appropriate to the product and the number of sample results which must conform with these limits for the product to be acceptable (n, c, m, and M, see Section 11.2). In this regard, it should be remembered that for certain food-borne pathogens such as *Staphylococcus aureus* or *Clostridium perfringens*, their mere presence does not necessarily indicate a hazard and specification of some numerical limits is necessary.

These last two points can present the greatest problem. In applying the microbiological criterion it is assumed that the analytical results obtained are an accurate reflection of the microbiological quality of the whole batch of food. How justified that extrapolation is will depend upon the accuracy and precison of the tests used and on how representative the samples were that were tested.

Micro-organisms are rarely distributed uniformly throughout a food, nor in fact are they usually distributed randomly. When micro-organisms are dispersed in a food material in the course of its production, some may die, some may be unable to grow and others may find themselves in microenvironments in which they can multiply. The resulting distribution, containing aggregates of cells, is described as a contagious distribution (Figure 11.1).

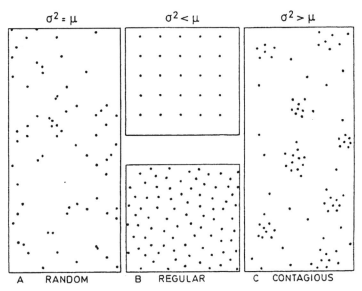

Figure 11.1   *Possible types of spatial distribution of micro-organisms in food* σ^2, *variance;*
    μ *mean*
    (Reproduced with permission from Jarvis (1989))

As the number of samples tested increases, so does our confidence in the result but so too does the cost of testing. To be sure of the quality of the batch or lot we would have to test it all, but since microbiological testing is destructive, this would result in almost absolute confidence in the product quality but none left to sell. A compromise must therefore be struck between what is practicable and what gives the best estimate of lot quality.

11.2   SAMPLING SCHEMES

The sampling scheme most commonly applied in the microbiological testing of foods is that of sampling for attributes. It makes no assumption about the distribution of micro-organisms within the batch of food and is therefore particularly suited to situations where we have no knowledge of this; for example with imported foods at their port of entry.

11.2.1   Two-class Attributes Plans

In an attributes sampling scheme analytical results are assigned into classes; in the simplest type, the two-class scheme, samples are classified as acceptable or defective depending on the test result. A sample is described as defective if it is shown to contain more than a specified number of organisms or, in cases where a presence or absence test is applied, the target organism is detected.

A two-class sampling scheme is defined by three numbers:

n – the number of sample units to be tested;
m – the count above which the sample is regarded as defective. This term would not appear in schemes employing a presence/absence test since a positive result is sufficient for the sample to be defective;
c – the maximum allowable number of sample units which may exceed m before the lot is rejected.

Such schemes do not make full use of the numerical data obtained but simply classify sample units according to the test result. For example, if m is 10^4 cfu g^{-1}, those samples giving counts of 10^2, 9×10^3, and 1.2×10^4 would be considered acceptable, acceptable, and defective respectively, despite the fact that the first sample had a count almost 2 log cycles lower than the second and the difference in count between the second and third samples is relatively small.

Using this approach, results from a number of sample units can be classified according to the proportion defective and the frequency of occurrence of defective units described by a binomial distribution. If p represents the proportion of defective sample units in the whole lot, *i.e.* the probability of a single sample unit being defective, and q represents the proportion of acceptable units (the probability of a sample unit being acceptable), then the probability distribution of the various possible outcomes is given by the expansion of the binomial:

$$(p + q)^n \tag{11.1}$$

Where n, the number of sample units examined, is small compared with the lot size.

Since, in a two-class plan, a sample can only be acceptable or defective then

$$p + q = 1 \tag{11.2}$$

The probability that an event will occur x times out of n tests is given by:

$$P_{(x)} = (n!/(n - x)! \, x!) \, p^x q^{(n-x)} \tag{11.3}$$

or if we substitute $(1-p)$ for q:

$$P_{(x)} = (n!/(n - x)! \, x!) \, p^x (1 - p)^{n-x} \tag{11.4}$$

If we have a sampling plan which does not permit any defective samples ($c = 0$), then by putting $x = 0$ into Equation (11.4) we obtain an expression for the probability of obtaining zero defective samples, *i.e.* the probability of accepting a lot containing a proportion p defective samples:

$$P_{acc} = (1 - p)^n \tag{11.5}$$

It follows that the probability of rejection is:

$$P_{rej} = 1 - (1 - p)^n \tag{11.6}$$

We can use this equation to determine how effective such relatively simple sampling schemes are. For example, Table 11.1 shows how the frequency (probability) of finding a defective sample changes with the level of defectives in the lot as a whole and with the number of samples taken. This could apply to a *Salmonella*-testing scheme where detection of the organism in a single sample is sufficient for the whole lot to be rejected. If the incidence of *Salmonella* is 1% ($p = 0.01$), a level

Table 11.1   *Acceptance and rejection thresholds in a sampling scheme*

Incidence of defectives[a] (%)	No. of samples tested					
	10		20		100	
	P_{acc}	P_{rej}	P_{acc}	P_{rej}	P_{acc}	P_{rej}
0.01	99.9	0.1	99.8	0.2	99.0	1.0
0.1	99.0	1.0	98.0	2.0	90.0	10.0
1	90.0	10.0	82.0	18.0	37.0	63.0
2	82.0	18.0	67.0	33.0	13.0	87.0
5	60.0	40.0	36.0	64.0	0.6	99.4
10	35.0	65.0	12.0	88.0	0.1	99.9

[a] *e.g.* Presence of *Salmonella* in a sample or surviving mesophiles in packs of an appertized food

corresponding to 40 positive 25 g samples in a lot of 1 tonne (4000 × 25g samples), then the lot would be accepted 90% of the time if 10 samples were tested on each occasion. When the number of samples taken at each testing is increased then the chances of rejection are increased, so that with 20 samples the lot would be accepted 82% of the time. If we went to testing 100 samples, we would accept the same lot only 37% of the time.

A statistically equivalent situation would be the testing of an appertized food for the presence of surviving organisms capable of spoiling the product, and where detection of one defective pack would mean rejection of the lot. Using Table 11.1 again, we can see that if there was a failure rate of one pack in a thousand ($p = 0.001$), even taking 100 packs for microbiological testing we would only reject a lot on one out of ten occasions. In fact we can calculate the number of packs it would be necessary to take in order to have a 95% probability of finding one defective sample ($P_{rej} = 0.95$). This is done simply by substituting in Equation (11.6) and solving for n. The answer, 2995, demonstrates why it is necessary to have alternatives to microbiological testing to control the quality of appertized foods.

The probabilities of acceptance or rejection associated with an attributes sampling plan can be calculated from the binomial distribution. For large batches of product these can also be represented graphically by what is known as an operating characteristic (OC) curve of the type shown in Figure 11.2. For each level of defectives in the lot, the probability of its acceptance or rejection using that plan can be read off the curve. Figure 11.3 demonstrates that as n increases for a given value of c, the stringency of the plan increases since a lot's overall quality must increase with n for it to have the same chance of being passed. If c is increased for a given value of n (Figure 11.4), so the plan becomes more lenient as lot quality can decrease but still retain the same chance of being accepted.

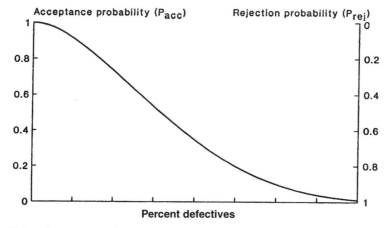

Figure 11.2 *An operating characteristic curve*

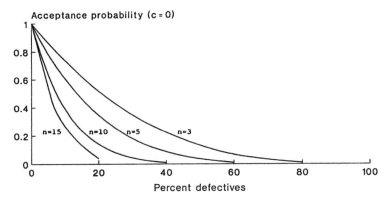

Figure 11.3 *Operating characteristic curves (increasing stringency).* $c = 0$, $n = 3, 5, 10, 15$

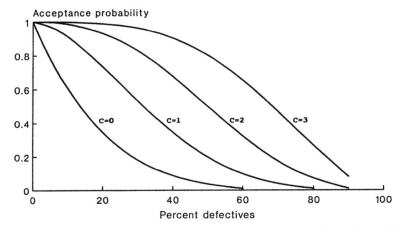

Figure 11.4 *Operating characteristic curves (decreasing stringency).* $n = 5$, $c = 0, 1, 2, 3$

The ideal OC curve would resemble Figure 11.5 with a vertical cut off at the maximum acceptable level of defectives. To achieve this would require testing an unacceptably high number of samples so the purchaser has to adopt a sampling plan which will accept lots defined as being of good quality most (*e.g.* 95%) of the time and has a high probability (*e.g.* 90%) of rejecting lots of poor quality (Figure 11.6). Two types of error are identifiable in this approach: the producer's risk that a lot of acceptable quality would be rejected (5%), and the consumer's or purchaser's risk that a lot of unacceptable quality would be accepted (10%). Lots of intermediate quality would be accepted at a frequency of between 10 and 95%.

11.2.2 Three-class Attributes Plans

Three-class attributes sampling plans introduce a further category and divide samples into three classes: acceptable, marginally acceptable, and

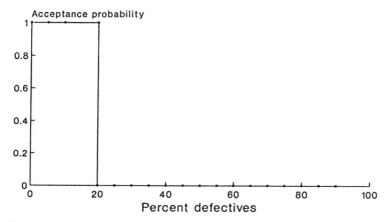

Figure 11.5 *An ideal operating characteristic curve*

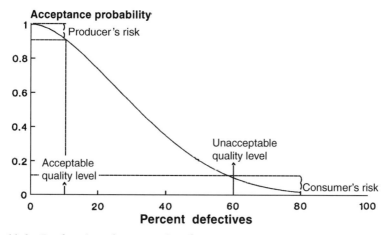

Figure 11.6 *Producer's and consumer's risk*

unacceptable. Use of this extra classification of marginally acceptable means that they are not used with presence or absence tests but only with microbiological count data. A three-class plan is defined by four numbers:

n – the number of samples to be taken from a lot;

M – a count which if exceeded by any of the test samples would lead to rejection of the lot;

m – a count which separates good quality from marginal quality and which most test samples should not exceed;

c – the maximum number of test samples which may fall into the marginally acceptable category before the lot is rejected.

As with the two-class plan increasing c or decreasing n increases the leniency of a three-class plan. OC curves for three-class plans can be derived from the trinomial distribution and describe a 3-dimensional

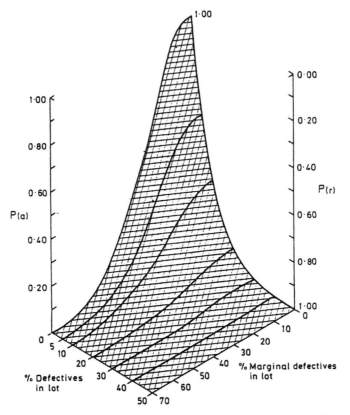

Figure 11.7 *Operating characteristic surface for a three-class plan.* n = 10, c = 2
(Reproduced with permission from Jarvis (1989)

surface with axes defining the probability of acceptance/rejection, the proportion of marginally defectives in a lot, and the proportion of defectives in a lot (Figure 11.7).

Microbiological criteria using attributes sampling plans for different foods have been produced by a number of organizations and some examples of these are given in Table 11.2. As is clear, the stringency of the sampling plan varies with the type of food and the organism being sought.

In 2006 a regulation came into force in the countries of the EU (2073/2005) setting down microbiological criteria for some foods based on attribute plans. This consolidated a number of previous regulations and criteria. It also introduced two types of criterion: food safety criteria and process hygiene criteria. If foods fail to meet food safety criteria they should not be placed on the market or withdrawn from sale. Failure to meet process hygiene criteria is less serious but would necessitate a review of food safety management procedures with a view to improving product quality. The criteria are not intended as quality control measures in

Table 11.2 *Attributes sampling plans*

Product	Organisms	Plan class	n	m	M	c	Source
Ice cream	APC	3	5	10^5	10^6	2	Canada
	coliforms	3	5	10	10^3	1	
Dried milk	APC	3	5	5×10^4	2×10^5	2	International Dairy Federation
	coliforms	3	5	10	100	1	
	Salmonella	2	15	0	–	0	
Frozen raw crustaceans	APC	3	5	10^6	10^7	3	ICMSF
	E. coli	3	5	11	500	3	
	Staph. aureus[a]	3	5	10^3	10^4	2	
	Salmonella[a]	2	5	0	–	0	
	V. parahaemolyticus[a]	3	5	10^2	10^3	1	
Frozen cooked crustaceans	APC	3	5	5×10^5	10^7	2	ICMSF
	E. coli	3	5	11	500	2	
	Staph. aureus	2	5	10^3	–	0	
	Salmonella[a]	2	10	0	–	0	
	V. parahaemolyticus[a]	3	5	10^2	10^3	1	
Frozen fruits and vegetables pH > 4.5	*E. coli*	3	5	10^2	10^3	2	ICMSF
Minced meat	APC	3	5	5×10^5	5×10^6	2	EU
	E. coli	3	5	50	500	2	
	Staph. aureus	3	5	100	5000	2	
	Salmonella (25 g samples)	2	5	–	–	0	

[a] Additional tests

m and M values are expressed as cfu g^{-1}, APC = Aerobic Plate Count

themselves and they do not (with a few exceptions) specify any minimum frequency of testing, increased routine testing or the introduction of a positive release system. The intention is that they serve as a means of ensuring food safety management systems such as HACCP (see 11.6) are functioning correctly.

11.2.3 Choosing a Plan Stringency

Two important principles governing the choice of plan stringency are presented in Table 11.3. As the severity of the hazard being tested for increases, so too must the stringency of the sampling plan. For example, spoilage can be regarded as more of a risk to the product than to the consumer and so tests for indicators of shelf-life such as aerobic plate counts will have the most lenient sampling plans. Even though such plans may quite frequently pass products which are defective, they can still be effective in the sense that regular rejection of say 1 in 5 batches of product would represent a significant economic loss to the producer and would be a strong incentive to improve quality.

Table 11.3 *ICMSF suggested sampling plans*

Degree of concern relative to utility and health hazard	Conditions in which food is expected to be handled and consumed after sampling, in the usual course of events[a]		
	Conditions reduce degree of concern	Conditions cause no change in concern	Conditions may increase concern
No direct health hazard Utility, e.g. shelf-life and spoilage	Increase shelf-life 3-class $n = 5$, $c = 3$	No change 3-class $n = 5$, $c = 2$	Reduce shelf-life 3-class $n = 5$, $c = 1$
Health hazard Low, indirect (indicator)	Reduce hazard 3-class $n = 5$, $c = 3$	No change 3-class $n = 5$, $c = 2$	Increase hazard 3-class $n = 5$, $c = 1$
Moderate, direct, limited spread *e.g. Staph. aureus C. perfringens*	3-class $n = 5$, $c = 2$	3-class $n = 5$, $c = 1$	3-class $n = 10$, $c = 1$
Moderate, direct, potentially extensive spread *e.g. Salmonella*	2-class $n = 5$, $c = 0$	2-class $n = 10$, $c = 0$	2-class $n = 20$, $c = 0$
Severe, direct *e.g. C. botulinum S. typhi*	2-class $n = 15$, $c = 0$	2-class $n = 30$, $c = 0$	2-class $n = 60$, $c = 0$

[a] More stringent sampling plans would generally be used for sensitive foods destined for susceptible populations
Adapted from ICMSF (1986)

Interpretation of the significance of high numbers of indicator organisms will depend on the indicator and the food involved. Enterobacteriaceae or coliform counts can provide an indication of the adequacy of process hygiene, though they are naturally present in substantial numbers on several raw foods. *Escherichia coli* is indicative of faecal contamination and the possible presence of enteric pathogens, although there is no direct relationship and interpretation is not clear-cut. Because of these uncertainties, indicator tests are of only moderate stringency.

When looking for known pathogens, more stringent sampling plans are appropriate and these become more demanding as the severity of the illness the pathogen causes increases.

The conditions under which the food is to be handled after sampling must also be accommodated in any plan. For example, a sampling plan for *E. coli* in raw meats can be quite lenient since the organism is not uncommon in raw meats and the product will presumably be cooked before consumption, thus reducing the hazard. A more stringent plan is required for the same organism if the subsequent handling of a food will produce no change in the hazard; for example, ice cream which is stored frozen until consumption. The most stringent plan would be required for products where subsequent handling could increase the hazard. This would be the case with dried milk where the product could be reconstituted and held at temperatures which would allow microbial growth to resume.

Plan stringency should also take account of whether the food is to be consumed by particularly vulnerable groups of the population such as infants, the very old, or the very sick.

11.2.4 Variables Acceptance Sampling

Very often we have no idea how micro-organisms are distributed within a batch of food and have no alternative but to use an attributes sampling scheme which makes no assumption on this question. In many cases though, studies have found that micro-organisms are distributed lognormally, that is to say the logarithms of the counts from different samples conform to a normal distibution. For example, a survey of nearly 1300 batches of frozen and dried foods found that, on average, only 7.8% of batches did not conform to a log-normal distribution. When this is the case it is possible to use a variables acceptance sampling procedure which achieves better discrimination by making full use of the numerical data obtained from testing rather than just assigning test results into classes as is done in sampling by attributes.

The shape of a normal distribution curve is determined by the parameters μ, the population mean which determines the maximum height of the curve, and σ, the standard deviation of the distribution which determines its spread. For any log count V in a log-normal distribution, a certain proportion of counts will lie above V determined by:

$$(V - \mu)/\sigma \, (= K) \tag{11.7}$$

where K is known as the standardized normal deviate. For example, when:

$K = 0 \, (V = \mu)$ then 50% of values will lie above V;
$K = 1 \, (V = \mu + \sigma)$ then 16% of values lie above V;
$K = 1.65 \, (V = \mu + 1.65\sigma)$ then 5% of values lie above V.

Rearranging, we get:

$$\mu + K\sigma = V \tag{11.8}$$

If V is chosen to represent a log count related to a safety or quality limit and K determines the acceptable proportion of samples in excess of V then a lot would be acceptable if:

$$\mu + K\sigma \leq V \tag{11.9}$$

and unacceptable if:

$$\mu + K\sigma > V \tag{11.10}$$

In practice V is likely to be very close to the logarithm of M, used in three-class attribute plans.

Since we do not know μ or σ, we must use estimates derived from our testing, $\bar{x}$, the mean log count, and s, the sample standard deviation. K is replaced with a value k_1, derived from standard Tables, which makes allowance for our imprecision in estimating K and chosen to give a desired lowest probability for rejection of a lot having an unacceptable proportion of counts greater than V. This gives us the condition for rejection:

$$\bar{x} + k_1 s > V \tag{11.11}$$

Some k_1 values are presented in Table 11.4. If we decrease the desired minimum probability of rejection for a given proportion exceeding V, *i.e.* decrease the stringency of the plan, then k_1 decreases. Application of a variables plan for control purposes will give a lower producer's risk than the equivalent attributes scheme.

It is possible to apply the variables plan as a guideline to Good Manufacturing Practice (GMP). In this case the criterion is:

$$\bar{x} + k_2 s < v \tag{11.12}$$

where k_2 is derived from Tables and gives a certain minimum probability of acceptance provided less than a certain proportion exceeds v, a lower limit value characteristic of product produced under conditions of Good Manufacturing Practice (Table 11.4). The value v will be very similar to

Table 11.4 k *Values for setting specifications for variables sampling*

Safety/quality $\bar{x} + k_1 s > V$

Probability of rejection	Proportion exceeding v	No. of replicate samples							
		3	4	5	6	7	8	9	10
0.95	0.05	7.7	5.1	4.2	3.7	3.4	3.2	3.0	2.9
	0.10	6.2	4.2	3.4	3.0	2.8	2.6	2.4	2.4
	0.30	3.3	2.3	1.9	1.6	1.5	1.4	1.3	1.3
	0.50	1.7	1.2	0.95	0.82	0.73	0.67	0.62	0.58
0.90	0.10	4.3	3.2	2.7	2.5	2.3	2.2	2.1	2.1
	0.25	2.6	2.0	1.7	1.5	1.4	1.4	1.3	1.3

GMP criterion $\bar{x} + k_2 s < v$

Probability of acceptance	Proportion exceeding v	3	4	5	6	7	8	9	10
0.95	0.10	0.33	0.44	0.52	0.57	0.62	0.66	0.69	0.71
	0.20	-0.13	0.02	0.11	0.17	0.22	0.26	0.29	0.31
	0.30	-0.58	-0.36	-0.24	-0.16	-0.10	-0.06	-0.02	0
0.90	0.10	0.53	0.62	0.68	0.72	0.75	0.78	0.81	0.83
	0.20	0.11	0.21	0.27	0.32	0.35	0.38	0.41	0.43
	0.30	-0.26	-0.13	-0.05	0.01	0.04	0.07	0.10	0.12

the logarithm of m used in the three-class attributes plan. If a lot were to fail a GMP criterion it would not lead to rejection of the lot but would alert the manufacturer to an apparent failure of GMP.

If we take a practical example, let us assume that the critical safety/quality limit is $10^7 \, \text{cfu g}^{-1}$ $(V=7)$, but that under conditions of GMP a level below 10^6 is usually attainable $(v=6)$. We are testing five samples from the lot and wish to be 95% certain to reject lots where more than 10% of samples exceed a count of 10^7. From Table 11.4, $k_1 = 3.4$ and our specification becomes:

$$\text{if } \bar{x} + 3.4s > 7 \text{ then reject.}$$

If we also wish to be 95% sure of accepting lots with less than 20% greater than our GMP limit then from Table 11.4, k_2 is 0.11 and our GMP criterion becomes:

$$\text{if } \bar{x} + 0.11s < 6 \text{ then accept.}$$

Three batches of product give the following log counts:

	A	B	C
	5.48	5.55	5.95
	5.23	5.30	6.12
	5.81	6.10	6.20
	5.97	5.75	5.65
	5.46	6.01	6.21
$\bar{x}$	5.59	5.74	6.03
s	0.30	0.33	0.23

Applying the safety/quality limit and GMP formulae, batches A and B are acceptable according to both criteria. Batch C, though acceptable with respect to the safety/quality limit, is not acceptable according to the GMP criterion and this may warrant further investigation.

11.3 QUALITY CONTROL USING MICROBIOLOGICAL CRITERIA

In the 1970s in Oregon in the United States standards were introduced governing the microbiological quality of ground meat in retail stores. After they had been in force for a few years an enquiry was conducted into their effect. The principal conclusion was that, although there was felt to have been a general improvement in sanitation and handling, the standards had produced no significant change in quality. There was no evidence that the bacterial load on ground meat or the risk of foodborne illness had been reduced and, on the debit side, significant costs had been incurred as a result of rejection of material not meeting the standard and through the expense of microbiological testing. It was also felt that consumers had been misled, since their expectation had been that the introduction of standards would lead to an improvement in quality. As a result of this enquiry, the standards were revoked.

An interesting comparison is provided by the Milk Marketing Board's scheme of paying English and Welsh farmers on the basis of the bacterial count of the milk they supply (see Chapter 5). In this case, feedback of the results to farmers resulted in a dramatic decrease in the recorded count of milk over a period of just four months.

The difference in these two experiences can be ascribed to a number of reasons. Firstly, microbiological testing of milk is more likely to give an accurate reflection of microbiological quality in the batch as a whole since it is easier to obtain truly representative samples of a liquid. Also, much is known on how to produce raw milk hygienically so that bacterial contamination is minimized, farmers had simply not been assiduous in the application of these procedures until financial penalties acted as an incentive. Another crucial difference is that the standards in Oregon had applied later in the supply chain, at the point of sale. Earlier stages in meat production such as conditions of slaughter, dressing and storage make a major contribution to the microbiological quality of meat and the standards had done nothing to improve these. The enquiry had noted that there had been an improvement of hygiene at the retail level but since this produced no significant reduction in count it clearly indicates that the problem lay elsewhere.

These two cases indicate two important features of microbiological quality control. Namely the ineffectiveness of retrospective systems of quality control and the importance of control at source.

A system of retrospective quality control based on testing samples of a product and accepting or rejecting a lot on the basis of test results suffers from a number of limitations. We have already discussed the inhomogeneous distribution of micro-organisms in food, the problems of representative sampling and the producer's and purchaser's risks associated with any sampling plan. To minimize these risks requires plans entailing the testing of large numbers of samples and these entail high costs as a result of both the amount of product required to be tested and the costs of laboratory resources. Even with representative samples there is the problem of the relative inaccuracy of traditional microbiological methods and their long elapsed times. If results of laboratory tests are required before a product can be released for sale (a positive release system), then the product's useful shelf-life is reduced. Finally, a major weakness of retrospective systems of quality control is that they provide little in the way of remedial information. They help identify that there is a problem but often give little information as to where it has arisen and what is required for its solution. If a product has high counts, is this due to poor quality raw materials, poor hygiene in the production process, poor conditions of storage, or some combination of all three?

The most effective way of controlling quality is through intervention at source, during the production process. On its own, any amount of testing will not improve product quality one jot, to do this requires action where the factors which determine quality operate, namely in the processing and supply chain itself.

11.4 CONTROL AT SOURCE

Control of processing . . . is of far greater importance than examination of the finished article.

Sir Graham Wilson

The traditional approach to control of microbiological quality at source has relied upon a combination of a well-trained workforce, rigorous inspection of facilities and supervision of operations, coupled with microbiological testing, not only of finished product, but also of ingredients, product in process, equipment, the environment, and personnel.

11.4.1 Training

Food handlers should be trained in the basic concepts and requirements of food and personal hygiene as well as those aspects particular to the specific food-processing operation. The level of training will vary depending on the type of operation and the precise job description of the employee, however some form of induction training with regular updating or refresher courses is an absolute minimum.

Training should give food handlers an understanding of the basic principles of hygiene, why it is necessary, and how to achieve it in practice. A core curriculum for any such course should emphasize:

(1) Micro-organisms as the main cause of food spoilage and food-borne illness and the characteristics of the common types of food poisoning.
(2) How to prevent food poisoning through the control of microbial growth, survival or contamination.
(3) Standards of personal hygiene required of food handlers. These are principally to avoid contamination of food with bacteria the food handler may harbour as part of the body's flora, *e.g Staph. aureus, Salmonella* or that they may bring in with them from the outside world, *e.g. Listeria, B. cereus.* Some do's and don'ts associated with good personal hygiene are listed in Table 11.5.

Table 11.5   *Some do's and don'ts of personal hygiene for food handlers*

DO

Wash your hands regularly throughout the day and especially:

–after going to the toilet;
–on entering a food room and before handling food or equipment;
–after handling raw foods;
–after combing or touching the hair;
–after eating, smoking, coughing or blowing the nose;
–after handling waste food, refuse or chemicals.

Keep fingernails short and clean.
Cover any cuts, spots or boils with a waterproof dressing.
Keep hair clean and covered to prevent hair/dandruff entering the food.
Always wear clean protective clothing (including footwear) in food processing areas.

DON'T
Do not smoke, chew gum, tobacco, betel nut, fingernails or anything else.
Do not taste food.
Do not spit, sneeze or cough over food.
Do not pick nose, ears or any other body site.
Do not wear jewellery when handling food.
Do not wear protective clothing outside the production areas.

(4) Principles of the handling and storage of foods such as the correct use of refrigerators and freezers, the importance of temperature monitoring, the need for stock rotation and the avoidance of cross-contamination.

(5) Correct cleaning procedures and the importance of the 'clean-as-you-go' philosophy.

(6) Knowledge of the common pests found in food premises and methods for their exclusion and control.

(7) An introduction to the requirements of current food legislation.

These topics should be illustrated and supplemented with material relevant to the specific type of food business and the foods being handled.

11.4.2   Facilities and Operations

The environment in which food processing is conducted is an important factor in determining product quality. The premises should be of sufficient size for the intended scale of operation and should be sited in areas which are free from problems such as a particular pest nuisance, objectionable odours, smoke or dust. The site should be well accessed by metalled roads and have supplies of power and potable water adequate to the intended purpose. Particular attention should also be paid to the provision of facilities for the efficient disposal of processing wastes.

Buildings must be of sound construction and kept in good repair to protect the raw materials, equipment, personnel and products within, and to prevent the ingress of pests. The grounds surrounding the plant should be well maintained with lawns cut regularly and a grass-free strip of gravel or tarmac around the buildings. Well-tended grounds will not only prove aesthetically pleasing but will help in the control of rodent pests. Landscaping features such as ponds are not advisable since they may encourage birds and insects.

It is important that the buildings provide a comfortable and pleasant working environment conducive to good hygienic practices. They should be well lit, well ventilated and of sufficient size to maintain the necessary separation between processes that could give rise to cross-contamination. Features such as control of temperature and relative humidity and a positive pressure of filtered air may be required in some process areas for the benefit of both personnel and product.

In processing areas, floors should be made of a durable material which is impervious, non-slip, washable, and free from cracks or crevices that may harbour contamination. Where appropriate, floors should be gently sloped to floor drains with trapped outlets. Internal walls should be smooth, impervious, easily cleaned and disinfected, and light coloured. The angle between floors and walls should be coved to facilitate cleaning. Ceilings should be light-coloured, easy to clean, and constructed to minimize condensation, mould growth and flaking. Pipework, light fittings and other services should be sited to avoid creating difficult-to-clean recesses or overhead condensation. A false ceiling separating processing areas from overhead services has sometimes been advocated though these are generally used only in particularly sensitive areas. Light fittings should be covered to protect food below in the event of a bulb or fluorescent tube shattering. Windows should have sills sloped away from the glass and, in some climates, should be covered with well-maintained fly screens. All entrances to the plant must be protected by close fitting, self-closing doors to prevent the ingress of birds and other pests. Air curtains may also be used to protect some work areas.

Toilets and changing facilities should be clean, comfortable, well lit and provide secure storage for employees' belongings. Toilets should not open directly on to food-processing areas and must be provided with hand-washing facilities supplied with hot water, soap and hand drying facilities. Ideally, taps and soap dispensers should be of the non-hand-operated type and single-use disposable towels or an air blower be provided for hand drying. Hand washing facilities should also be available elsewhere in the plant wherever the process demands.

The overall layout of the plant should ensure a smooth flow-through from raw materials reception and storage to product storage and dispatch. Areas may be designated as 'high risk' or 'low risk' depending on the sensitivity of the materials being handled and the processes used. High-and low-risk areas of a production process should be physically separated, should use different sets of equipment and utensils, and workers should be prevented from passing from one area to the other without changing their protective clothing and washing their hands. The principal situation where such a separation would be required is between an area dealing with raw foods, particularly meat, and one handling the cooked or ready-to-eat product.

It should hardly need emphasizing that the same rules governing access, behaviour and the wearing of protective clothing also apply to management, visitors and anyone requiring to visit the processing areas.

11.4.3 Equipment

Equipment and its failings can be a source of product contamination and some notable examples are presented in Table 11.6. The main objectives of the design of hygienic food-processing equipment are to produce equipment that performs a prescribed task efficiently and economically while protecting the food under process from contamination. There is general agreement on the basic principles of hygienic design, as outlined by a number of bodies. Those given below are taken from the Institute of Food Science and Technology (UK) publication, 'Good Manufacturing Practice: A Guide to its Responsible Management' with slight modification.

(1) All surfaces in contact with food should be inert to the food under conditions of use and must not yield substances that might migrate to or be absorbed by the food.

(2) All surfaces in contact with the food should be microbiologically cleanable, smooth and non-porous so that particles are not caught in microscopic surface crevices, becoming difficult to dislodge and a potential source of contamination.

(3) All surfaces in contact with food must be visible for inspection, or the equipment must be readily dismantled for inspection, or it must be demonstrated that routine cleaning procedures eliminate the possibility of contamination.

(4) All surfaces in contact with food must be readily accessible for manual cleaning, or if clean-in-place techniques are used, it should be demonstrated that the results achieved without disassembly are equivalent to those obtained with disassembly and manual cleaning.

(5) All interior surfaces in contact with food should be so arranged that the equipment is self-emptying or self-draining. In the design of equipment it is important to avoid dead space or other

Table 11.6 *Examples of equipment-related spoilage or foodborne illness*

Equipment	Problem	Consequences	Correction
Grain silo	Areas of high moisture	Mouldy grain[a]	Proper ventilation and grain turnover
Can reformer	Holes in cans of salmon	Botulism	Proper maintenance of equipment
Gelatin injector	Welds difficult to clean	Salmonellosis from meat pies	Smooth weld
Wood smoke sticks	Bacteria surviving cleaning	Spoilage of sausage	Replace wood with metal
Heat exchanger (cooling side)	Cracked cooling unit permitting entrance of contaminated water	Salmonellosis from milk powder	Replace heat exchanger
Pump	Worn gasket	Spoilage of mayonnaise	Replace gaskets more frequently
Deaerator	Not properly cleaned or located in processing scheme	Contamination of pasteurized milk, enterotoxigenic cheese	Properly clean deaerator and move upstream of pasteurizer
Commercial oven	Poor heat distribution	Areas of under-cooking, rapid spoilage potential foodborne illness	Correct heat distribution in oven, monitor temperature to detect failure

[a] Moulds can produce a range of mycotoxins
From Shapton and Shapton (1991)

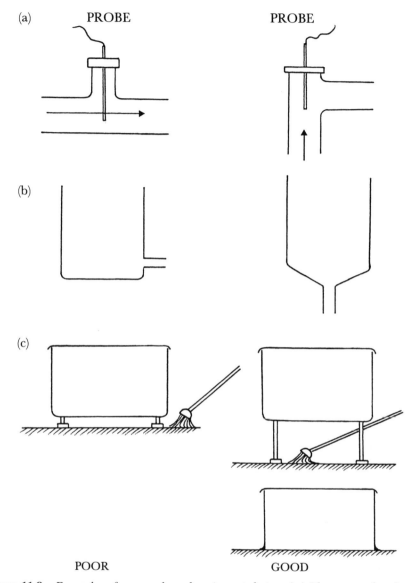

Figure 11.8 *Examples of poor and good equipment design. (a) Placement of probes, (b) avoidance of dead space, (c) avoidance of cleaning problems*

conditions which trap food and may allow microbial growth to take place (Figure 11.8).

(6) Equipment must be so designed to protect the contents from external contamination and should not contaminate the product from leaking glands, lubricant drips and the like; or through inappropriate modifications or adaptations.

(7) Exterior surfaces of equipment not in contact with food should be so arranged to prevent the harbouring of soils, micro-organisms or

pests in and on equipment, floors, walls and supports. For example, equipment should fit either flush with the floor or be raised sufficiently to allow the floor underneath to be readily cleaned.

(8) Where appropriate, equipment should be fitted with devices which monitor and record its performance by measuring factors such as temperature/time, flow, pH, weight.

11.4.4 Cleaning and Disinfection

In the course of its use, food processing equipment will become soiled with food residues. These may impair its performance by, for instance, impeding heat transfer, and can act as a source of microbiological contamination. Hygienic processing of food therefore requires that both premises and equipment are cleaned frequently and thoroughly to restore them to the desired degree of cleanliness. Cleaning should be treated as an integral part of the production process and not regarded as an end-of-shift chore liable to be hurried or superficial.

What appears to be clean visually can still harbour large numbers of viable micro-organisms which may contaminate the product. Cleaning operations in food processing have, therefore, two purposes:

(i) physical cleaning to remove 'soil' adhering to surfaces which can protect micro-organisms and serve as a source of nutrients; and

(ii) microbiological cleaning, also called sanitizing or disinfection, to reduce to acceptable levels the numbers of adhering micro-organisms which survive physical cleaning.

These are best accomplished as distinct operations in a two-stage cleaning process (Figure 11.9), although combined detergent/sanitizers are sometimes used for simplicity and where soiling is very light.

In a general cleaning/disinfecting procedure, gross debris should first be removed by brushing or scraping, possibly combined with a pre-rinse of clean, potable (drinking quality) water. This should be followed by a more thorough cleaning which requires the application of a detergent solution. The detailed composition of the detergent will depend on the nature of the soil to be removed, but a main component is likely to be a surfactant; a compound whose molecules contain both polar (hydrophilic) and nonpolar (hydrophobic) portions. Its purpose in detergent formulations is to reduce the surface tension of the aqueous phase, to improve its penetrating and wetting ability and contribute to other useful detergent properties such as emulsification, dispersion and suspension. There are three main types of surfactant,

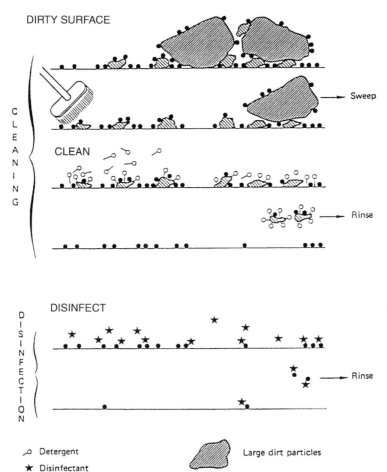

Figure 11.9 *Two-stage cleaning.*
(Reproduced with permission from WHO document VPH/83.42, copyright retained)

classified according to the nature of the hydrophilic portion of the molecule:

(i) *anionic* – in these, which include soaps, alkyl and alkylbenzene sulfonates and alcohol sulfates, the hydrophilic portion is a negatively charged ion produced in solution. They are incompatible with the use of quaternary ammonium compounds (QUATs) which are positively charged;

(ii) *non-ionic* – made by condensing ethylene oxide on to the polar end of a fatty acid, fatty alcohol or alkyl phenol;

(iii) *cationic* – quaternary ammonium compounds (QUATs) which have a positive charge in solution and are used mainly for their bacteriostatic and bacteriocidal activity rather than their cleaning properties.

Detergent preparations also often include alkalis such as sodium hydroxide, sodium silicates, or sodium carbonate which assist in solubilizing organic material such as fats and proteins. Acids are used in other formulations designed to remove the tenacious mineral scales such as milkstone which build up on surfaces, particularly heated ones, after repeated use.

Phosphates have a number of useful functions in detergents though their subsequent environmental impact can pose problems. Detergent performance is improved by sequestering agents which chelate calcium and magnesium ions and prevent the formation of precipitates. Polyphosphates are often used for this purpose, although ethylenediamine tetraacetic acid (EDTA) and gluconic acid are alternatives which have the advantages of heat stability and compatibility with QUATs. Polyphosphates also inhibit the redeposition of soil; a role for which sodium carboxymethyl cellulose is sometimes included.

Several other factors contribute to an effective cleaning procedure in addition to the physico-chemical activity of the detergent solution. Heat generally improves the efficiency of cleaning, particularly with fat-containing soils, although the temperature used must be compatible with the detergent, the soil type, and the processing surface being cleaned. Mechanical energy in the form of shear forces created by turbulence, scrubbing or some other form of agitation considerably assists in the cleaning process. For smaller items of equipment this can be done manually but for larger areas and pieces of equipment some form of power cleaning is necessary. This may involve the use of a high pressure low volume (HPLV) jet of water or detergent solution. HPLV systems operate at pressures in the range 40–100 bar ($kg\,cm^{-2}$) with flow rates between 5 and 90 $l\,min^{-1}$ and are best suited for cleaning equipment where it is necessary to direct a powerful jet into relatively inaccessible areas (though hygienic design should minimize these). They are also used to rinse off detergents applied to equipment in the form of gels or foams; systems which give longer contact times between detergent and soil than would be obtained simply by spraying with an aqueous detergent solution. Gel or foam cleaning is particularly suited to use with more recalcitrant soils and on non-horizontal equipment surfaces where conventional detergent solutions would quickly run off.

Low pressure/high volume (LPHV) cleaning ($\approx 5\,bar$; $\approx 500\,l\,min^{-1}$) is suitable for areas with a low level of soiling with water soluble residues or washing light debris to a floor drain.

It is desirable that equipment is not left wet after cleaning since micro-organisms will be able to grow in any residual water film. This is best achieved by provision of sufficient drainage points and natural air drying, although drying with single-use tissues may be required in some circumstances.

Many micro-organisms will be removed along with the soil in the course of cleaning, but many may remain on an apparently clean surface. It is therefore necessary to disinfect equipment after cleaning. A most efficient means of doing this is through the application of moist heat, which has a distinct advantage over chemical disinfectants in that its efficacy is not impaired by residual organic matter. It does however require careful control to ensure that the required temperature is maintained long enough for it to be effective. This is most appropriate in enclosed systems and is not always practicable in other areas for which chemical disinfectants are the method of choice.

Six types of chemical disinfectant are most commonly used in food processing:

(1) chlorine and chlorine compounds
(2) iodophors
(3) quaternary ammonium compounds (QUATs)
(4) biguanides
(5) acid anionic surfactants
(6) amphoteric surfactants

Hydrogen peroxide and peracetic acid are also used in some applications such as the disinfection of packing materials.

Chemical disinfectants do not act specifically on a single aspect of a microbial cell's metabolism but have a more broadly based inhibitory effect. In the case of chlorine, iodophors and peracetic acid, they act as non-specific oxidizing agents oxidizing proteins and other key molecules within the cell, while others such as QUATs and amphoterics act as surfactants, disrupting the cell membrane's integrity. For this reason, development of microbial resistance requires quite complex cellular changes. This has been noted in capsulated Gram-negative bacteria where changes in the composition of the cell membrane have resulted in resistance to QUATs and amphoterics. Development of resistance by some pseudomonads to these agents can, however, be prevented by addition of a sequestering agent which is believed to interfere with calcium and magnesium binding in the outer membrane and capsule, making the cell more vulnerable. Acquisition of resistance to oxidizing disinfectants has not been observed.

The main considerations in choosing a chemical disinfectant for use in the food industry are:

(1) Its microbiological performance – the numbers and types of organisms to be killed.
(2) How toxic is it and what is its effect on the food?
(3) What is its effect on plant – does it stain or corrode equipment?
(4) Does it pose any hazard to staff using it?
(5) Is it adversely affected by residual soil?
(6) What are the optimal conditions for its use, *i.e.* temperature, contact time, pH, water hardness?
(7) How expensive is it?

Some of these characteristics are summarized in Table 11.7.

All disinfectants are deactivated to some extent by organic matter. This is why they are best used after thorough cleaning has removed most of the soil.

Chlorine in the form of hypochlorite solution is the cheapest effective disinfectant with a broad range of antimicrobial activity which includes spores. The active species is hypochlorous acid (HOCl) which is present in aqueous solutions at pH 5–8. It is corrosive to many metals including stainless steel although this can be minimized by using it at low concentrations, at alkaline pH, at low temperature and with short contact times. For most purposes an exposure of 15 minutes to a solution containing $100\,mg\,l^{-1}$ available chlorine at room temperature is sufficient.

Organochlorine disinfectants such as the chloramines are generally weaker antimicrobials but are more stable and less corrosive than hypochlorite allowing longer contact times to be used.

In iodophors, iodine is dissolved in water by complexing it with a non-ionic surfactant. Phosphoric acid is often included since the best bactericidal activity is observed under acidic conditions. To disinfect clean surfaces a solution containing $50\,mg\,l^{-1}$ available iodine at a pH < 4 is usually required. The amber colour of iodophors in solution has two useful functions: it provides a crude visual indication of the strength of the solution and it will stain organic and mineral soils yellow indicating where equipment has been inadequately cleaned. However, they can also stain plastics and can taint some foods.

QUATs are highly stable with a long shelf-life in concentrated form. They are non-corrosive and can therefore be used at higher temperatures and with longer contact times than other disinfectants. However, at low concentrations ($<50\,mg\,l^{-1}$) and low temperatures they are less effective against Gram-negative bacteria. This is not usually a problem under normal conditions of use ($150–250\,mg\,l^{-1}$; $>40\,°C$; contact time $> 2\,min$),

Table 11.7 *Characteristics of common disinfectants used in the food industry*

	Steam	Chlorine	Iodophors	QUATs	Amphoterics	Acid anionic	Peracetic acid
Activity against:							
Vegetative bacteria							
Gram-positive	++	++	++	++	++	++	++
Gram-negative	++	++	++	++	++	++	+++
Yeasts	++	++	++	+++	+++	++	+++
Moulds	++	++	++	++	++	++	+++
Bacterial spores	+	++	+++	0	0	+	+++
Adversely affected by soil	0	++	+++	+	+	+	++
Corrosive	0	++	++	0	0	0	0
Possibility of tainting with poor rinsing	0	+	0	0	0	+	+
Active at neutral pH	+	+	0	++	++	0	0
Affected by water hardness	0	0	0	+	+	+	0
Instability in hot water	0	++	++	0	0	0	+

although incorrect usage could result in a build up of QUAT-resistant bacteria on equipment. Because of their surfactant properties, QUATs (and amphoterics) adhere to food-processing surfaces even after rinsing. This can be an advantage; in one study in a poultry plant, levels of bacteria on plant were shown to continue decreasing for nine hours after disinfection as a result of the effect of residual QUAT. In some areas though, it can be a problem; residual QUAT or amphoteric may inhibit starter culture activity in cheese and yoghurt production and can also affect head retention on beer.

Biguanides are similar to QUATs but have greater activity against Gram-negatives, although development of resistance has been noted here too.

Amphoterics are surfactants with a mixed anionic and cationic character which are far less affected by changes in pH than other disinfectants. Their high foaming characteristics make them unsuitable for some uses.

In modern food processing much equipment cleaning is automated in the form of cleaning-in-place (CIP) systems. These are most readily applied to cleaning and disinfecting plant which handles liquid foods and have therefore found widest application in the brewing and dairy industries, although they are now appearing in meat processing plants. A CIP system is a closed section of plant which can be cleaned by draining the product followed by circulation of a sequence of solutions and water rinses that clean and then disinfect the plant leaving it ready for resumed production. Though the initial capital investment is high, CIP offers a number of advantages. Its running costs are lower than traditional cleaning procedures since labour costs are low and it gives optimal use of detergents, disinfectants, water and steam. As it does not involve the dismantling of plant prior to cleaning, CIP minimizes unproductive 'down time' and the risk of equipment damage during disassembly. It is also safer since personnel are no longer required to perform the sometimes hazardous operations of climbing up on to, or into, equipment and, provided the system is correctly formulated, it gives a consistent result with little chance of human error.

To ensure cleaning and disinfecting procedures are achieving the desired result some form of assessment is necessary. The inadequacy of visual inspection of equipment to determine its microbiological status has already been alluded to. It is however worth noting that, with few exceptions, if a surface is visually dirty it is also likely to be microbiologically dirty. Culturing micro-organisms removed from a cleaned surface by swabbing, rinsing or a contact-transfer technique will give an indication of the level of contamination, but only after sufficient time has elapsed for the organisms to produce visible growth. Recently the use of ATP bioluminescence has found increasing use in this area. It provides a rapid measure of the hygienic status of a surface without having to distinguish between microbial and non-microbial ATP since high levels of ATP, whatever the origin, will indicate inadequate cleaning and disinfection (see Section 10.5.3).

Cleaning dry process areas presents an entirely different problem from that discussed above. By their very nature these areas are inimical to microbial growth due to the absence of water. To introduce moisture in the name of hygiene could have exactly the opposite effect and give rise to microbiological problems. Cleaning in dry process areas should therefore be mechanical, using vacuum cleaners, wipes and brooms.

11.5 CODES OF GOOD MANUFACTURING PRACTICE

The features of control at source outlined above are often enshrined in official regulations or codes of Good Manufacturing Practice (GMP). GMP is defined as those procedures in a food-processing plant which consistently yield products of acceptable microbiological quality suitably monitored by laboratory and in-line tests. A code of GMP must define details of the process that are necessary to achieve this goal such as times, temperatures, *etc.*, details of equipment, plant layout, disinfection (sanitation) and hygiene practices and laboratory tests.

Codes of GMP have been produced by a variety of organizations including national regulatory bodies, international organizations such as the Codex Alimentarius Commission as well as trade associations and professional bodies. They can be used by manufacturers as the basis for producing good quality product but may also be used by inspectors from regulatory bodies.

While they can be very useful, a frequent limitation is that in their desire to be widely applicable they tend to be imprecise. This leads to the use of phrases such as 'appropriate cleaning procedures', without specifying what these may be; 'cleaning as frequently as possible', without specifying a required frequency; 'undesirable organisms', without specifying which organisms. They also often fail to identify which are the most important requirements affecting food quality and which are of lesser importance. As a result, someone conducting, supervising or inspecting an operation is left uncertain as to what specifically is required to ensure that the operation is conducted in compliance with GMP.

This sort of information is often only available based on a detailed analysis of an individual processing operation.

11.6 THE HAZARD ANALYSIS AND CRITICAL CONTROL POINT (HACCP) CONCEPT

In the food industry today approaches based on Good Manufacturing Practice are being largely superseded by application of the Hazard

Analysis Critical Control Point (HACCP) concept. This has improved on traditional practices by introducing a more systematic, rule-based approach for applying our knowledge of food microbiology to the control of microbiological quality. The same system can also be adopted with physical and chemical factors affecting food safety or acceptability, but here we will confine ourselves to microbiological hazards. It should also be remembered that HACCP is primarily a preventative approach to quality assurance and as such it is not just a tool to control quality during processing but can be used to design quality into new products during their development.

HACCP was originally developed as part of the United States space programme by the Pillsbury Company, the National Aeronautics and Space Administration (NASA) and the US Army Natick Laboratories who used it to apply the same zero defects philosophy to food for astronauts as to other items of their equipment. It is based on an engineering system known as the Failure Modes Analysis Scheme which examines a product and all its components asking the question 'What can go wrong?'.

In 1973 it was adopted by the US Food and Drug Administration for the inspection of low-acid canned food. It has since been more and more widely applied to all aspects of food production, food processing and food service, and to all scales of operation from large industrial concerns, through to cottage industries and even domestic food preparation.

The meaning of the terms *hazard* and *risk* in the HACCP system differs from their common everyday usage as synonyms. In HACCP, a *hazard* is a source of danger; defined as a biological, chemical or physical property with the potential to cause an adverse health effect. Individual hazards can be assessed in terms of their severity and risk. Clearly botulism is a far more severe hazard than say *Staphylococcus aureus* food poisoning. *Risk* is an estimate of the likely occurrence of a hazard so, although *C. botulinum* is a more severe hazard, epidemiological evidence shows that the risk it poses is generally very low.

Before HACCP can be applied, it is essential that good manufacturing and hygienic practices are already in place. Factors such as hygienically designed plant and premises, effective cleaning regimes, employee hygiene, pest control, *etc.* provide the necessary foundation on which a successful HACCP system can be built. When HACCP regulations were introduced in the United States to cover fish and fishery products (1995) and meat and poultry (1996), this requirement for prerequisites based on good manufacturing practice was built into the regulations in the form of prescribed Sanitation Standard Operating Procedures (SSOPs).

HACCP itself has evolved since its first formulation and has been the subject of considerable international discussion and debate. In recent years however national and international bodies seem to have settled on an agreed definition based on seven essential principles of a HACCP system:

(1) Conduct a hazard analysis.
(2) Determine the Critical Control Points (CCPs).
(3) Establish critical limits.
(4) Establish a system to monitor control of the CCP.
(5) Establish corrective action to be taken when monitoring indicates that a particular CCP is not under control.
(6) Establish procedures to verify that the HACCP system is working effectively.
(7) Establish documentation concerning all procedures and records appropriate to these principles and their application.

To apply these principles in practice it is necessary to go through a series of steps outlined in Table 11.8.

A HACCP study is best conducted by a multidisciplinary team comprising a microbiologist, a process supervisor, an engineer and a quality assurance manager, all of whom will be able to bring their own particular expertise and experience to bear on the task in hand. Involvement of production personnel will also ensure identification with the plan by those who will have to implement it. It is important to decide on the terms of reference or scope of the HACCP plan. Experience suggests that best results are obtained when the study's terms of reference specify particular microbial hazards for consideration since this will allow the team to define specific controls. The choice of hazard considered will depend on whether there is epidemiological evidence linking a particular micro-organism with the food in question. In the absence of such evidence, factors such as the

Table 11.8   *Steps in the application of HACCP*

1	Assemble the HACCP team	
2	Describe the product	
3	Identify intended use	
4	Construct flow diagram	
5	On-site confirmation of flow diagram	
6	List all potential hazards	**Principle 1**
	Conduct a hazard analysis	
	Determine control measures	
7	Determine CCPs	**Principle 2**
8	Establish critical limit for each CCP	**Principle 3**
9	Establish a monitoring system for each CCP	**Principle 4**
10	Establish corrective action for deviations that may occur	**Principle 5**
11	Estabish verification procedures	**Principle 6**
12	Estabish documentation and record keeping	**Principle 7**

product's physical and chemical characteristics and the way it is eventually used by the consumer must provide the basis for selection.

The HACCP team produces a full description of the product, its composition and intended use, and conducts a detailed evaluation of the entire process to produce a flow diagram. This must cover all process steps under the manufacturer's control but may also extend beyond this, from before the raw materials enter the plant to the product's eventual consumption. If the eventual consumers include a high proportion of a particularly vulnerable group of the population such as infants, the elderly or sick this too should be identified.

The flow diagram must contain details of all raw materials, all processing, holding and packaging stages, a complete time–temperature history, and details of factors such as pH and a_w that will influence microbial growth and survival. Additional information covering plant layout, design and capacity of process equipment and storage facilities, cleaning and sanitation procedures will also be necessary to assess the possible risks of contamination.

Once completed, it is important that the accuracy of the final document is verified in a separate assessment during which the process is inspected on-site using the flow diagram as a guide.

11.6.1 Hazard Analysis

Hazard Analysis determines which hazards could pose a realistic threat to the safety of those consuming the product and must therefore be controlled by the production process. It is best approached in a systematic way by working through a list of raw materials, ingredients and steps in processing, packaging, distribution and storage, listing alongside each the hazards that might reasonably be expected to occur. It must identify:

(i) raw materials or ingredients that may contain micro-organisms or metabolites of concern, the likely occurrence of these hazards and the severity of their adverse health effects;

(ii) the potential for contamination at different stages in processing:

(iii) intermediates and products whose physical and chemical characteristics permit microbial growth and/or survival, or the production and persistence of toxic metabolites; and

(iv) measures that will control hazards such as process steps which are lethal or bacteriostatic.

Clearly, the expertise of the food microbiologist plays a key role at this stage; for example, helping the team distinguish between raw materials/ingredients that are microbiologically sensitive, *e.g.* meat, eggs, nuts and those which are not, *e.g.* sugar, vinegar, *etc.* The use of quantitative tools such as predictive models to calculate the potential for growth or the extent of survival at each step can also provide the hazard analysis with valuable information.

11.6.2 Identification of Critical Control Points (CCPs)

Once the hazard analysis has produced a list of the potential hazards, where they could occur, and measures that would control them, critical control points (CCPs) are identified. A CCP is defined as a location, step or procedure at which some degree of control can be exercised over a microbial hazard; that is, the hazard can be either prevented, eliminated, or reduced to acceptable levels. Loss of control at a CCP would result in an unacceptable risk to the consumer or product.

A raw material could be a CCP if it is likely to contain a microbial hazard and subsequent processing, including correct consumer use, will not guarantee its control. Specific processing steps such as cooking, chilling, freezing, or some feature of formulation may be CCPs, as could aspects of plant layout, cleaning and disinfection procedures, or employee hygiene. Many are self-evident, but decision trees can be used to help in their identification (Figure 11.10).

The questions in the decision tree should be asked for each hazard at each step in the process. Though it is necessary to consider hazards individually it will emerge that some points in a process are CCPs for more than one hazard. For example, using the decision tree, pasteurization will be a critical control point for VTEC, *L. monocytogenes, Salmonella* and *Campylobacter* in the processing of milk. Though chill storage of the milk prior to pasteurization is clearly beneficial, since it will reduce or prevent the growth of these and other organisms, it will not be a CCP for these particular hazards since they are eliminated by subsequent pasteurization. Chilling would, however, be a CCP with regard to *Staph. aureus* because by preventing growth of the organism, it will prevent production of heat resistant toxin that could survive pasteurization.

If a hazard is identified at a step where control is necessary for safety, and there is no control measure at that or any subsequent step, then the product or process should be modified to include a control measure.

11.6.3 Establishment of CCP Critical Limits

For each of the CCPs identified, criteria must be specified that will indicate that the process is under control at that point. These will usually

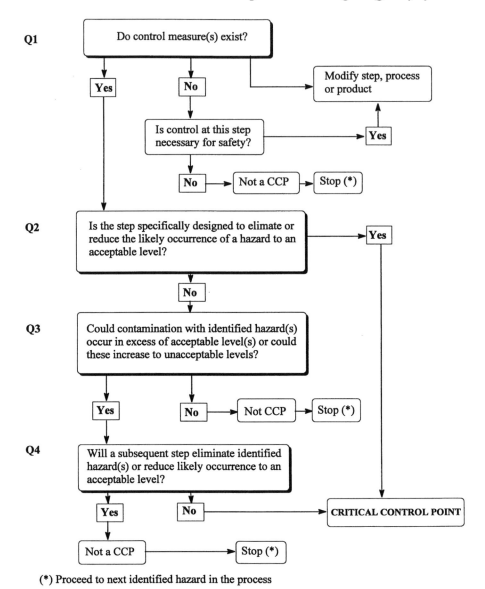

(*) Proceed to next identified hazard in the process

Figure 11.10 *Example of decision tree to identify CCPs (answer questions in sequence)*

take the form of critical limits (with tolerances where appropriate) necessary to achieve control of the hazard. Criteria may include:

(i) *physical parameters* such as temperature/time, humidity, quantity of product in a pack, dimensions of can seams, or depth of product in trays to be chilled;
(ii) *chemical parameters* such as pH in fermented or acidified foods, a_w in intermediate-moisture foods, salt concentra-

tion, available chlorine in can cooling water, or level of preservative;

(iii) *sensory information* such as texture, appearance, or odour; or,

(iv) *management factors* such as the correct labelling of products with instructions for use and handling, or efficient stock rotation.

Critical limits can be derived from a number of sources such as in-house expertise, published data, expert advice, mathematical models or from experiments conducted specifically to provide this information.

11.6.4 Monitoring Procedures for CCPs

Crucial to the application of criteria at CCPs is the introduction of monitoring procedures to confirm and record that control is maintained. It is important to remember that the assurance given by monitoring procedures will only be as good as the methods used and these too must be regularly tested and calibrated.

To achieve the on-line control of a processing operation, monitoring procedures should wherever possible be continuous and give 'real time' measurement of the status of a CCP. In some cases, the availability of appropriate monitoring procedures could govern the choice of criteria. If continuous monitoring is not possible then it should be of a frequency sufficient to guarantee detection of deviations from critical limits, and those limits should be set taking into account the errors involved in periodic sampling.

The long elapsed times involved in obtaining microbiological data means that microbiological criteria are not generally used for routine monitoring of CCPs, other than perhaps the testing of incoming raw materials. Microbiological testing does however play an important part in verification.

Records should be kept of the performance of CCPs. These will assist in process verification and can also be analysed for trends which could lead to a loss of process control in the future. Early recognition of such a trend would allow pre-emptive remedial action to be taken.

11.6.5 Protocols for CCP Deviations

When routine monitoring indicates that a CCP is out of control there should be clearly described procedures for its restoration, who is responsible for taking action and for recording the action taken. In addition to measures to restore the process, it should also prescribe what should be done with product produced while the CCP was out of control.

11.6.6   Verification

Verification is the process of checking that a HACCP plan is being applied correctly and working effectively. It is an essential feature of quality control based on HACCP and is used both when a system is first introduced and to review existing systems. Verification uses supplementary information to that gathered in the normal operation of the system and this can include extensive microbiological testing. To verify that criteria or critical limits applied at CCPs are satisfactory will often require microbiological and other, more searching, forms of testing. For example, microbial levels may need to be determined on equipment where day-to-day monitoring of cleaning procedures is based on visual inspection alone, or the precise lethality of a prescribed heat process may have to be measured. In normal operation, only limited end product testing is required because of the safeguards built into the process itself, but more detailed qualitative and quantitative microbiological analyses of final product and product-in-process may be required in a verification programme. Supplementary testing must be accompanied by a detailed on-site review of the original HACCP plan and the processing records.

One of the great strengths of the HACCP approach is its specificity to individual production facilities. Differences in layout, equipment and/or ingredients between plants producing the same product could mean that different CCPs are identified. Similarly, small changes in any aspect of the same production process could lead to verification identifying new CCPs or weaknesses in existing criteria or monitoring procedures.

11.6.7   Record Keeping

The HACCP scheme should be fully documented and kept on file. Documentation should include details of the HACCP team and their responsibilities; material from the Hazard Analysis such as the product description and process flow diagram; details of the CCPs – the hazards associated with them and critical limits; monitoring systems and corrective action; procedures for record keeping and for verification of the HACCP system. This should be accompanied by associated process records obtained during operation of the scheme. It will also include material such as documentation to verify suppliers' compliance with the processor's requirements, records from all monitored CCPs, validation records and employee training records.

Because of its highly specific and detailed nature, it is not possible to present here a full HACCP system for particular food products. But by way of illustration, Figure 11.11 shows the flow diagram of a process for the production of a yoghurt flavoured by the addition of fruit or nut puree with critical control points identified.

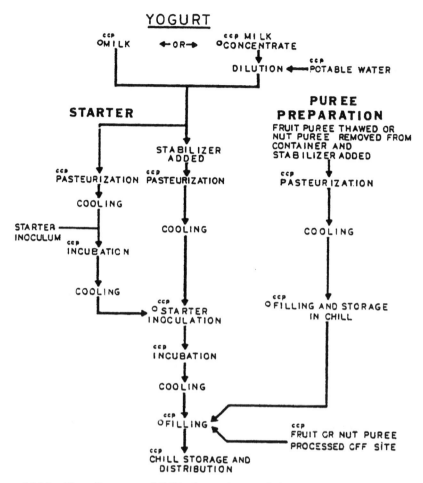

YOGURT

Figure 11.11 *Flow diagram and CCPs for yoghurt with fruit or nut puree. CCP, critical control point; O, major contamination source*
(Reproduced with permission from Shapton and Shapton (1991))

Here the microbiological safety hazards are the presence of pathogens or their toxins. The final product has a pH of 3.9–4.2 and, stored at chill temperatures, it will not permit pathogen growth and will in fact have a moderate lethal effect depending on the pathogen considered. The properties of the product will however have no effect on preformed toxins introduced during processing.

Control of pathogens in the product is obtained by pasteurization of the milk and by ensuring a satisfactory fermentation to give pH < 4.3 so CCPs will be located at points critical to achieving these goals. For example, incoming milk must be tested for antibiotic residues which may inhibit starter-culture activity, and the time and temperature of heat treatment, and factors governing the fermentation such as temperature and starter composition, must all be strictly controlled and monitored.

The possibility that the fruit or nut puree may contain pre-formed toxins is a matter which is under the control of the supplier. The pH of the fruit puree is likely to control any possibility of growth and toxin production by bacterial pathogens, although mycotoxins might be a concern. To control yeasts which could reduce the product's shelf-life, it may be necessary to specify the heat process given to the fruit puree and to store it at chill temperatures prior to use. Nut puree requires more stringent control because of its higher pH. The supplier should provide evidence that it has received a botulinum cook and that the nuts used in its preparation were of good quality and free from aflatoxin.

The US Department of Agriculture has produced a generic HACCP analysis of the production of raw beef (Figure 11.12). Such documents can provide a useful guide to a HACCP team but care must be taken that unique factors applying to the operation under study are not overlooked.

Clearly the full rigours of the HACCP approach are disproportionate to the needs and capabilities of many small food businesses. Since these make up a substantial part of the food industry in most countries, particularly in the food service sector, simplified HACCP-based programmes have been developed for such cases. One example of this is the "Safer Food Better Business" scheme developed by the UK's Food Standards Agency.

11.7 QUALITY SYSTEMS: BS 5750 AND ISO 9000 SERIES

In line with the thinking described above, there has been a change in the way standards are being used in quality management. They have moved away from being specific and prescriptive to being more conceptual in approach. This is best illustrated by the British Standard BS 5750 and its International Standards Organization (ISO) equivalent, the ISO 9000 series, on Quality Systems which are applicable to any processing or productive activity.

A quality system is a means of ensuring that products of a defined quality are produced consistently and it represents an organized commitment to quality. Quality systems work by requiring documented evidence at all stages, from product research and development, through raw materials purchase, to supply to the customer, that quality is rigorously controlled. In the food industry which is increasingly pursuing certification under such standards, HACCP documentation can play an essential role in this as evidence of a commitment to quality.

Factories can achieve approval under BS 5750 from a certifying body such as the British Standards Institution. Quality assessors study the company's 'quality manual' to ensure it meets all the requirements of the standard and then make a detailed on-site assessment of actual practices

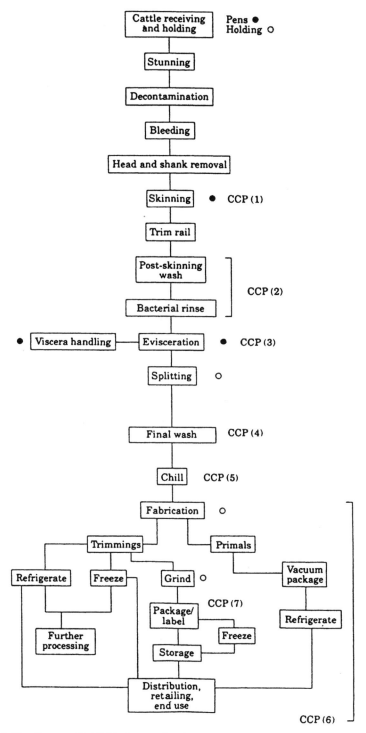

Figure 11.12 *Generic HACCP for raw beef*
(Reproduced with permission from Food Microbiology, 1993, **10**, 449–488. Academic Press)

to verify that prescribed procedures are understood and followed. Following certification, regular follow-up visits are made to ensure continued conformance with the standard.

The value to a company of seeking this outside endorsement of their quality systems is that it provides objective evidence to potential customers of the company's commitment to quality and can therefore pay substantial commercial dividends.

11.8   RISK ANALYSIS

Regulatory authorities established by governments are charged with the task of protecting the public from unsafe food and to do this they must be able to assess foodborne risks and implement strategies for their control. In the past, governments have adopted a variety of approaches to achieve this, largely subjective and based on local interests and conditions. Increasingly, however, there is a move to more systematic and unified approaches to the problem. In part, this has been driven by perceived weaknesses in existing systems, increasing concerns about the safety of food and the need for cost effective strategies to prevent and, where necessary, to reduce the risk from foodborne hazards. A major impetus behind the introduction of transparent, science-based approaches to risk management has however been the needs of international commerce.

Food and food products are important items of international trade and the loss of export markets can have a serious economic impact on producing countries. It is therefore important to be sure that when one country rejects imported food on the grounds that it is unsafe this is being done for sound scientific reasons rather than simply to erect a trade barrier to protect domestic producers or to penalise the exporting country. For this reason world trade agreements under the General Agreement on Tariffs and Trade (GATT), now the World Trade Organization (WTO), have recognized that so-called Sanitary and Phytosanitary (SPS) and Technical Barriers to Trade should be transparent and based on sound scientific principles. In particular:

> Members shall ensure that their SPS measures are based on an assessment...of the risks to human, animal or plant life or health, taking into account risk assessment techniques developed by the relevant international organisations.

Risk assessment is the scientific component of an overall system known as risk analysis (Figure 11.13). Risk assessment should provide an estimate, preferably quantitative, of the probability of occurrence and the severity of adverse health effects resulting from human exposure to

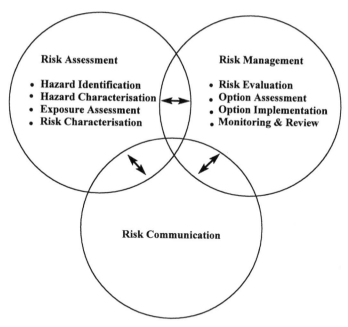

Figure 11.13 *Risk Analysis*
(Adapted form Lammerding, J. Food protection 1997, 1420–1425)

foodborne hazards, known as the risk estimate. There are four steps in risk assessment.

(1) *Hazard identification* is similar to the hazard analysis stage in HACCP and must identify the agents that are hazards to health and that may be present in a particular food. These will be the focus of subsequent stages in the risk assessment process. As with HACCP, these hazards may be chemical, physical or biological, though in this case we are concerned with micro-organisms and their toxins.

(2) *Exposure assessment* is a qualitative/quantitative evaluation of the likely intake of a hazardous agent. This must employ information on consumption patterns, *i.e.* the amounts of a particular food consumed by individuals, taking into account any variation with factors such as age, socio-economic status, religion, *etc*. It must also estimate the level of the microbial hazard in the food at the time of consumption – information that could be obtained from survey data and the use of predictive models describing the growth/survival of the organisms under likely conditions of storage and processing prior to consumption.

(3) *Hazard characterization* is the qualitative and/or quantitative evaluation of the adverse effects associated with the particular

hazard. Dose–response models which relate the number of micro-organisms ingested and the likelihood and severity of clinical illness are a valuable tool, but development of such models for foodborne pathogens is still in its early stages. There are very limited scientific data available compared with chemical hazards and there are additional difficulties arising from the complexity of micro-organisms. Some dose–response data have been obtained from studies conducted with volunteers, but these can be a crude model of the real situation since the volunteers all tend to be healthy adults. Other complicating factors can be the wide var-iation in virulence between different strains of the same organism, variation of virulence with the organism's physiological state and the important role played by the food vehicle in modulating the ability of the organism to cause infection. Alternative approaches have used information from disease outbreaks to generate more realistic dose–response models, but this is an area where consid-erable developments are likely over the next few years.

(4) *Risk characterization* integrates the results from the previous three stages to give an estimate, including attendant uncertain-ties, of the probability and severity of illness in a given popula-tion. The accuracy of the estimate can be assessed by comparison with independent epidemiological data where available.

Risk management (Figure 11.13) is the process of deciding, in collabo-ration with risk assessors, which risk assessments should be undertaken and then weighing policy alternatives to accept, minimize or reduce assessed risks. Risk managers have to decide what level of risk is acceptable (zero risk is an unachievable objective), assess the costs and benefits of different control options and if required select and implement appropriate controls, including regulatory measures. Management also includes the subsequent evaluation of the effectiveness of the measures taken and their review, if necessary.

The final component in risk analysis is risk communication – the interactive exchange of information and opinions between risk assessors, risk managers, consumers and other interested parties. This is an integral part of risk analysis and has a number of goals including the promotion of awareness, understanding, consistency and transparency.

Following an assessment of risk and its management options, it may be possible to define a food safety objective – a statement of the frequency or the maximum concentration of a microbiological hazard in a food that is considered to give an acceptable level of consumer protection. This can then be used by the food industry as a goal to be delivered through the application of good hygienic practices and HACCP. With this we reach something of a high point in the development of food microbiology to date. Management systems of the type described help ensure effective decision making and the integrated application of our increasing knowledge of foodborne micro-organisms, their detection and control. This in its turn can help deliver a varied, safe and reliable food supply, making a significant contribution to the overall quality of our lives.

CHAPTER 12

Further Reading

Information Sources on the Internet

The Internet is a huge and valuable resource with many sites of interest
to food microbiologists. Here we have listed a few, many of which will
lead you on to other related sites.

Food safety Resources on the Internet: http://bubl.ac.uk/link/f/
foodsafety.htm

Food Standards Agency (UK): http://www.food.gov.uk/

International Life Sciences Institute (ILSI) http://europe.ilsi.org

Health Protection Agency (UK): http://www.hpa.org.uk/

Department for the Environment, Food and Rural Affairs (DEFRA)
(UK): http://www.defra.gov.uk/

Center for Food Safety & Applied Nutrition, USA: http://vm.cfsan.fda.gov/
—particularly useful is the 'Bad Bug Book': http://vm.cfsan.fda.gov/
~mow/intro.html

Mycotoxicology Newsletter: an international forum for mycotoxins,
especially useful for summaries of recent symposia:
http://www.mycotoxicology.org/mtnl2c.html

Institute of Food Science and Technology, UK: http://www.ifst.org/
—particularly the hot topics section: http://www.ifst.org/hottop.htm

World Health Organisation (WHO), Geneva: http://www.who.ch/
—particularly the food safety programme: http://www.who.int/fsf/

Canadian Food Inspection Agency: http://www.cfia-acia.agr.ca/

Foodnet, Canada: http://foodnet.fic.ca/

National Center for Food Safety and Technology (NCFST) is a consor-
tium of industry, academia and government organized to address the
complex issues raised by emerging food technologies: http://www.
iit.edu/~ncfs/

CSIRO, Australia: http://www.dfst.csiro.au/

Australian Office of Food Safety: http://www.dpie.gov.au/ocvo/ofs.html

The following is a selective list of further reading on topics covered in this volume. It is intended primarily to give guidance into the broader literature. Although divided by chapter, the coverage of many of the references will overlap several topic areas.

Chapter 1

WHO, 'WHO Commission on Health and Environment. Report of the Panel on Food and Agriculture', 1992, WHO/EHE/92.2, 191pp.
This discusses many important global issues and gives some indication of the broader relevance of food microbiology.

Chapter 2

R.G. Board, D. Jones, R.G. Kroll and G.L. Pettipher, (eds.), 'Ecosystems: Microbes: Food', Society for Applied Bacteriology Symposium Series Number 21, 1992, 178pp.
M.E. Lacey and J.S. West, 'The Air Spora', Springer, The Netherlands, 2006, 156 pp.
D.E.S. Stewart-Tull, P.J. Dennis and A.F. Godfree, (eds.), 'Aquatic Microbiology', Society for Applied Bacteriology Symposium Series Number 28, 1999, 271pp.

Chapter 3

ICMSF, 'Microbial Ecology of Foods. Volume 1 Factors affecting life and death of microorganisms', Academic Press, New York, 1980, 332pp.
Though some of the ICMSF books have been published for some time now, they still represent a magnificent collection of relevant material, well written and authoritative.
B.M. Lund, T.C. Baird-Parker and G.W. Gould (eds) The Microbiological safety and quality of food. Vol. 1. Aspen Publishers Inc. Gaithersburg 2000, 972pp.
An excellent book covering factors affecting growth, preservation technologies and the microbiology of specific food commodities.
L. Leistner and G.W. Gould, Hurdle Technologies. Kluwer Academic/Plenum Publishers New York 2002, 194 pp.
D.A.A. Mossel and M. Ingram, The physiology of the microbial spoilage of foods, *J. Appl. Bacteriol.*, 1955, **18**, 232–268.
This article is included for its seminal importance in the development of modern food microbiology.
L.R. Beuchat, (ed.), 'Food and Beverage Mycology', 2nd Edn., Van Nostrand Reinhold, New York, 1987, 661pp.
A collection of excellent essays on all aspects of food mycology, particularly the chapter on water activity.

S. Notermans, P. in't Veld, T. Wijtzes and G.C. Mead, A user's guide to microbial challenge testing for ensuring the safety and stability of food products, *Food Microbiol.*, 1993, **10**, 145–158.

T.A. McMeekin, J.N. Olley, T. Ross and D.A. Ratkowsky, 'Predictive Microbiology: Theory and Application', Research Studies Press Ltd., Taunton, England, 1993, 340pp.

Chapter 4

Advisory Committee on the Microbiological Safety of Food, 'Report on Vacuum Packaging and Associated Processes', HMSO, London, 1992, 69pp.

The ACMSF has produced a number of reports that can be very profitably read. Though naturally centred on the UK situation, they contain much that is of relevance for a far wider audience. Their reports can be accessed through the Food Standards Agency website.

Two excellent books which develop much of the material covered in this Chapter are:

P.M. Davidson and A.L. Branen, (eds.), 'Antimicrobials in Foods', 2nd Edn., Marcel Dekker Inc., New York, 1993, 647pp.

G.W. Gould, (ed.), 'Mechanisms of action of food preservation procedures', Elsevier Applied Science Publishers, London, 1989, 441pp.

The two classic texts on heat processed foods are:

C.R. Stumbo, 'Thermobacteriology in Food Processing', Academic Press, New York and London, 1965, 236pp.

A.C. Hersom and E.D. Hulland, 'Canned Foods', Churchill Livingstone, Edinburgh, London and New York, 7th Edn., 1980, 380pp.

H. Burton, 'UHT Processing of Milk and Milk Products', Elsevier Applied Science Publishers, London, 1988.

H. Burton, Thirty five years on—a story of UHT research and development, *Chem. Ind.*, August 1985, 546–553.

J.M. Farber, Microbiological aspects of modified-atmosphere packaging technology – a review, *J. Food Prot.*, 1991, **54**, 58–70.

G.W. Gould, A.D. Russell and D.E.S. Stewart-Tull, (eds.), 'Fundamental and Applied Aspects of Bacterial Spores', Society for Applied Bacteriology Symposium Series Number 23, 1994, 138pp.

R.T. Parry, (ed.), 'Principles and Applications of Modified Atmosphere Packaging of Food', Blackie, 1993, 305pp.

C.C. Seow, (ed.), 'Food Preservation by Moisture Control', Elsevier Applied Science Publishers, London, 1988, 277pp.

Chapter 5

C. de W. Blackburn (ed.), Food spoilage microorganisms, Woodhead Publishing, Cambridge, UK 2006, 712pp.

M.H. Brown, (ed.), 'Meat Microbiology', Elsever Applied Science Publishers, London, 1982.

ICMSF, 'Micro-organisms in Foods. Vol. 6, Microbial ecology of food commodities', Blackwell Academic & Professional, London 1998, 615pp.

J.I. Pitt and A.D. Hocking, 'Fungi and Food Spoilage', 2nd Edn., Blackie Academic, London, 1997, 593pp.

An excellent guide to the identification of many fungi found associated with foods and with chapters dealing with the mould spoilage of specific commodities.

R.K. Robinson (ed.) Dairy microbiology handbook, 3rd edn., Wiley Interscience 2002, 765pp.

ICMSF, 'Microbial Ecology of Foods. Volume 2 Food Commodities', Academic Press, New York, 1980, 664pp.

Chapters 6, 7 and 8

There is a very substantial, generally high quality literature on food-borne illness from which we can recommend:

T.J. Mitchell, A.F. Godfree and D.E.S. Stewart-Tull, (eds.), 'Toxins', Society for Applied Bacteriology Symposium Series Number 27, 1998, 160pp.

ICMSF, 'Micro-organisms in Foods. Vol. 5, Microbiological specifications of food pathogens', Blackwell Academic & Professional, London, 1996, 513pp.

M.P. Doyle, L.R. Beuchat and T.J. Montville (eds) Food microbiology: fundamentals and frontiers. ASM Press, Washington D.C. 2001, 872 pp.

D.O. Cliver and H.R. Riemann (eds), Foodborne diseases, Academic Press, San Diego, 2002, 411pp.

M. Griffiths (ed), Understanding pathogen behaviour, Woodhead Publishing, Cambridge, UK 2005, 611pp.

Y. Motarjemi and M. Adams (eds), Emerging foodborne pathogens, Woodhead Publishing, Cambridge, UK 2006, 634pp.

A.H.W. Hauschild and K.L. Dodds,(eds.), 'Clostridium botulinum: Ecology and Control in Foods', Marcel Dekker, New York, 1993, 412pp.

M. Richmond, 'The Microbiological Safety of Food Part 1. Report of the Committee on the Microbiological Safety of Food', HMSO, London, 1990, 147pp.

M. Richmond, 'The Microbiological Safety of Food Part II. Report of the Committee on the Microbiological Safety of Food', HMSO, London, 1990, 220pp.

The two parts of the Richmond report are primarily of relevance to the UK but will nonetheless contain much of interest to a broader readership.

K.K. Sinha and D. Bhatnagar, (eds) 'Mycotoxins in Agriculture and Food Safety', Marcel Dekker, New York, 1998, 511 pp.

L.S. Jackson, J.W. DeVries and L.B. Bullerman, (eds) 'Fumonisins in Food', Plenum, New York. 1996, 399 pp.

J.W. DeVries, M.W. Trucksess and L.S. Jackson, (eds) 'Mycotoxins and Food Safety', Kluwer Academic, New York. 2002, 295 pp.

N. Magan and M. Olsen, (eds) 'Mycotoxins in Food', Woodhead Publishing, Cambridge, UK, 2004, 471 pp.

A.E. Desjardins, 'Fusarium Mycotoxins: Chemistry, Genetics and Biology', APS Press, Minnesota, 2006, 260 pp.

A.H. Varnam and M.G. Evans, 'Foodborne Pathogens, an Illustrated Text', Wolfe, London, 1991, 557pp.

W.M. Waites and J.P. Arbuthnott, (eds.), 'Foodborne Illness', Edward Arnold, London, 1991, 146pp.

Chapter 9

The proceedings of four major conferences on the lactic acid bacteria have been published and contain a wealth of excellent and authoritative reviews. The three most recent are:

FEMS Microbiol. Rev., 1987, **46**, pp.201–379.
FEMS Microbiol. Rev., 1990, **87**, pp.3–188.
FEMS Microbiol. Rev., 1993, **12**, pp.3–271.

A.P.J. Trinci, 'Mycoprotein: a twenty year overnight success story', *Mycological Research,* **1992**, 96, pp.1–33.

R.G. Board, D. Jones and B. Jarvis, (eds.), 'Microbial Fermentations: Beverages, Foods and Feeds', Society for Applied Bacteriology Symposium Series, Number 24, 1995, 145pp.

There are a number of texts which can provide further background information on specific fermented foods:

B.J.B. Wood, (ed.) 'Microbiology of Fermented Foods', 2 Volumes, 2nd edn., Blackie Academic & Professional, London, 1998, 852 pp.

F.G. Priest and I. Campbell, (eds.), 'Brewing Microbiology', Elsevier Applied Science, London, 1987, 275pp.

D.E. Briggs, C.A. Boulton, P.A. Brookes and R. Stevens, Brewing science and practice. Woodhead Publishing, Cambridge, UK 2004, 881pp.

S. Salminen, and A. von Wright, (eds.), 'Lactic acid bacteria: microbiology and functional aspects', 2nd edn., Marcel Dekker Inc., New York, 1998.

K.H. Steinkraus, (ed.), 'Handbook of Indigenous Fermented Foods', Marcel Dekker Inc., New York, 1983, 671pp.

Chapter 10

Some of the references cited here give further details on techniques described in Chapter 10, while others illustrate the collaborative efforts made to assess and standardize traditional cultural techiques.

J.E.L. Corry, G.D.W. Curtis, R.M. Baird (eds), Culture media for food microbiology, Progress in Industrial Microbiology vol. 34, Elsevier, Amsterdam, 1995, 491pp.

C.H. Collins, P.M. Lyne, J.M. Grange and J.O. Falkingham III, Microbiological methods 8th edn, Arnold, London, 2004, 456pp.

L.R. Beuchat and T. Deak, (eds.), Advances in Methods for Detecting, Enumerating and Identifying Yeasts in Foods, *Int. J. Food Microbiol.*, 1993, **19**, 1–86.

L. Gram, Evaluation of the bacteriological quality of seafood, *Int. J. Food Microbiol.*, 1992, **16**, pp. 25–39.

W.F. Harrigan, 'Laboratory Methods in Food Microbiology', 3rd edn., Academic Press, San Diego, 1998, 532 pp.

ICMSF, 'Microorganisms in Foods I: Their significance and methods of enumeration', 2nd Edn., University of Toronto Press, Toronto, 1978, 434pp.

P.D. Patel, (ed.), 'Rapid Analysis Techniques in Food Microbiology', Blackie, London, 1994, 294pp.

R.G. Kroll, A. Gilmour and M. Sussman, 'New Techniques in Food and Beverage Microbiology', SAB Technical Series No. 31, Blackwell, Oxford, 1993, 302pp.

Though out of date on many of the methods it describes, this book contains a useful discussion on the significance of micro-organisms in foods.

R.A. Samson, A.D. Hocking, J.I. Pitt and A.D. King, 'Modern methods in Food Mycology', Elsevier, Amsterdam, 1992.

The report of the 2nd International Workshop on Standardization of Methods for the Mycological examination of Foods containing details of recommended methods and media.

R.A. Samson, E.S. Hoekstra, J.C. Frisvad and O. Filtenborg, 'Introduction to Food-Borne Fungi', 5th Edn., Cenntraalbureau voor Schimmelcultures, Baarn, 1996, 322pp.

A very strongly recommended introduction to the most common fungi found associated with foods.

G.M. Wyatt, 'Immunoassays for food poisoning bacteria and their toxins', Chapman and Hall, London, 1992, 129pp.

T.A. McMeekin, Detecting pathogens in food, Woodhead Publishing, Cambridge, UK 2003, 370pp.

C. Bell, P. Neaves and A.P. Williams, Food Microbiology and Laboratory Practice, Blackwell Science, Oxford, 2005, 324pp.

Chapter 11

The following are some key references from a large and increasing literature on the microbiological quality control of food:

ICMSF, 'Micro-organisms in Foods. Vol. 2, Sampling for Microbiological Analysis: Principles and Specific Applications', 2nd Edn., Blackwell Scientific Publications, Oxford, 1986, 293pp.

ICMSF, 'Micro-organisms in Foods. Vol. 4, Application of the hazard analysis critical control point (HACCP) system to ensure microbiological safety and quality', Blackwell Scientific Publications, Oxford, 1988, 357pp.

ICMSF, Micro-organisms in Foods. Vol 7. Microbiological testing in food safety management, Kluwer Academic/Plenum Publishers, New York, 2002, 362pp.

B. Jarvis, 'Progress in Industrial Microbiology', Vol. 21, Statistical aspects of the microbiological analysis of foods, Elsevier, Amsterdam, 1989, 179pp.

D.A. Shapton and N.F.Shapton, (eds.), 'Principles and practices for the safe processing of food', Butterworth-Heinemann, Oxford, 1991, 457pp.

IFST, 'Good manufactoring practice: a guide to its responsible management', 4th Edn., Institute of Food Science and Technology (UK), London, 1998, 172pp.

IFST, 'Food hygiene training: a guide to its responsible management', Institute of Food Science and Technology (UK), London, 1992, 28pp.

National Advisory Committee on Microbiological Criteria for Foods (USA), 'Hazard analysis and critical control point principles and application guidelines', *J. Food Protection,* **61**, 1246–1259.

National Advisory Committee on Microbiological Criteria for Foods (USA), Generic HACCP for raw beef, *Food Microbiol.,* 1993, **10**, 449–488.

A.T. Jackson, Cleaning of food processing plant, *Dev. Food Preserv.,* 1985, **3**, 95–126.

WHO/FAO, 'Application of Risk Analysis to Food Standards Issues', 1995, WHO/FNU/FOS/95.3, 39pp.

A.M. Lammerding and G.M. Paoli, 'Quantitative Risk Assessment: an emerging tool for emerging foodborne pathogens', *Emerging Infectious Diseases,* 1997, **3**, 483–487.

WHO, 'HACCP Introducing the hazard analysis and critical control point system', 1997, WHO/FSF/FOS/97.2, 21pp.

M. Brown and M. Stringer, Microbiological risk assessment in food processing, Woodhead Publishing, Cambridge, 2002, 301 pp.

Subject Index